AF454465

Advances in Cereal Crops Breeding

Advances in Cereal Crops Breeding

Editor

Igor G. Loskutov

MDPI • Basel • Beijing • Wuhan • Barcelona • Belgrade • Manchester • Tokyo • Cluj • Tianjin

Editor
Igor G. Loskutov
N. I. Vavilov Institute of Plant
Genetic Resources (VIR)
Russia

Editorial Office
MDPI
St. Alban-Anlage 66
4052 Basel, Switzerland

This is a reprint of articles from the Special Issue published online in the open access journal *Plants* (ISSN 2223-7747) (available at: https://www.mdpi.com/journal/plants/special_issues/Cereal_Breeding_Advance).

For citation purposes, cite each article independently as indicated on the article page online and as indicated below:

LastName, A.A.; LastName, B.B.; LastName, C.C. Article Title. *Journal Name* **Year**, *Volume Number*, Page Range.

ISBN 978-3-0365-2650-8 (Hbk)
ISBN 978-3-0365-2651-5 (PDF)

Cover image courtesy of Igor G. Loskutov

Contents

About the Editor

Igor G. Loskutov was born in 1956. In 1978, he graduated from Saint Petersburg State University with a diploma of Agrochemistry and Soil Science. In 1985, he completed his Ph. D (plant breeding) and in 2003, his D. Sc. (botany and plant breeding) from N.I. Vavilov Institute of Plant Genetic Resources (VIR).

He has vast experience in different areas of Plant Genetic Resources: the theoretical, practical and legislative aspects of collecting, evaluation and storage of plant genetic resources; botany, systematic, evolution, distribution and diversity of cultivated and wild oats; genomic, genetics and breeding of the main characters and properties of cereals.

At present, he is Head of the Department of Genetic Resources of Oat, Barley and Rye, N.I. Vavilov Institute of Plant Genetic Resources (VIR), Russia; Professor of Department of Agrochemistry, Biology Faculty, Sankt-Petersburg State University, Russia.

He was supervisor of 8 Bachelors and 3 Masters of Science Dissertations in St-Petersburg State University and supervisor of 10 PhD Dissertations in Vavilov Institute of Plant Genetic Resources.

He has published 437 papers in Soviet/Russian national and international peer-reviewed journals such as *Euphytica, J. Bot., Genome, J. Agric. Food Chem., Gen. Res. Crop Evol., Rus. J. Gen.: Appl. Res., Rus. J. Gen., Agronomy, Plants, Molecules*, etc., and from different journals and book of Springer. He is editorial board member of several International Journals. He has six monographs. He has six Russian patents. Prof. Loskutov also played an instrumental role in different prestigious Russian and internal collaborative research projects with USA, Germany, Italy, Switzerland, Sweden, France, China, etc.

As a Visiting Professor, he made some lectures in Universities and Research Institutes in USA, Germany, Italy, China, Brazil, Sweden, Tajikistan, Finland, France, UK, Turkey, Israel, etc.

Preface to "Advances in Cereal Crops Breeding"

This Special Issue presents some advances in the results of cereal crop breeding. These studies address only some of the bottlenecks in the breeding of specific crops. At the same time, the advances in the modern breeding of grain crops are multifaceted and diverse, occurring in different countries and different continents of the world. At present, in the breeding of agricultural crops, the factor of climate change and the associated changing conditions for the cultivation of many crops important for humanity are becoming increasingly important. Climate change can lead to an excess and a sharp shortage of precipitation, coupled with an increase in temperature, which will affect edaphic factors of plant growth and development, expressed in the salinization/acidification or drying out of the soil. On the other hand, this can lead to the emergence of new diseases and stronger epiphytoties of already known diseases, or to a greater spread of agricultural pests. The above factors ultimately affect the productivity and quality of the products obtained, on which the food security of each country depends. In future, breeders of the world will be assisted in solving many of these problems, along with the traditional ones, using recently developed "omix" genotyping technologies.

Igor G. Loskutov
Editor

Editorial

Advances in Cereal Crops Breeding

Igor G. Loskutov

Federal Research Center the N.I. Vavilov All-Russian Institute of Plant Genetic Resources (VIR), St. Petersburg 190000, Russia; i.loskutov@vir.nw.ru

Citation: Loskutov, I.G. Advances in Cereal Crops Breeding. *Plants* **2021**, *10*, 1705. https://doi.org/10.3390/plants10081705

Received: 9 August 2021
Accepted: 16 August 2021
Published: 19 August 2021

Cereals are the main food and feed crops on our planet, with wheat, rice, and maize occupying three-quarters of the total acreage. The vast majority of plant breeders and plant geneticists around the world are engaged in cereal breeding. The genetic resources for crop genepools, including breeding and research materials, landraces, and wild crop relatives, which collectively are the pillars of modern plant breeding, are maintained ex situ in gene banks. The main challenges or bottlenecks in the advanced breeding techniques currently used in cereals are connected with concerns related to climate change, with breeding programs aiming to increase yield and tolerance to biotic and abiotic stresses (e.g., yield potential and resistance to main diseases and pets, as well as increased drought, heat tolerance, and nutrient efficiency). In the last few years, a trend has occurred in cereal crop breeding aimed at combining high agronomic and biochemical parameters in a single cultivar. Currently, traditional genetic and innovative molecular genetic methods are widely used in the breeding of grain crops. The success of biotechnology approaches has expanded the breeding possibilities and allowed interspecies and intergenus hybrids to be obtained. The development of molecular biology and genomics has completely overcome the barriers limiting the breeding of living organisms, while methods for genome editing of agricultural crops are still being improved to achieve higher levels of accuracy. Studies aimed at finding genes and quantitative traits loci (QTLs) that affect the main breeding traits and at identifying the desired allelic variants are currently relevant. In the field of genetic sequencing, genotyping by sequencing, also called GBS, is a method used to discover single-nucleotide polymorphisms (SNP) in order to perform genotyping studies, such as genome-wide association studies (GWAS).

The acquisition of large-scale phenotypic data has become one of the major bottlenecks hindering crop breeding and functional genomics studies. Nevertheless, recent technological advances have provided potential solutions to relieve such bottlenecks and to explore advanced methods for large-scale phenotyping, data acquisition, and data processing in the coming years. The phenomics data generated are already beginning to be used to identify genes and QTL through QTL mapping, association mapping, and genome-wide association studies (GWAS), in order to achieve crop improvements through genomics-assisted breeding (GAB). There is no doubt that accurate high-throughput phenotyping platforms will accelerate improvements in plant genetics.

This Special Issue on 'Advances in Cereal Crops Breeding' comprises 9 papers covering a wide array of aspects, ranging from the expression-level investigation of genes in terms of salinity stress adaptations and their relationships with proteomics in rice, the use of genetic analysis to assess the general combining ability (GCA) and specific combining ability (SCA) in promising hybrids of maize, the use of DNA markers based on PCR in rice, the identification of quantitative trait loci (QTLs) in wheat and simple sequence repeats (SSR) in rice, the use of single-nucleotide polymorphisms (SNP) in a genome-wide association study (GWAS) in cereals, and Nanopore direct RNA sequencing of related with LTR RNA retrotransposon in triticale prior to genomic selection of heterotic maize hybrids.

In order to better understand the mechanisms involved in salinity stress adaptations in rice, two contrasting rice cultivars were compared in a recent study—Luna Suvarna, a salt-tolerant cultivar, and IR64, a salt-sensitive cultivar. The expression-level investigation of auxin signaling pathway genes revealed increases in the transcript levels of several auxin homeostasis genes in Luna Suvarna compared with IR64 under salinity stress. Furthermore, protein profiling showed 18 proteins that were differentially regulated between the roots of two cultivars, some of which were salinity-stress-responsive proteins found exclusively in the proteome of Luna Suvarna roots, revealing the critical role of these proteins in imparting salinity stress tolerance. The results show that Luna Suvarna involves a combination of morphological and molecular traits of the root system that could prime the plant to better tolerate salinity stress [1].

The tolerance of rice to salinity stress involves diverse and complementary mechanisms, such as the regulation of genome expression, activation of specific ion transport systems to manage excess sodium at the cell or plant level, and anatomical changes that mitigate sodium penetration into the inner tissues of the plant. The identification of salinity tolerance QTLs associated with different mechanisms involved in salinity tolerance requires the greatest possible genetic diversity to be explored. In the investigation of genotyped rice landraces, SNP markers were used, with the aim of identifying new QTLs involved in salinity stress tolerance via a genome-wide association study (GWAS). Twenty-one identified QTLs colocalized with known QTLs. Several genes within these QTLs have functions related to salinity stress tolerance and are mainly involved in gene regulation, signal transduction, and hormone signaling. This study provides promising QTLs for breeding programs to enhance salinity tolerance and identifies candidate genes that should be further functionally studied to better understand salinity tolerance mechanisms in rice [2].

In addition to water flooding and salinity, rice growers in some parts of the world are also facing drought; thus, developing new rice genotypes tolerant to water scarcity is one of the best strategies to maximize yield potential and achieve water savings. In a recent study, rice genotypes were characterized for grain and agronomic parameters under normal and drought stress conditions and genetic differentiation was determined via specific DNA markers related to drought tolerance using simple sequence repeats (SSR) and cultivar grouping, establishing their genetic relationships with different traits. All genotypes were grouped into two major clusters with 66% similarity based on Jaccard's similarity index. As a result of the study, genotypes were identified that could be included as appropriate materials for developing a drought-tolerant breeding program. Genetic diversity is needed to grow new rice cultivars that combine drought tolerance with high grain yields, which is essential to maintaining food security [3].

Recent studies on the tolerance to biotic and abiotic stressors in rice hybrids with donor lines of the genes of interest showed the effectiveness of such hybrids. As a result of the studies carried out using molecular marking based on PCR in combination with traditional breeding, early-maturing rice lines with genes resistant to salinity (SalTol) and flooding (Sub1A) were obtained, which are suitable for cultivation in southern Russia. The development of resistant rice varieties and their introduction into production will allow us to avoid the epiphytotic development of the disease, preserving the biological productivity of rice and resulting in environmentally friendly agricultural products [4].

The combining ability and genetic diversity of plants are important prerequisites for the development of outstanding hybrids that are tolerant to high plant density. A recent study was carried out to assess general combining ability (GCA) and specific combining ability (SCA), identify promising hybrids, estimate genetic diversity among the inbred lines, and correlate genetic distance (GD) to hybrid performance and SCA across different plant densities. As a result, no significant correlation was found between GD and either hybrid performance or SCA for grain yield and other traits, proving to be of no predictive value. Nevertheless, SCA could be used to predict hybrid performance across all plant densities. Overall, this study presents useful information regarding the inheritance of maize grain yield and other important traits under high plant density [5].

In addition to studying the productivity of plants and genes associated with general adaptability, the genetic improvement of root systems is of interest as an efficient approach to improve the yield potential and nitrogen use efficiency (NUE) of crops. QMrl-7B is a major stable quantitative trait locus (QTL) controlling the maximum root length in wheat. Two types of near isogenic lines (A-NILs with superior and B-NILs with inferior alleles) were used to specify the effects of QMrl-7B on root, grain output, and nitrogen-related traits under both low-nitrogen (LN) and high-nitrogen (HN) environments. The QMrl-7B A-NILs manifested larger root systems compared to the B-NILs, which is favorable to N uptake and accumulation, and eventually enhanced grain production. This study provides valuable information for the genetic improvement of root traits and breeding of elite wheat varieties with high yield potential [6].

Traditional plant breeding approaches supplemented with SNP markers used for genome-wide associative studies (GWASs) and genetic editing, as well as high-throughput chemotyping techniques, are exploited to speed up the breeding of desired genotypes. To enrich cereal grains with functional components, the new breeding programs need a source of genes in order to improve the contents of the beneficial components. The sources of these valuable genes are plant genetic resources deposited in genebanks, including landraces, rare crop species, and even wild relatives of cultivated plants. Correlations between the contents of certain bioactive compounds and the resistance to diseases or tolerance to certain abiotic stressors suggest that breeding programs aimed at increasing the levels of health-benefiting components in cereal grain might at the same time allow the development of cultivars adapted to unfavorable environmental conditions [7].

Using Nanopore long-term forward RNA sequencing, functionally important but unexplored RNA molecules have been identified, including long non-coding RNAs (lncRNAs), as they are often associated with repeat-rich regions of genomes and transposon-derived transcripts expressed during early stages of seed development in triticale. Detailed analysis of the protein-coding potential of the RTE-RNAs showed that 75% of them carry open reading frames (ORFs) for a diverse set of GAG proteins, the main components of virus-like particles of LTR retrotransposons. This demonstrated experimentally that certain RTE-RNAs originate from autonomous LTR retrotransposons, with ongoing transposition activity during early stages of triticale seed development. Overall, these results provide a framework for further exploration of the newly discovered lncRNAs and RTE-RNAs in functional and genome-wide association studies in triticale and wheat. The results also demonstrate that Nanopore direct RNA sequencing is an indispensable tool for the elucidation of lncRNA and retrotransposon transcripts [8].

Genomic selection (GS) shows great promise in terms of strongly increasing rates of genetic improvement in plant breeding programs. It allows for comparative larger gains from selection by estimating all marker effects simultaneously, while the subsequent selection of genetically superior individuals is based on their genomic estimated breeding value (GEBV) instead of using a few significant markers, as is the case in classical marker-assisted selection (MAS). GS is ideal for complex traits with lower heritability and complex genetic architectures.

Genomic selection (GS) can accelerate variety improvement when the training set (TS) size and its relationship with the breeding set (BS) are optimized for the prediction accuracies (PAs) of genomic prediction (GP) models. Sixteen GP algorithms were run on phenotypic best linear unbiased predictors (BLUPs) and best linear unbiased estimators (BLUEs) of resistance to both fall armyworm (FAW) and maize weevil (MW) in a tropical maize panel. Random-based training sets (RBTS) and pedigree-based training sets (PBTSs) were designed to study biotic resistance. For PBTS, the FAW resistance PAs were generally higher than those for RBTS, except for one dataset. GP models generally showed similar PAs across individual traits, whilst the TS designation was determinant, since a positive correlation between TS size and PAs was observed for RBTS, while for the PBTS, this correlation was negative. The resulting population could be of interest in future breeding activities targeted at improving insect resistance in maize and could be potentially useful for GS of complex traits with low to moderate heritability. This study has pioneered the use of GS for maize resistance to insect pests [9].

Advances in cereals breeding to develop new improved cultivars are some of the most important factors in agricultural production, playing an essential role in ensuring sustainable agriculture. Along with classical breeding goals, innovative, modern plant breeding methodologies are applied here to create new cultivars of crops for current and future agriculture applications. This endeavor includes the development of cultivars for stress cultivation conditions to achieve sustainable agricultural production, increased food quality, and increased security, and to supply raw materials for innovative industrial products and to meet the needs of mankind.

Funding: This research received no external funding.

Conflicts of Interest: The authors declare no conflict of interest.

References

1. Saini, S.; Kaur, N.; Marothia, D.; Singh, B.; Singh, V.; Gantet, P.; Pati, P. Morphological Analysis, Protein Profiling and Expression Analysis of Auxin Homeostasis Genes of Roots of Two Contrasting Cultivars of Rice Provide Inputs on Mechanisms Involved in Rice Adaptation towards Salinity Stress. *Plants* **2021**, *10*, 1544. [CrossRef]
2. Le, T.; Gathignol, F.; Vu, H.; Nguyen, K.; Tran, L.; Vu, H.; Dinh, T.; Lazennec, F.; Pham, X.; Véry, A.-A.; et al. Genome-Wide Association Mapping of Salinity Tolerance at the Seedling Stage in a Panel of Vietnamese Landraces Reveals New Valuable QTLs for Salinity Stress Tolerance Breeding in Rice. *Plants* **2021**, *10*, 1088. [CrossRef]
3. Gaballah, M.M.; Metwally, A.M.; Skalicky, M.; Hassan, M.M.; Brestic, M.; El Sabagh, A.; Fayed, A.M. Genetic Diversity of Selected Rice Genotypes under Water Stress Conditions. *Plants* **2021**, *10*, 27. [CrossRef]
4. Dubina, E.; Kostylev, P.; Ruban, M.; Lesnyak, S.; Krasnova, E.; Azarin, K. Rice Breeding in Russia Using Genetic Markers. *Plants* **2020**, *9*, 1580. [CrossRef]
5. Kamara, M.M.; Rehan, M.; Ibrahim, K.M.; Alsohim, A.S.; Elsharkawy, M.M.; Kheir, A.M.S.; Hafez, E.M.; El-Esawi, M.A. Genetic Diversity and Combining Ability of White Maize Inbred Lines under Different Plant Densities. *Plants* **2020**, *9*, 1140. [CrossRef]
6. Liu, J.; Zhang, Q.; Meng, D.; Ren, X.; Li, H.; Su, Z.; Zhang, N.; Zhi, L.; Ji, J.; Li, J.; et al. *QMrl-7B* Enhances Root System, Biomass, Nitrogen Accumulation and Yield in Bread Wheat. *Plants* **2021**, *10*, 764. [CrossRef]
7. Loskutov, I.G.; Khlestkina, E.K. Wheat, Barley, and Oat Breeding for Health Benefit Components in Grain. *Plants* **2021**, *10*, 86. [CrossRef]
8. Kirov, I.; Dudnikov, M.; Merkulov, P.; Shingaliev, A.; Omarov, M.; Kolganova, E.; Sigaeva, A.; Karlov, G.; Soloviev, A. Nanopore RNA Sequencing Revealed Long Non-Coding and LTR Retrotransposon-Related RNAs Expressed at Early Stages of Triticale SEED Development. *Plants* **2020**, *9*, 1794. [CrossRef]
9. Badji, A.; Machida, L.; Kwemoi, D.B.; Kumi, F.; Okii, D.; Mwila, N.; Agbahoungba, S.; Ibanda, A.; Bararyenya, A.; Nghituwamhata, S.N.; et al. Factors Influencing Genomic Prediction Accuracies of Tropical Maize Resistance to Fall Armyworm and Weevils. *Plants* **2021**, *10*, 29. [CrossRef]

Review

Wheat, Barley, and Oat Breeding for Health Benefit Components in Grain

Igor G. Loskutov * and Elena K. Khlestkina

Federal Research Center the N.I. Vavilov All-Russian Institute of Plant Genetic Resources (VIR), St. Petersburg 190000, Russia; e.khlestkina@vir.nw.ru

* Correspondence: i.loskutov@vir.nw.ru

Abstract: Cereal grains provide half of the calories consumed by humans. In addition, they contain important compounds beneficial for health. During the last years, a broad spectrum of new cereal grain-derived products for dietary purposes emerged on the global food market. Special breeding programs aimed at cultivars utilizable for these new products have been launched for both the main sources of staple foods (such as rice, wheat, and maize) and other cereal crops (oat, barley, sorghum, millet, etc.). The breeding paradigm has been switched from traditional grain quality indicators (for example, high breadmaking quality and protein content for common wheat or content of protein, lysine, and starch for barley and oat) to more specialized ones (high content of bioactive compounds, vitamins, dietary fibers, and oils, etc.). To enrich cereal grain with functional components while growing plants in contrast to the post-harvesting improvement of staple foods with natural and synthetic additives, the new breeding programs need a source of genes for the improvement of the content of health benefit components in grain. The current review aims to consider current trends and achievements in wheat, barley, and oat breeding for health-benefiting components. The sources of these valuable genes are plant genetic resources deposited in genebanks: landraces, rare crop species, or even wild relatives of cultivated plants. Traditional plant breeding approaches supplemented with marker-assisted selection and genetic editing, as well as high-throughput chemotyping techniques, are exploited to speed up the breeding for the desired genotypes. Biochemical and genetic bases for the enrichment of the grain of modern cereal crop cultivars with micronutrients, oils, phenolics, and other compounds are discussed, and certain cases of contributions to special health-improving diets are summarized. Correlations between the content of certain bioactive compounds and the resistance to diseases or tolerance to certain abiotic stressors suggest that breeding programs aimed at raising the levels of health-benefiting components in cereal grain might at the same time match the task of developing cultivars adapted to unfavorable environmental conditions.

Keywords: barley; breeding; marker-assisted selection; genes; genetic resources; genome editing; health benefits; metabolomics; oat; QTL; wheat

Citation: Loskutov, I.G.; Khlestkina, E.K. Wheat, Barley, and Oat Breeding for Health Benefit Components in Grain. *Plants* **2021**, *10*, 86. https://doi.org/10.3390/plants10010086

Received: 3 December 2020
Accepted: 30 December 2020
Published: 3 January 2021

1. Introduction

Cereal crops are the main food and feed sources worldwide, supplying more than half of the calories consumed by humans [1]. An overwhelming majority of plant breeders and geneticists work on no other crops but cereals. Breeding methods depend on the biological features of a crop and on the genetic research standards, traditions, economic objectives, and levels of agricultural technologies in the country where plant breeding is underway. The general breeding trend of the past decades, however, was finding solutions to the problem of higher yields in cereal crops; furthermore, special attention was paid in many countries to increasing plant resistance against diseases and various abiotic stressors. The concentration of all efforts on these two targets and none other resulted in a certain decline in the genetic diversity in those plant characters that are associated with the biochemical composition of cereal grain [2]. In the last few years, cereal crop breeding generated a trend aimed at combining high biochemical and agronomic parameters in one cultivar [3–5]. In addition to protein, cereal grains are rich in other chemical compounds, such as fats with their good assimilability by the organism and a well-balanced composition of chemical constituents, including fatty acids [6–10], vitamins of the B, A, E, and F groups, organic compounds of iron, calcium, phosphorus, manganese, copper, molybdenum, and other trace elements [3], and diverse biologically active compounds–polysaccharides, phenolic compounds, carotenoids, tocopherols, avenanthramides, etc.

In recent years, the world food market has seen the emergence of a wide range of new cereal crop products designed for dietetic purposes. Currently, available data confirm the importance of biochemical composition in cereal crop grains since it underpins their dietetic, prophylactic, and curative effect on the human organism [11]. Cereals are rich in protein, starch, oils, vitamins, micronutrients, and various antioxidants. The research that examines the potential of a number of cereal crops for prophylactic or medicinal uses has been expanding from year to year [12–16]. In addition to determining types of bioactivity for different grain components, an important challenge is to concentrate further efforts of researchers on disclosing the mechanisms of their effect [17].

It is admitted that breeding techniques can help to increase the percentage of individual constituents in the grain to a very high level. An important role in promoting this breeding trend is played by the achievements in modern genetics of cereal crops and traits associated with the quality and dietary value of their products. New breeding programs imply that the developed high-yielding cultivars will combine maximum contents of the abovementioned components and optimal correlations among them with other grain quality indicators and resistance to biotic stressors. Marker-assisted selection techniques are used more and more often to accelerate the development of cultivars enriched in useful grain components [4,18]. There are examples of the works employing genetic editing technologies for these purposes [19–21]. The current review aims to consider current trends and achievements in wheat, barley, and oat breeding for health-benefiting components.

2. Major Dietary Components in Grain and Breeding Programs for Health Benefit
2.1. Micronutrients

The long-standing problem of micronutrient deficiencies in human diets is the most significant for public healthcare worldwide. It is especially true for cereal-based diets: They are poor in both the number of micronutrients and their bioavailability for the organism since breeding of these major food and feed crops primarily aims at developing higher-yielding varieties to meet global demand. Due to dilution effects, an increase in grain mass sometimes causes a reduction in micronutrient contents. In most countries, people eat meals produced from cereal crops with low micronutrient content; it is a serious global problem invoked by the uniformity of different diets and may lead to significant health deteriorations [22,23]. Iron-deficiency anemia is one of the most widespread health disorders provoked by the worldwide deficit in micronutrients [24], while zinc deficiency in food is faced on average by one-third of the world's population [25]. Increasing the content of these trace elements in wheat by breeding techniques is considered one of the ways to enhance the consumption of micronutrients with food [26].

It has been noticed that cereal crop cultivars can be enriched in the desired micronutrients through the application of agricultural practices or by plant breeding [22,27–30]. Such procedures, however, might lead to an increase in micronutrient content in leaves but not in grain [31]. Methods combining breeding and agrochemical approaches were proposed to solve this problem: They helped accumulate micronutrients in the edible parts of plants [27–29,32]. There are considerable variations in the concentration of micronutrients in seeds or kernels of most crops [3,32]. Genetic variability in the micronutrient content is often observed to be less expressed in fruit and more in leaves. Nevertheless, screening large collections of staple cereal crops reveals extensive diversity of micronutrient concentrations in their grains [26,32,33]. Increased content of most micronutrients was observed in local varieties and landraces of wheat and other cereals, compared with improved commercial cultivars [34].

The content of micronutrients in grain was analyzed in 65 commercial Russian cultivars of four major cereal crops: wheat, barley, rye, and oat. Statistically significant variations were found in the content of all studied trace elements (Fe, Zn, and Mn). The highest levels were registered for barley and oat cultivars. Among barley genotypes, the content of Fe, Zn, and Mn varied with a 3-to 5.5-fold difference between the extremes (Table 1). Oat cultivars manifested a 7-fold difference between the extremes in the Zn content and nearly 3-fold in Mn [3].

Table 1. Average values and ranges for the content of micronutrients (Fe, Mn, Zn) in caryopses of cereal crops [3].

Crops	Content, mg/kg		
	Fe	Mn	Zn
Winter soft wheat (*Triticum aestivum* L.)	**21.8** (19−4)	**4.3** (3.3−4.9)	**17.1** (13−21)
Spring soft wheat (*T. aestivum*)	**17.5** (15−22)	**3.3** (2.4−4.1)	**19.2** (14−22)
Soft wheat (mean)	**19.7** (15−24)	**3.8** (2.4−4.9)	**18.2** (13−22)
Winter and spring rye (*Secale cereale* L.)	**20.3** (14−30)	**4.2** (2.6−7.0)	**18.4** (15−24)
Spring barley (*Hordeum vulgare* L.)	**33.2** (24−79)	**10.1** (7−21)	**10.6** (6−33)
Oats (*Avena sativa* L.)	**26.7** (19−37)	**6.1** (3.5−9.9)	**26.3** (10−70)

A detailed study of a set of commercial oat cultivars of different geographical origin in the context of their micronutrient content and biochemical parameters showed that genotypic differences in the Fe and Zn levels in grain were small (1.9–2.7 times), but in Mn, they were relatively high (10.5 times). A 1.8-fold difference was observed between the lowest (10.9%) and the highest (19.3%) protein content levels in oat grain [3]. A wide range of variation in oil content (2.7–8.1%) was found in all studied oat accessions. The amounts of protein, oil, oleic acid, and Zn in grain demonstrated statistically significant positive correlations among themselves [3]. The identified oat cultivars with high nutritive value will be included in breeding programs and used directly in high-quality food production.

Molecular-genetic research on 335 spring barley accessions was conducted for more effective utilization of the micronutrient diversity in cereal crop breeding. A genome-wide association study (GWAS) was employed for mapping quantitative trait loci (QTL) linked to the content of macro- and micronutrients in grain (Fe, Zn, Ba, Ca, Cu, K, Mg, Mn, Na, P, S, Si, and Sr). The analyses of the tested populations helped to identify specific QTL for each of the studied indicators and map them on chromosomes. The QTL identified are valuable for the future development of barley cultivars with increased content of nutrients, especially Zn and Fe [35].

2.2. β-glucans

A physiologically important dietary component in the grain is (1,3;1,4)-β-D-glucan, or the non-starchy water-soluble polysaccharide β-glucan. This component is reported to be typical of some species of the Poaceae family: its content varies within 3–11% in barley, 1–2% in rye, and <1% in wheat, while in other cereals, it is present only in trace amounts [36]. At the same time, the content and composition of dietary fibers in various cereal crop species are genetically determined. It means, as opined by many scientists, that it is possible to produce new lines of such crops with different correlations between the levels of β-glucan polysaccharides and arabinoxylans that would be optimal for various uses [37–39]. Studying of the β-glucan content in oat and barley cultivars is associated with their uses for dietetic and medical purposes [37,38].

The β-glucans are not evenly distributed within a grain: its larger amount is found in the endosperm cell walls, aleurone, and subaleurone layers, and its content varies from 1.8 to 7% [40,41]. The concentration of β-glucans in oat grain and their degree of polymerization depend not only on the cultivar but also on the conditions of cultivation, grain processing, and post-harvest storage [42].

The presence in the grain of a higher amount of β-glucans, which are dietary soluble fiber (or soluble non-starch polysaccharide), determines the viscosity of oat and barley broths, which have a beneficial effect on important functions of the human gastrointestinal tract, so they are widely used in the food industry for dietetic and curative purposes [36,43]. Among numerous products of barley and oat biosynthesis, probably the most valuable for the human organism is soluble cellulose fibers and β-glucans first of all (also arabinoxylan, xyloglucan, and some other secondary cellulose components), as they can reduce the level of cholesterol in the blood and noticeably mitigate the risk of cardiovascular diseases [38,44,45]. Multiple evidence of the beneficial role played by β-glucans impelled the U.S. Food and Drug Administration (FDA) to make an official statement that soluble dietary fibers extracted from whole oat grain to produce flakes, bran, or flour helped to reduce the risks of cardiovascular diseases [46]. Insoluble fractions of dietary fiber are partly cellulose, xylose, and arabinose [39]. Insoluble dietary fiber has general gastrointestinal effects and, in most cases, has an impact on weight loss. There is convincing evidence that β-glucans contained in oat grain are partially responsible for decreasing the levels of glucose in the human blood and of cholesterol in serum [12]; it is associated with its physicochemical and rheological characteristics, such as molecular weight, solubility in water, and a viscosity [42,47].

Genetic diversity of barley and oats in the content of β-glucans in their grain was evaluated in the framework of two European Union (EU) programs. The HEALTHGRAIN Diversity Screen project resulted in finding significant differences in the content of β-glucans and antioxidants in the grain of five tested oat cultivars [48]. The AVEQ project (*Avena* genetic resources for quality in human consumption) analyzed 658 oat cultivars and confirmed the contribution of both genetic and environmental aspects to the formation of the tested character [49]. It is worth mentioning that, compared with cultivated and other wild di- and tetraploid oat species, higher contents of β-glucans and other antioxidants were found in the hexaploid (wild) *A. fatua*, *A. occidentalis*, and (cultivated) *A. byzantina*, and diploid (wild I) *A. atlantica* [38,39,49–51].

Measuring the content of β-glucans in oat grain in large and diverse sets of cultivars and species showed that its values were significantly dispersed [37,38,49]. Naked oat forms demonstrated a higher total content of the analyzed polysaccharide than hulled ones, but the latter contained more insoluble β-glucans in their grain [52–54]. Computer modeling helped to provide a ranking of the factors affecting the β-glucan content in hulled and naked oat cultivars during their cultivation. The analysis showed that the selection of the cultivar is the most important parameter of the model for determining the final β-glucan accumulation in grain, among the other factors [55]. There are contradictory data concerning the results of comparative studies on naked and hulled barley as well. Some authors failed to disclose significant differences between these two forms of the crop [56,57], while others found that naked barleys contained more β-glucans than hulled ones [43,58]. Meanwhile, the group of Tibetan naked barleys was reported to have the highest content of β-glucans in their grain [59].

In the meantime, the amount of β-glucans in oat grain is associated with protein and fat accumulation, grain volume weight, and grain productivity [60,61]. The content of these polysaccharides depended on meteorological conditions and agricultural practices used in oat cultivation [61]. The content of β-glucans in barley grain is determined by both the genotype and the growing conditions [43,59,62]; some authors insist that it is the genotype that plays a decisive role [63,64], while others give preference to the environmental conditions [65,66]. When 33 barley cultivars and lines were tested in two arid areas in the United States, it was shown that the variability in the content of β-glucans in grain was determined by the genotype for 51% [64] to 66% [67]. At the same time, the protein content in grain depended on environmental conditions for 69%, whereas yield size and the grain volume weight for 83 and 70%, respectively [64]. The study of 9 barley cultivars and 10 oat ones showed that cultivar-specific differences in the β-glucan content persisted across the years [63]. The content of β-glucans in the grain is also influenced by plant development phases. It was reported that the content of β-glucans gradually increased in the process of grain formation, and in the maturation phase, it either reached the plateau or decreased [57]. At present, there are contradictory data concerning the linkage of β-glucan accumulation in barley grain with 1000 grain weight, protein content, or starch content [56,62]. Some authors did not find any interplay between these characters, while others reported a positive correlation. When the content of β-glucans was measured in the grain of six-row and two-row barley cultivars, no differences between these two cultivar groups were reported [43].

The 1700 oat lines with mutations induced by TILLING of high-frequency mutagenesis have been produced for breeding purposes with molecular-based, high precision selection methods from cv. 'Belinda' (Sweden) to evaluate the variability of β-glucans content in this crop [68]. Their assessment resulted in identifying 10 lines with β-glucan concentrations in their grain higher than 6.7% and 10 lines with the content of β-glucans less than 3.6% (β-glucan concentration in cv. 'Belinda' was 4.9%). The maximum range of variation in the content of these polysaccharides was from 1.8 to 7.5% [69]. The comparatively recent identification of genes participating in the biosynthesis of β-glucans in cereals [70] and their first genetic map open new opportunities for genetic improvement of grain quality indicators and resulting food products, which is very important for human health [71].

Three markers (Adh8, ABG019, and Bmy2) significantly linked to β-glucan content regulation were identified in barley grain, and a group of *HvCslF* genes was mapped: At least two of them were in the region of barley chromosome 2H explained by the QTL for (1,3;1,4)-β-glucan near the Bmy2 marker [72]. A genome-wide association study (GWAS) employing oat germplasm of worldwide origin from the American Gene Bank was aimed at the identification of QTL linked to β-glucan content in grain and resulted in finding three independent markers closely associated with the target character. A comparison of these results with the data obtained for rice showed that one of the described markers, localized on rice chromosome 7, was adjacent to the *CslF* gene family responsible for β-glucan synthesis in grain. Thus, GWAS in oat can be a successful QTL detection technique with the future development of higher-density markers [73].

By now, the GWAS approach has already started to be used to analyze the association between the genotype and the content of β-glucans and fatty acids in oats. Researchers have identified four loci contributing to changes in the fatty acid composition and content in oat grain. However, genome regions conducive to changing the content of proteins, oils, saccharic and uronic acids, which, in their turn, produce a direct effect on grain quality, remain unexplored [74]. Furthermore, positive correlations were demonstrated in barley between 1000 grain weight and tocol concentration, between dietary fiber content and phenolic compounds, and between husk weight and total antioxidants in hulled barley [38,50].

2.3. Antioxidants

Cereal crop grains are known to have high nutritive value and contain diverse chemical compounds with antioxidant properties. Research efforts have been undertaken in recent years to study the content of antioxidants in the grain of various cultivated cereals [50,75–79].

Starting in the mid-1930s, oat flour has been used as a natural antioxidant. Later, more in-depth research was done to assess the antioxidative properties of oat flour versus those of chemical antioxidants. It was ascertained that adding sterols extracted from oat to heated soybean oil significantly decelerated its oxidation compared with the reference. At present, along with the extensive utilization of synthetic antioxidants, oat flour has found its stable niche as a natural ingredient in eco-friendly food products [7].

A comparison of bakery products made from wheat that synthesized such antioxidant compounds as anthocyanins with those from an anthocyanin-free wheat line demonstrated that the presence of anthocyanins increased the shelf life of bakery products and their resistance to molding under provocative conditions [80]. Cereal crops contain secondary metabolites with antioxidant activity belonging to three groups: phenolic compounds, carotenoids, and tocopherols [81].

2.4. Phenolic Compounds and Avenanthramides

Oat and barley grains contain a considerable amount of various phenolic compounds exhibiting biological activity, including antioxidative, anti-inflammatory, and antiproliferative (preventive activity against cancerous and cardiac diseases) effects [50]. One of the most abundant and powerful antioxidants found in nature, the flavonoid quercetin, has been found in wheat. It is characterized by numerous biological effects, including antithrombotic activity [82].

Many published studies testify that a major part of phenolic compounds in grain occurs in a bound form: Their content in oat and wheat grains reaches 75% [83,84]. Phenolic acids, like most flavonoids in cereal crops, are concentrated in structures bound to the cell wall: 93% of the total flavonoid content in wheat and 61% in oats [83]. The highest level of total flavonoids is characteristic of maize grain, followed by wheat, oats, and rice [83]. Phenolic acids are the most widespread phenolic compounds in oats, especially ferulic acid (250 mg/kg), which is present mainly inbound forms linked through ester or ether bonds to cell wall components but also exists in the free form [85].

Bioactive chemical compounds are unevenly distributed within the grain. Grains of four naked barley cultivars were divided into five layers to measure the total phenolic content and total antioxidant activity. The total content of soluble phenolic compounds was observed to decrease from the outer layer (2.8–7.7 µg/g) towards the inner endosperm structures (0.87–1.35 µg/g) [78,86]. It has been proven that most antioxidants contained in whole grain are located in the bran and germ fractions of the grain. For example, whole-grain wheat flour was found to contain in its bran/germ fraction 83% of the total phenolic content in grain and 79% of total flavonoids [87].

In the study of molecular mechanisms of 'melanin-like' black seed pigments known to be strong antioxidants, comparative transcriptome analysis of two near-isogenic lines differing by the allelic state of the *Blp* (black lemma and pericarp) locus revealed that black seed color is related to the increased level of ferulic acid and other phenolic compounds [88]. The melanic nature of the purified black pigments was confirmed by a series of solubility tests and Fourier transform infrared spectroscopy, while intracellular pigmented structures were described to appear in chloroplast-derived plastids designated "melanoplasts" [89]. The most frequently mentioned flavonoids of cereal crops are the flavonols kaempferol and quercetin, the flavanone naringenin and its glycosylated forms, catechin, and epicatechin in barley [90–93].

Pigmentation of the grain's outer coating can be analyzed as an important indicator of antioxidant activity. A barley cultivar with purple grain contained 11 anthocyanins, while only one anthocyanin was observed in black and yellow barley grains. The purple barley bran extract had the highest total antioxidant activity [94]. Another study of naked barley demonstrated the presence of higher antioxidant activity in pigmented grains compared with non-pigmented ones [78]. A study of naked and hulled oats showed that naked oat cultivars had significantly higher values of total antioxidant activity. Among hulled oat cultivars, these values were higher in dark-hulled forms compared with white-hulled oats [50].

Differences between naked and hulled oats and barleys, generated a perfect model interesting for comparative analyses: the mutant barley line for the *Nud* gene (nakedness), derived by gene editing from cv. 'Golden Promise' [21]. Using this model will help to distinguish the pleiotropic effects of the *Nud* gene on the grain's biochemical composition from the influence of closely linked genes.

Analyzing grain extracts of wheat lines with different combinations of the *Ba* (*Blue aleurone*) and *Pp* (*Purple pericarp*) genes on the genetic background of elite cultivars demonstrated a higher diversity of flavonoid compounds in the carriers of dominant alleles of *Ba* and *Pp* genes. Comparing the products made from the grain of a purple-grained line with those from an anthocyanin-free isogenic line revealed significant differences, which was also true for the samples that had passed a full processing cycle, including baking at elevated temperatures [80,95]. The analysis of anthocyanin extracts obtained under conditions simulating those of food digestion by a human organism showed that ingesting 100 g of bread crisps or biscuits made from flour with added purple wheat grain bran raised the assimilation of anthocyanins to 1.03 and 0.83 mg, respectively, i.e., 100 g of bran would supply the organism with up to 3.32 g of anthocyanins. Besides, purple-grained wheat matched or even exceeded the reference line in the quality and taste of its products [95].

Recently, new high-yielding wheat cultivars, resistant to fungal diseases and having high anthocyanin content in grain have been developed [4]. The efficiency of the breeding strategy lasting only three years from the first cross until the state cultivar competitive testing has been demonstrated. The strategy is based on marker-assisted selection (MAS) [4]. MAS also demonstrated its efficacy in creating barley with certain alleles of anthocyanin regulatory genes [18]. For breeding blue-grained wheat, besides molecular markers, FISH or C-banding are needed since the *Ba* gene is alien for wheat and can be inherited from wheat lines with either 4B or 4D chromosome substituted by the *Thinopyrum ponticum* chromosome 4 [96,97]. Unlike bread wheat, barley has its own *Ba* gene. Recent findings of regulatory features of anthocyanin biosynthesis in barley [98] are useful for both MAS-based and genetic editing-based breeding strategies.

Interestingly, 30 years ago, the purple- and blue-grain characters were regarded as having "a limited practical use from a scientific point" [99]. Since that time, some studies demonstrating the health benefit of plant anthocyanins, including those from wheat grain [16], have been carried out, denying the old point of view and proving these traits to be economically important. Commercial cultivars of wheat with increased anthocyanin content have been released in Canada, China, Japan, and several European countries [100,101].

The class of phenolics with antioxidative effect and bioactivity includes avenanthramides (AVA), a class of hydroxycinnamoyl anthranilate alkaloids contained only in oats. Twenty-five components of these compounds were detected in kernels, and twenty in hulls [102]. The most widespread in oats are AVA-A (2p), AVA-B (2f), and AVA-C (2c) [9,103,104]. There is documented evidence that avenanthramides demonstrate antioxidant, anti-inflammatory, antiatherogenic, and antiproliferative activity [105–107].

It has been shown that oat cultivars differed in the AVA content in grain. The cultivated diploid species *A. strigosa* had a very high AVA content reaching 4.1 g/kg, and the hexaploid *A. byzantina* contained 3.0 g/kg. Contrariwise, wild oat species with different ploidy levels were characterized by relatively low AVA content values (240–1585 mg/kg) [108]. Analyzing a representative set of cultivated and wild oat species revealed an even wider diversity of the AVA content in grain [109]. A conclusion has been made that wild oat species are an important source of diversity for breeding programs, which dictates the necessity of further studies into the pattern of AVA content and composition variability across the genus *Avena* L. Wild oat species might incorporate a unique AVA composition, promising for crosses with cultivated oats.

2.5. Tocols

The health benefits of oats are also associated with the presence of several antioxidant compounds known as tocols, specifically tocopherols and tocotrienols. The fat-soluble vitamin E contains tocopherols and tocotrienols [110], which make the oil more resistant to oxidation. Both tocopherols and tocotrienols have several isomeric forms designated as α, β, γ, and δ [111]. All in all, vitamin E can comprise eight isomers, with prevailing α-isomers (70–85%) and δ-isomers not exceeding 1%. The total tocopherol content in oat cultivars can reach 2.6–3.2 mg/100 g, which is many times lower than in barley [101]. Tocopherols are mainly present in the germ fraction of grain, while tocotrienols are found in the pericarp and endosperm. Tocotrienols prevail in oats, barley, and wheat; their concentrations vary from 40 to 60 μg/g depending on the crop [112].

Eight isomers of tocols have been found in barley grain oil (four tocopherols and four tocotrienols). They play an exceptionally important role, regulating cholesterol in human blood. Tocols also demonstrate very high activity as antioxidants, blocking harmful peroxidation of lipids in cell membranes [101]. Tocols (16–94 mg/kg) consist of a polar chromanol ring linked to an isoprenoid-derived hydrocarbon chain. They are powerful scavengers of free radicals, also demonstrating an ability to inhibit the proliferation of some cancer cells [108].

Furthermore, positive correlations were demonstrated in barley between 1000 grain weight and tocol concentration, between dietary fiber content and phenolic compounds, and between husk weight and total antioxidants in hulled barley [38,50]. Presently, molecular-genetic studies of this type of antioxidant are based on simple-sequence repeats (SSR) markers. It is worth mentioning that the naked barley with the *Waxy* gene and zero amylase content in starch has higher contents of both β-glucans and tocols [113].

2.6. Sterols

Sterols are important components of vegetable oils. Their content in oat grain varies, according to different sources, from 0.1% to 9.3% of the total fatty acid content. This indicator often depends not only on the oat genotype but also on the extraction technique. Cultivars of rye, wheat, barley, and oats grown in the same year and same location were compared, the highest plant sterol content was observed in rye (mean content 95.5 mg/100 g, wb), whereas the total sterol contents (mg/100 g, wb) of wheat, barley, and oats were 69.0, 76.1, and 44.7, respectively [114]. Among the six components of sterol content, the main one is sitosterol, whose content reaches 70% of the total sterol content; additionally, about 20% are allocated to campesterol and stigmasterol [7,101]. The content of sterols in oats can reach 447 mg/kg and include, in addition to the abovementioned, D-5 and D-7 avenasterols [114] and phytic acid (5.6–8.7 mg/g); the latter manifests antioxidant activity due to its ability to chelate metal ions, thus making them catalytically inactive and inhibiting the metal-mediated formation of free radicals. However, this chelating activity reduced the bioavailability of major minerals [110].

2.7. Carotenoids

Carotenoids (yellow, orange, and red pigments) relating to isoprenoids are among the most widespread plant antioxidants. Carotenoid content in oat grain can reach 1.8 μg/g [86]; besides, lutein is considered the main xanthophyll in wheat, barley, and oat grains, and zeaxanthin is the secondary one [115].

Comparative investigation of four groups of wheat genotypes (spelt wheat, landraces, old cultivars, and primitive wheat) for carotenoid content and composition in grain revealed a high level of variation among the genotypes and the groups in the content of carotenoids. Lutein contributed 70–90% of the carotenoids in the grain [116]. In durum wheat, which is used for the production of pasta, carotenoid content is also an important technological and market indicator. In semolina and pasta, a yellow color is desirable, and it depends on the carotenoid accumulation in kernels. Genetic dissection of the carotenoid content character showed quantitative trait loci (QTL) on all wheat chromosomes [117]. The major QTL, responsible for 60% of heritability, is located on the long arms of chromosomes 7A and 7B. Variability in these QTL is explained by allelic variations of the phytoene synthase (PSY) genes. Molecular markers for MAS-based breeding programs aimed at the enrichment of durum wheat grain with carotenoid content are available [117].

2.8. Other Antioxidant Compounds

Oat is the only cereal grass that contains saponins, steroidal glycosides known as avenocosides A and B (65.5 and 377.5 mg/kg, respectively), which exhibit anticancer activity at the expense of diverse, complex mechanisms, including inhibition of neoplasm cell growth through cell cycle arrest and, inter alia, stimulation of cancer cell apoptosis [13]. Oat also accommodates two classes of saponins: avenocosides (steroid-linked saccharides) and avenacides (triterpenoid-linked saccharides), which were shown to drop the cholesterol level, stimulate the immune system, and demonstrate anticancer properties [14]. Targeted breeding for increased content of these compounds in oat lines has not yet been attempted, but interline and interspecies differences in this indicator have already been identified [118]. Grains of five Finnish barley cultivars grown in 2006–2008 were analyzed for their total content of folic acid. It was noted that the external and germ-containing grain layers had the highest levels of this compound (up to 1710 ng/g) [77,79].

3. Assessment of Cereal Crop Genetic Resources According to the Diversity and Concentration of Health-Friendly Dietary Grain Components

Secondary metabolites associated with quality traits in the released and processed products are presently identified using metabolomic profiling or chemotyping. Such an approach enables researchers to evaluate plant genetic resources according to these traits, including varieties of cultivated species and populations of wild ones. Chemotyping the grain of cultivated and wild *Avena* L. spp. showed that the range of variability in the metabolomic profile of improved cultivars was significantly narrower than that of wild species. Metabolites, the content of which may have been reduced in the process of domestication and breeding in comparison to wild oats, are identified [2]. Presumably, it might be connected with the selection during oat domestication and a decline of metabolome diversity while "domestication syndrome" traits were shaped [119]. The diversity of metabolomic profiles may be lost in the process of selection when highly specialized single-line intensive-type cultivars are developed because this process is always accompanied by a decrease in genetic polymorphism in a breeding object compared with the metagenome of numerous ecotypes, local varieties, and natural races of dozens of wild species [2,119]. A study of naked and hulled oat forms disclosed differences in their metabolites, which serves as an additional justification of the differentiation between these subspecies of common oat [2]. Landraces, which are plant varieties selected and grown regionally but not officially tested and released as registered varieties, are a source of special genetic characteristics derived by (many years of) adaptation to the respective territory. Such local varieties are often more resistant to biotic and abiotic stresses typical for their environment. In addition, such varieties may be a source for special phytochemicals (also known as bioactives) considered as health-beneficial, while the content of these compounds may be lower in commercial cultivar [2,120].

The bands of secondary metabolites in oat accessions exposed to *Fusarium* infection were analyzed, and correlations between metabolites and resistance were disclosed. High-protein oat forms with increased content of certain secondary metabolites demonstrated less damage from *Fusarium*, accumulated fewer toxins, and were more adaptable to the biotic stress [121].

Matthews et al. [122] used metabolite profiling to compare 45 lines of tetraploid and hexaploid wheat. The extracts were analyzed by the ultraperformance liquid chromatography coupled with time-of-flight mass spectrometry (UPLC-TOF-MS). Two different species of bread and durum wheat formed two distinct groups differing in sterols, fatty acids, and phospholipids, while *T. aestivum* L. split into two groups (corresponding to hard and soft bread wheat) according to differences in heterocyclic amines and polyketides. This and similar studies underpin the use of chemotyping in breeding both for desired agronomic traits and for higher contents of health-benefiting compounds in cereal grain.

Information obtained with the molecular metabolomic approach on mQTL (metabolite quantitative trait loci) and mGWAS (metabolome-based GWAS) ensures a new level for qualitative and quantitative characterization of secondary metabolites interesting for breeding. Such analyses can provide knowledge about the interactions among metabolites themselves and between them and important breeding indicators. It may lead to the development of more rational models linking a certain metabolite with such characters as plant productivity or end-product quality. Even more promising is the possibility to examine the interplay between quantitative variation in metabolites and changes in the plant phenotype [123].

Due to the genetic potential of grain crops through the directed formation of the properties and structure of the kernel in the process of ontogenesis, when developing new cultivars, it is possible to attend to the target component composition of the final product. Wider application of chemotyping, chemical research methods, metabolomic analysis of grain quality, and searching for high content of rare beneficial (dietary or curative) components will result in the release of new crop cultivars, thus promoting next-generation breeding trends and technologies [50].

4. The Effect of Dietary Components in Grain on Life Functions of Plants Themselves

Content of all biochemical components in the grain of cereal crops there are variations in the composition of it. These variations arise from differences between environments, variation in the genotype of the crop, and interactions between biotic and abiotic factors and genotype. Biotic and abiotic factors change depending on climate change, soil, and various stressors affecting plants. The genotypic variation includes the differences between individual genotypes.

4.1. Biotic Stress Resistance

Generally, an explanation why grain in the soil is not affected by microorganisms despite the environmental conditions favorable for infection was given by the presence of antimicrobial flavonoid compounds in extracts from barley and wheat grains soaked in water [124]. Higher disease resistance of plants with enhanced flavonoid biosynthesis has been described in rye, barley, and wheat [125]. In vitro infection of developing barley caryopses of wild type and proanthocyanidin-free mutants with fungal pathogens *Fusarium poae*, *F. culmorum*, and *F. graminearum* revealed mutants to be more sensitive to *Fusarium* attack than wild-type plants [126].

Considering the available data on interactions between compounds with antioxidant properties in cereal crop kernels and *Fusarium* spp., it seems appropriate to suppose that some of the former could significantly contribute to the grain's protection mechanism against toxicogenic fungi and mycotoxin accumulation. It has been proven that the crucial role in Fusarium Head Blight (FHB) resistance is played by five main classes of antioxidant metabolites: phenolic acids, flavonoids, carotenoids, tocopherols, and benzoxazinoids [127].

Cereal crop diseases caused by pathogenic and toxicogenic species of the *Fusarium* genus (FHB) inflict serious economic losses worldwide. Therefore, the development of sustainable strategies to prevent FHB contamination and mycotoxin accumulation has become a target of intensive research in recent years, and the use of FHB-resistant genotypes has been chosen as one of the prioritized trends in breeding practice [121,128,129]. Even now, however, the knowledge of complex mechanisms regulating resistance in cereal crops is still insufficient, and selecting resistant genotypes remains a difficult task for breeders. It has been established that, in addition to their fungicidal properties, a number of antioxidant secondary metabolites in cereals can regulate mycotoxin production by various pathogenic fungi [127].

The first weighty general argument in favor of phenolic compounds, carotenoids, and tocopherols is their ability to suppress reactive oxygen species (ROS), thus protecting biological cells. Besides, tocopherols and carotenoids can entrap free radicals of lipid peroxides and, therefore, arrest lipid peroxidation chain development [130]. Cinnamic acid derivatives, such as sinapic, caffeic, *p*-coumaric, chlorogenic, and ferulic acids, are effective inhibitors of *F. graminearum* and *F. culmorum* development, while benzoic acid derivatives, except syringic acid, produce an antiactivating effect [131,132]. There is an opinion that cereal crop metabolites with antioxidant activity suppress toxigenic action of a fungal infection. Numerous research works demonstrated the efficiency of phenolic compounds [133,134], carotenoids [135], tocopherols, and even benzoxazinoids [136] in restraining the growth and mycotoxin production of toxigenic *Fusarium* fungi. Finally, phenolic compounds partaking in plant structure enforcement are known to contribute to building a physical barrier against pathogenic infection. There is a positive interrelation between the content of phenolic acids, both free and bound to the cell wall, and FHB resistance in wheat [137]. A high level of FHB resistance in barley with the black-pigmented grain is supposedly associated with increased content of phenolic compounds [133].

High-protein oat forms were observed to be less affected by *Fusarium* head blight and accumulate fewer toxins; they are more adaptable to biotic stress. A relationship was identified between FHB resistance and accumulation of pipecolic acid, monoacylglycerols, tyrosine, galactinol, certain phytosterols, saccharides, and adenosine [121].

There were, however, many unproven assumptions on the participation of metabolites in the FHB resistance mechanism in cereals. Although the genetic architecture that supports secondary metabolite synthesis and regulation in cereal crops is exceptionally intricate, such proof may be retrieved in the process of comprehensive genetic and functional genomic studies [127].

Accumulation of avenanthramides in oats is also associated with the penetration of a fungal infection. Avenanthramides are mostly contained in oat grain, but under an attack by crown rust or leaf blotch, they start to synthesize in leaves as a means of protection against disease agents [110]. The fact that the amount of avenanthramides in grain significantly increases during imbibition [138], plant development [139], steeping [140], and storage [141] is also related to plant protection against potential susceptibility to pathogenic flora.

4.2. Abiotic Stress Resistance

Polyphenolic compounds in grain may protect seeds from unfavorable abiotic environmental conditions. Some of these compounds may act as sunscreens against potentially damaging UV-B radiation [142]. This may explain the presence of a purple grain color and other parts of the plant in tetraploid wheat *T. aethiopicum* Jakubz. [143] adapted to intensive solar UV-B radiation in highland areas in Ethiopia. Studies of near-isogenic wheat lines differing in the anthocyanin content in the pericarp and coleoptile under various stress conditions showed that both pericarp and coleoptile anthocyanins protected seedlings from osmotic stress [144], while protection of seedlings under a moderate irradiation dose (pre-treatment of dry seeds with 50 Gy before sowing) or moderate Cd toxicity (25 μM CdCl$_2$) was due to the coleoptile's anthocyanins only [145,146]. Flavonoid substances can prevent negative effects of excessive moisture, such as pre-harvest seed sprouting by reducing the permeability of seed coat to water [147], inhibiting α-amylase (an enzyme whose activity is directly related to seed germination of grain) [148], or inactivating dehydrogenase required for the initial phase of respiration in ripening grain and young shoots [149].

Avenanthramide accumulation in oat grain is affected by weather and geographic conditions under which the studied material is cultivated [109,150–153]. Changes in the concentration of avenanthramides in response to salinity stress in CBF3 transgenic oat demonstrated that these compounds might have a potential role in enhancing abiotic stress tolerance in oats [154]. Havrlentova et al. [155] suggested that oats with higher β-D-glucan content may have thicker and, therefore, more insulating cell walls, better adapted to heat stress conditions. The same conclusion between higher content of β-D-glucan and greater cell wall thickness has been reported in barley [156]. Sterol might be important for cold acclimation of wheat [157,158] and oat [159]. Thus, breeding programs aimed at an increase in the content of health benefit components in cereal grain are at the same time eligible to solve the task of cultivar adaptability to unfavorable environmental conditions.

5. Conclusions

Each of the abovementioned natural components (dietary or curative) is promising for use as a food additive or an ingredient of pharmaceutical and cosmetic products. They are expected to play an ever-growing role in food industries, expanding the assortment of healthy food for the population. The demand for such products has already instigated plant breeders to launch new breeding programs aimed at the development of cereal crop cultivars with higher contents of bioactive components in grain. Such programs have often been based on molecular breeding techniques from the very beginning. Screening promising cultivars and hybrids for the content of antioxidants and other bioactive compounds in the grain is required to expand and promote this breeding trend. It also seems expedient to apply simple, undamaging and, as a rule, indirect techniques of plant genotype assessment for the levels of antioxidants in the grain to increase the performance and efficiency of such screening, employing the entire genetic diversity of cereal crops for identification of contrasting initial sources for breeding food and feed cultivars. The results obtained in the process of studying already existing cereal cultivars and the achievements of plant breeding in releasing new high-yielding and high-quality cultivars enable producers to use them in the development of a wide assortment of health-friendly dietary products contributing to the physical fitness of the human organism.

Author Contributions: Conceptualization, I.G.L., E.K.K.; writing, I.G.L., E.K.K. All authors have read and agreed to the published version of the manuscript.

Funding: This research received no external funding.

Acknowledgments: The article was made with the support of the Ministry of Science and Higher Education of the Russian Federation under agreement № 075-15-2020-911 date 16.11.2020 on providing a grant in the form of subsidies from the Federal budget of the Russian Federation. The grant was provided for state support for the creation and development of a World-class Scientific Center "Agrotechnologies for the Future".

Conflicts of Interest: The authors declare no conflict of interest.

References

1. FAO Save and Grow. Available online: http://www.fao.org/ag/save-and-grow/MRW/index_en.html (accessed on 12 December 2020).
2. Loskutov, I.G.; Shelenga, T.V.; Rodionov, A.V.; Khoreva, V.I.; Blinova, E.V.; Konarev, A.V.; Gnutikov, A.A.; Konarev, A.V. Application of metabolomic analysis in exploration of plant genetic resources. *Proc. Latv. Acad. Sci.* **2019**, *73*, 494–501. [CrossRef]
3. Bityutskii, N.; Loskutov, I.; Yakkonen, K.; Konarev, A.; Shelenga, T.; Khoreva, V.; Blinova, E.; Rymin, A. Screening of *Avena sativa* cultivars for iron, zinc, manganese, protein and oil contents and fatty acid composition in whole grains. *Cereal Res. Commun.* **2019**, *48*. [CrossRef]
4. Gordeeva, E.; Shamanin, V.; Schoeva, O.; Khlestkina, E. The strategy for marker-assisted breeding of anthocyanin-rich spring bread wheat (*Triticum aestivum* L.) cultivars in West Siberia. *Agronomy* **2020**, *10*, 1603. [CrossRef]
5. Morgounov, A.; Karaduman, Y.; Akin, B.; Aydogan, S.; Baenziger, S.; Bhatta, M.; Chudinov, V.; Dreisigacker, S.; Govindan, V.; Güler, S.; et al. Yield and quality in purple grain wheat isogenic lines. *Agronomy* **2020**, *10*, 86. [CrossRef]
6. Leonova, S.; Shelenga, T.; Hamberg, M.; Konarev, A.V.; Loskutov, I.; Carlsson, A.S. Analysis of oil composition in cultivars and wild species of oat (*Avena sativa*). *J. Agric. Food Chem.* **2008**, *56*, 7983–7991. [CrossRef]
7. Marshall, H.G. *Oats Science and Technology (Agronomy)*; Sorrells, M.E., Ed.; American Society of Agronomy: Madison, WI, USA, 1992; p. 846.
8. Olson, R.A. *Nutritional Quality of Cereals Grains: Genetic and Agronomic Improvement*; Frey, K.J., Ed.; American Society of Agronomy: Madison, WI, USA, 1987.
9. Peterson, D.M. Oat–a Multifunctional Grain. In Proceedings of the 7th International Oats Conference, Helsinki, Finland, 28 July 2004; pp. 21–26.
10. Rasmusson, D.C. *Nutritional Quality of Barley*; American Society of Agronomy: Madison, WI, USA, 1985.
11. Malaguti, M.; Dinelli, G.; Leoncini, E.; Bregola, V.; Bosi, S.; Cicero, A.F.G.; Hrelia, S. Bioactive peptides in cereals and legumes: Agronomical, biochemical and clinical aspects. *Int. J. Mol. Sci.* **2014**, *15*, 21120–21135. [CrossRef]
12. Zou, Y.; Liao, D.; Huang, H.; Li, T.; Chi, H. A systematic review and meta-analysis of beta-glucan consumption on glycemic control in hypercholesterolemic individuals. *Int. J. Food Sci. Nutr.* **2015**, *66*, 355–362. [CrossRef]
13. Yang, J.; Wang, P.; Wu, W.; Zhao, Y.; Idehen, E.; Sang, S. Steroidal saponins in oat bran. *J. Agric. Food Chem.* **2016**, *64*, 1549–1556. [CrossRef]
14. Sang, S.; Chu, Y. Whole grain oats, more than just a fiber: Role of unique phytochemicals. *Mol. Nutr. Food Res.* **2017**, *61*, 1600715. [CrossRef]
15. Kumar, K.; Chauhan, D.; Kumar, S. Barley: A potential source of functional food ingredients. In Proceedings of the Conference: National Seminar on Technological Interventions in Food Processing and Preservation, Jaipur, India, 17 November 2017.
16. Tikhonova, M.; Shoeva, O.; Tenditnik, M.; Ovsyukova, M.; Akopyan, A.; Dubrovina, N.; Amstislavskaya, T.G.; Khlestkina, E. Evaluating the effects of grain of isogenic wheat lines differing in the content of anthocyanins in mouse models of neurodegenerative disorders. *Nutrients* **2020**, *12*, 3877. [CrossRef]
17. Martınez-Villaluenga, C.; Penas, E. Health benefits of oat: Current evidence and molecular mechanisms. *Curr. Opin. Food Sci.* **2017**, *14*, 26–31. [CrossRef]

18. Gordeeva, E.I.; Glagoleva, A.Y.; Kukoeva, T.V.; Khlestkina, E.K.; Shoeva, O.Y. Purple-grained barley (*Hordeum vulgare* L.): Marker-assisted development of NILs for investigating peculiarities of the anthocyanin biosynthesis regulatory network. *BMC Plant Biol.* **2019**, *19*, 52. [CrossRef] [PubMed]

19. Sánchez-León, S.; Gil-Humanes, J.; Ozuna, C.V.; Giménez, M.J.; Sousa, C.; Voytas, D.F.; Barro, F. Low-gluten, nontransgenic wheat engineered with CRISPR/Cas9. *Plant Biotechnol. J.* **2018**, *16*, 902–910. [CrossRef] [PubMed]

20. Jouanin, A.; Schaart, J.G.; Boyd, L.A.; Cockram, J.; Leigh, F.J.; Bates, R.; Wallington, E.J.; Visser, R.G.F.; Smulders, M.J.M. Outlook for coeliac disease patients: Towards bread wheat with hypoimmunogenic gluten by gene editing of α- and γ-gliadin gene families. *BMC Plant Biol.* **2019**, *19*, 333. [CrossRef] [PubMed]

21. Gerasimova, S.; Hertig, C.; Korotkova, A.; Kolosovskaya, E.; Otto, I.; Hiekel, S.; Kochetov, A.; Khlestkina, E.; Kumlehn, J. Conversion of hulled into naked barley by Cas endonuclease-mediated knockout of the NUD gene. *BMC Plant Biol.* **2020**, *20*, 255. [CrossRef] [PubMed]

22. Kennedy, G.; Nantel, G.; Shetty, P. The scourge of "hidden hunger": Global dimensions of micronutrient deficiencies. *Food Nutr. Agric.* **2003**, *32*, 8–16.

23. Bhullar, N.K.; Gruissem, W. Nutritional enhancement of rice for human health: The contribution of biotechnology. *Biotechnol. Adv.* **2013**, *31*, 50–57. [CrossRef]

24. WHO. *The World Health Report: 2002: Reducing Risks, Promoting Healthy Life*; World Health Organization: Geneva, Switzerland, 2002.

25. Hotz, C.; Brown, K.H. Assessment of the risk of zinc deficiency in populations and options for its control. *Food Nutr. Bull.* **2004**, *25*, 94–204.

26. Morgounov, A.; Gomez-Becerra, H.F.; Abugalieva, A.; Dzhunusova, M.; Yessimbekova, M.; Muminjanov, H.; Zelenskiy, Y.; Ozturk, L.; Cakmak, I. Iron and zinc grain density in common wheat grown in Central Asia. *Euphytica* **2007**, *155*, 193–203. [CrossRef]

27. Bouis, H.E. Micronutrient fortification of plants through plant breeding: Can it improve nutrition in man at low cost? *Proc. Nutr. Soc.* **2003**, *62*, 403–411. [CrossRef]

28. Graham, R.D.; Welch, R.M.; Bouis, H.E. Addressing micronutrient malnutrition through enhancing the nutritional quality of staple foods: Principles, perspectives and knowledge gaps. *Adv. Agron.* **2001**, *70*, 77–142.

29. Kutman, U.B.; Yidiz, B.; Ozturk, L.; Cakmak, I. Biofortification of durum wheat with zinc through soil and foliar application of nitrogen. *Cereal Chem.* **2010**, *87*, 1–9. [CrossRef]

30. Rengel, Z.; Graham, R.D. Importance of seed Zn content for wheat growth on Zn-deficient soil. *Plant Soil* **1995**, *173*, 259–266. [CrossRef]

31. Frossard, E.; Bucher, M.; Machler, F.; Mozafar, A.; Hurell, R. Potential for increasing the content and bioavailability of Fe, Zn and Ca in plants for human nutrition. *J. Sci. Food Agric.* **2000**, *80*, 861–879. [CrossRef]

32. White, P.; Broadley, M.R. Biofortifying crops with essential mineral elements. *Trends Plant Sci.* **2005**, *10*, 586–593. [CrossRef]

33. White, P.; Broadley, M.R. Physiological limits to Zn biofortification of edible crops. *Front. Plant Sci.* **2011**, *2*, 1–11. [CrossRef]

34. Frankin, S.; Kunta, S.; Abbo, S.; Sela, H.; Goldberg, B.Z.; Bonfil, D.J.; Levy, A.A.; Ragolsky, A.-N.; Nashef, K.; Roychowdhury, R. In Focus: The 1st International Conference of Wheat Landraces for Healthy Food. *J. Sci. Food Agric.* **2018**, *100*. [CrossRef]

35. Gyawali, S.; Otte, M.L.; Chao, S.; Jilal, A.; Jacob, D.L.; Amezrou, R.; Verma, R.P.S. Genome wide association studies (GWAS) of element contents in grain with a special focus on zinc and iron in a world collection of barley (*Hordeum vulgare* L.). *J. Cereal Sci.* **2017**, *77*, 266–274. [CrossRef]

36. Wood, P.J.; Beer, M.U. Functional oat products. In *Functional Foods: Biochemical and Processing Aspects*; Mazza, G., Ed.; Technomic Publishing Company Inc.: Lancaster, PA, USA, 1998; pp. 1–37.

37. Toole, G.A.; Gall, G.L.; Colquhoun, I.J.; Drea, S.; Opanowicz, M.; Bed, Z.; Shewry, P.R.; Mills, E.N.C. Spectroscopic analysis of diversity in the spatial distribution of arabinoxylan structures in endosperm cell walls of cereal species in the health grain diversity collection. *J. Cereal Sci.* **2012**, *56*, 134–141. [CrossRef]

38. Loskutov, I.G.; Polonskiy, V.I. Content of β-glucans in oat grain as a perspective direction of breeding for health products and fodder. *Agric. Biol.* **2017**, *52*, 646–657. [CrossRef]

39. Welch, R.W.; Leggett, J.M.; Lloyd, J.D. Variation in the kernel (1 3)(1 4)-β-D-glucan content of oat cultivars and wild Avena species and its relationship to other characteristics. *J. Cereal Sci.* **1991**, *13*, 173–178. [CrossRef]

40. Havrlentová, M.; Hlinková, A.; Ofajová, A.; Kováčik, P.; Dvončová, D.; Deáková, A. Effect of fertilization on b-D-glucan content in oat grain (*Avena sativa* L.). *Agriculture* **2013**, *59*, 111–119. [CrossRef]

41. Decker, E.A.; Rose, D.J.; Stewart, D. Processing of oats and the impact of processing operations on nutrition and health benefits. *Br. J. Nutr.* **2014**, *112*, 58–64. [CrossRef] [PubMed]

42. Clemens, R.; Van Klinken, B.J.W. The future of oats in the food and health continuum. *Br. J. Nutr.* **2014**, *112*, 75–79. [CrossRef] [PubMed]

43. Fastnaught, C.E.; Berglund, P.T.; Holm, E.T.; Fox, G.J. Genetic and environmental variation in β-glucan content and quality parameters of barley for food. *Crop. Sci.* **1996**, *36*, 941–946. [CrossRef]

44. Dzyubenko, N.; Zwer, P.; Loskutov, I. Abstracts of Oral and Poster Presentation. In Proceedings of the 10th International Oat Conference: Innovation for Food and Health "OATS 2016", St Petersburg, Russia, 11–15 July 2016.

45. Tiwari, U.; Cummins, E. Meta-analysis of the effect of beta-glucan intake on blood cholesterol and glucose levels. *Nutrition* **2011**, *27*, 1008–1016. [CrossRef] [PubMed]

46. Joyce, S.A.; Kamil, A.; Fleige, L.; Gahan, C.G. The cholesterol-lowering effect of oats and oat beta glucan: Modes of action and potential role of bile acids and the microbiome. *Front. Nutr.* **2019**, *6*, 171. [CrossRef]

47. Wood, P.J. Oat and rye beta-glucan: Properties and function. *Cereal Chem.* **2010**, *87*, 315–330. [CrossRef]

48. Piironen, V.; Anna-Maija, L.; Laura, N.; Li, L.; Mariann, R.; Anna, F.; Danuta, B.; Kurt, G.; Courtin, C.; Delcour, J.; et al. Phytochemical and fiber components in oat varieties in the Healthgrain diversity screen. *J. Agr. Food Chem.* **2008**, *56*, 9777–9784. [CrossRef]

49. Redaelli, R.; Frate, V.D.; Bellato, S.; Terracciano, G.; Ciccoritti, R.; Germeier, C.U.; Stefanis, E.D.; Sgrulletta, D. Genetic and environmental variability in total and soluble β-glucan in European oat genotypes. *J. Cereal Sci.* **2013**, *57*, 193–199. [CrossRef]

50. Polonskiy, V.; Loskutov, I.; Sumina, A. Biological role and health benefits of antioxidant compounds in cereals. *Biol. Commun.* **2020**, *65*, 53–67. [CrossRef]

51. Welch, R.W.; Brown, J.C.W.; Leggett, J.M. Interspecific and intraspecific variation in grain and groat characteristics of wild oat (Avena) species: Very high groat (1 3), (1 4)-β-D-glucan in an *Avena atlantica* genotype. *J. Cereal Sci.* **2000**, *31*, 273–279. [CrossRef]

52. Saastamoinen, M.; Plaami, S.; Kumpulainen, J. Genetic and environmental variation in β-glucan content of oats cultivated or tested in Finland. *J. Cereal Sci.* **1992**, *16*, 279–290. [CrossRef]

53. Havrlentová, M.; Bieliková, M.; Mendel, L.; Kraic, J.; Hozlár, P. The correlation of (1–3) (1–4)-β-d-glucan with some qualitative parameters in the oat grain. *Agriculture* **2008**, *54*, 65–71.

54. Gajdošová, A.; Petruláková, Z.; Havrlentová, M.; Červená, V.; Hozová, B.; Šturdík, E.; Kogan, G. The content of water-soluble and water-insoluble β-d-glucans in selected oats and barley varieties. *Carbohydr. Polym.* **2007**, *70*, 46–52. [CrossRef]

55. Tiwari, U.; Cummins, E. Simulation of the factors affecting β-glucan levels during the cultivation of oats. *J. Cereal Sci.* **2009**, *50*, 175–183. [CrossRef]

56. Griffey, C.; Brooks, W.; Kurantz, M.; Thomason, W.; Taylor, F.; Obert, D.; Moreau, R.; Flores, R.; Sohn, M.; Hicks, K.; et al. Grain composition of Virginia winter barley and implications for use in feed, food, and biofuels production. *J. Cereal Sci.* **2010**, *51*, 41–49. [CrossRef]

57. Knutsen, S.H.; Holtekjilen, A.K. Preparation and analysis of dietary fibre constituents in whole grain from hulled and hull-less barley. *Food Chem.* **2007**, *102*, 707–715. [CrossRef]

58. Huth, M.; Dongowski, G.; Gebhart, E.; Flamme, W. Functional properties of dietary fibre enriched exudates from barley. *J. Cereal Sci.* **2002**, *32*, 115–117. [CrossRef]

59. Zhang, G.; Wang, J.; Chen, J. Analysis of β-glucan content in barley cultivars from different locations of China. *Food Chem.* **2002**, *79*, 251–254. [CrossRef]

60. Cox, T.S.; Frey, K.J. Complementarity of genes for high groat-protein percentage from *Avena sativa* L. and *A. sterilis* L. *Crop. Sci.* **1985**, *25*, 106–109. [CrossRef]

61. Peterson, D.M.; Wesenberg, D.M.; Burrup, D.E. β-Glucan content and its relationship to agronomic characteristics in elite oat germplasm. *Crop. Sci.* **1995**, *35*, 965–970. [CrossRef]

62. Yalcin, E.; Celik, S.; Akar, T.; Sayim, I.; Koksel, H. Effects of genotype and environment on β-glucan and dietary fibre contents of hull-less barleys grown in Turkey. *Food Chem.* **2007**, *101*, 171–176. [CrossRef]

63. Lee, C.J.; Horsley, R.D.; Manthey, F.A.; Schwarz, P.B. Comparison of β-glucan content of barley and oat. *Cereal Chem.* **1997**, *74*, 571–575. [CrossRef]

64. Hang, A.; Obert, D.; Gironella, A.I.N.; Burton, C.S. Barley amylase and β-glucan: Their relationships to protein, agronomic traits, and environmental factors. *Crop. Sci.* **2007**, *47*, 1754–1760. [CrossRef]

65. Zhang, G.; Chen, J.; Wang, J.; Ding, S. Cultivar and environmental effects on (1–3, 1–4)-β-glucan and protein content in malting barley. *J. Cereal Sci.* **2001**, *34*, 295–301. [CrossRef]

66. Rey, J.I.; Hayes, P.M.; Petrie, S.E.; Corey, A.; Flowers, M.; Ohm, J.B.; Ong, C.; Rhinhart, K.; Ross, A.S. Production of dryland barley for human food: Quality and agronomic performance. *Crop. Sci.* **2009**, *49*, 347–355. [CrossRef]

67. Aman, P.; Graham, H.; Tilley, A. Content and solubility of mixed-linked (1–3;1–4)-β-glucan in barley and oats during kernel development and storage. *J. Cereal Sci.* **1989**, *10*, 45–50. [CrossRef]

68. Chawade, A.; Sikora, P.; Brautigam, M.; Larsson, M.; Vivekanand, V.; Nakash, M.A.; Chen, T.; Olsson, O. Development and characterization of an oat TILLING-population and identification of mutations in lignin and β-glucan biosynthesis genes. *BMC Plant Biol.* **2010**, *10*, 86–99. [CrossRef]

69. Sikora, P.; Tosh, S.M.; Brummer, Y.; Olsson, O. Identification of high β-glucan oat lines and localization and chemical characterization of their seed kernel β-glucans. *Food Chem.* **2013**, *137*, 83–91. [CrossRef]

70. Collins, H.M.; Burton, R.A.; Topping, D.L.; Liao, M.-L.; Bacic, A.; Fincher, G.B. Variability in fine structures of noncellulosic cell wall polysaccharides from cereal grains: Potential importance in human health and nutrition. *Cereal Chem.* **2010**, *87*, 272–282. [CrossRef]

71. Song, G.; Huo, P.; Wu, B.; Zhang, Z. A genetic linkage map of hexaploid naked oat constructed with SSR markers. *Crop. J.* **2015**, *3*, 353–357. [CrossRef]

72. Burton, R.A.; Wilson, S.M.; Hrmova, M.; Harvey, A.J.; Shirley, N.J.; Medhurst, A.; Fincher, G.B. Cellulose synthase-like *CslF* genes mediate the synthesis of cell wall (1–3;1–4)-β-D-glucans. *Science* **2006**, *311*, 1940–1942. [CrossRef] [PubMed]

73. Newell, M.A.; Asoro, F.G.; Scott, V.P.; White, P.J.; Beavis, W.D.; Jannink, J.-L. Genome-wide association study for oat (*Avena sativa* L.) beta-glucan concentration using germplasm of worldwide origin. *Appl. Genet.* **2012**, *125*, 1687–1696. [CrossRef]

74. Carlson, M.O.; Montilla-Bascon, G.; Hoekenga, O.A.; Tinker, N.A.; Poland, J.; Baseggio, M.; Sorrells, M.E.; Jannink, J.-L.; Gore, M.A.; Yeats, T.H.; et al. Multivariate genome-wide association analyses reveal the genetic basis of seed fatty acid composition in oat (*Avena sativa* L.). *G3 GenesGenomesGenet* **2019**, *9*, 2963–2975. [CrossRef]
75. Andersson, A.A.M.; Lampi, A.M.; Nyström, L.; Piironen, V.; Li, L.; Ward, J.L.; Gebruers, K.; Courtin, C.M.; Delcour, J.A.; Boros, D.; et al. Phytochemical and dietary fiber components in barley varieties in the healthgrain diversity screen. *J. Agric. Food Chem.* **2008**, *56*, 9767–9776. [CrossRef]
76. Do, T.D.T.; Cozzolino, D.; Muhlhausler, B.; Box, A.; Able, A.J. Antioxidant capacity and vitamin E in barley: Effect of genotype and storage. *Food Chem.* **2015**, *187*, 65–74. [CrossRef]
77. Edelmann, M.; Kariluoto, S.; Nyström, L.; Piironen, V. Folate in barley grain and fractions. *J. Cereal Sci.* **2013**, *58*, 37–44. [CrossRef]
78. Gong, L.X.; Jin, C.; Wu, L.J.; Wu, X.Q.; Zhang, Y. Tibetan hullless Barley (*Hordeum vulgare* L.) as a potential source of antioxidants. *Cereal Chem.* **2012**, *89*, 290–295. [CrossRef]
79. Edelmann, M.; Kariluoto, S.; Nyström, L.; Piironen, V. Folate in oats and its milling fractions. *Food Chem.* **2012**, *135*, 1938–1947. [CrossRef]
80. Khlestkina, E.K.; Usenko, N.I.; Gordeeva, E.I.; Stabrovskaya, O.I.; Sharfunova, I.B.; Otmakhova, Y.S. Evaluation of wheat products with high flavonoid content: Justification of importance of marker-assisted development and production of flavonoid-rich wheat cultivars. *Vavilovskii Zhurnal Genetiki Selektsii* **2017**, *21*, 545–553. [CrossRef]
81. Boutigny, A.-L.; Richard-Forget, F.; Barreau, C. Natural mechanisms for cereal resistance to the accumulation of Fusarium trichothecenes. *Eur. J. Plant Pathol.* **2008**, *121*, 411–423. [CrossRef]
82. Branković, G.; Dragičević, V.; Dodig, D.; Zorić, M.; Knežević, D.; Žilić, S.; Denčić, S.; Šurlan, G. Genotype × environment interaction for antioxidants and phytic acid contents in bread and durum wheat as influenced by climate. *Chilean J. Agric. Res.* **2015**, *75*, 139–146.
83. Adom, K.K.; Liu, R.H. Antioxidant activity of grains. *J. Agric. Food Chem.* **2002**, *50*, 6182–6187. [CrossRef] [PubMed]
84. Boz, H. Ferulic acid in cereals-a review. *Czech J. Food Sci.* **2015**, *33*, 1–7. [CrossRef]
85. Mattila, P.; Pihlava, J.M.; Hellstrom, J. Contents of phenolic acids, alkyl- and alkenylresorcinols, and avenanthramides in commercial grain products. *J. Agric. Food Chem.* **2005**, *53*, 8290–8295. [CrossRef]
86. Ndolo, V.U.; Beta, T. Comparative studies on composition and distribution of phenolic acids in cereal grain botanical fractions. *Cereal Chem.* **2014**, *91*, 522–530. [CrossRef]
87. Liu, R.H. Whole grain phytochemicals and health. *J. Cereal Sci.* **2007**, *46*, 207–219. [CrossRef]
88. Glagoleva, A.Y.; Shmakov, N.V.; Shoeva, O.Y.; Vasiliev, G.V.; Shatskaya, N.V.; Börner, A.; Afonnikov, D.A.; Khlestkina, E.K. Metabolic pathways and genes identified by RNA-seq analysis of barley near-isogenic lines differing by allelic state of the black lemma and pericarp (*Blp*) gene. *BMC Plant Biol.* **2017**, *17*, 182. [CrossRef]
89. Shoeva, O.Y.; Mursalimov, S.R.; Gracheva, N.V.; Glagoleva, A.Y.; Börner, A.; Khlestkina, E.K. Melanin formation in barley grain occurs within plastids of pericarp and husk cells. *Nat. Sci. Rep.* **2020**, *10*, 179. [CrossRef]
90. Bollina, V.; Kumaraswamy, G.K.; Kushalappa, A.C.; Choo, T.M.; Dion, Y.; Rioux, S.; Faubert, D.; Hamzehzarghani, H. Mass spectrometry-based metabolomics application to identify quantitative resistance-related metabolites in barley against Fusarium Head blight. *Mol. Plant Pathol.* **2010**, *11*, 769–782. [CrossRef] [PubMed]
91. Bollina, V.; Kushalappa, A.C.; Choo, T.M.; Dion, Y.; Rioux, S. Identification of metabolites related to mechanisms of resistance in barley against *Fusarium graminearum*, based on mass spectrometry. *Plant Mol. Biol.* **2011**, *77*, 355–370. [CrossRef] [PubMed]
92. Zilic, S.; Sukalovic, V.H.T.; Dodig, D.; Maksimovic, V.; Maksimovic, M.; Basic, Z. Antioxidant activity of small grain cereals caused by phenolics and lipid soluble antioxidants. *J. Cereal Sci.* **2011**, *54*, 417–424. [CrossRef]
93. Dykes, L.; Rooney, L.W. Phenolic compounds in cereal grains and their health benefits. *Cereal Foods World* **2007**, *52*, 105–111. [CrossRef]
94. Lee, C.; Han, D.; Kim, B.; Bae, N.; Baik, B.K. Antioxidant and anti-hypertensive activity of anthocyaninrich extracts from hulless pigmented barley cultivars. *Int. J. Food Sci. Technol.* **2013**, *48*, 984–991. [CrossRef]
95. Usenko, N.I.; Khlestkina, E.K.; Asavasanti, S.; Gordeeva, E.I.; Yudina, R.S.; Otmakhova, Y.S. Possibilities of enriching food products with anthocyanins by using new forms of cereals. *Foods Raw Mater.* **2018**, *6*, 128–135. [CrossRef]
96. Arbuzova, V.S.; Badaeva, E.D.; Efremova, T.T.; Osadchaya, T.S.; Trubacheeva, N.V.; Dobrovolskaya, O.B. A cytogenetic study of the blue-grain line of the common wheat cultivar Saratovskaya 29. *Rus. J. Genet.* **2012**, *4*, 785–791. [CrossRef]
97. Gordeeva, E.; Badaeva, E.; Yudina, R.; Shchukina, L.; Shoeva, O.; Khlestkina, E. Marker-assisted development of a blue-grained substitution line carrying the *Thinopyrum ponticum* Chromosome 4Th(4D) in the spring bread wheat saratovskaya 29 background. *Agronomy* **2019**, *9*, 723. [CrossRef]
98. Strygina, K.V.; Börner, A.; Khlestkina, E.K. Identification and characterization of regulatory network components for anthocyanin synthesis in barley aleurone. *BMC Plant Biol.* **2017**, *17*, 184. [CrossRef]
99. Zeven, A.C. Wheats with purple and blue grains: A review. *Euphytica* **1991**, *56*, 243–258. [CrossRef]
100. Syed Jaafar, S.N.; Baron, J.; Siebenhandl-Ehn, S.; Rosenau, T.; Böhmdorfer, S.; Grausgrube, H. Increased anthocyanin content in purple pericarp × blue aleurone wheat crosses. *Plant Breed.* **2013**, *132*, 546–552. [CrossRef]
101. King, C. New Possibilities with Purple Wheat. Available online: https://www.topcropmanager.com/new-possibilities-with-purple-wheat-20050/ (accessed on 29 October 2020).

102. Welch, R.W. The chemical composition of oats. In *The Oat Crop: Production and Utilization*; Welch, R.W., Ed.; Chapman & Hall: London, UK, 1995; pp. 279–320.
103. Collins, F.W. Oat phenolics: Avenanthramides, novel substituted N-cinnamoylanthranilate alkaloids from oat groats and hulls. *J. Agric. Food Chem.* **1989**, *37*, 60–66. [CrossRef]
104. Jastrebova, J.; Skoglund, M.; Dimberg, L.H. Selective and sensitive LC–MS determination of avenanthramides in oats. *Chromatographia* **2006**, *63*, 419–423. [CrossRef]
105. Fu, J.; Zhu, Y.; Yerke, A.; Wise, M.L.; Johnson, J.; Chu, Y.; Sang, S. Oat avenanthramides induce heme oxygenase-1 expression via Nrf2-mediated signaling in HK-2 cells. *Mol. Nutr. Food Res.* **2015**, *59*, 2471–2479. [CrossRef] [PubMed]
106. Pellegrini, G.G.; Morales, C.C.; Wallace, T.C.; Plotkin, L.I.; Bellido, T. Avenanthramides prevent osteoblast and osteocyte apoptosis and induce osteoclast apoptosis in vitro in an Nrf2-independent manner. *Nutrients* **2016**, *8*. [CrossRef]
107. Tripathi, V.; Mohd, A.S.; Ashraf, T. Avenanthramides of oats: Medicinal importance and future perspectives. *Pharmacogn. Rev.* **2018**, *12*, 66–71. [CrossRef]
108. Redaelli, R.; Dimberg, L.; Germeier, C.U.; Berardo, N.; Locatelli, S.; Guerrini, L. Variability of tocopherols, tocotrienols and avenanthramides contents in European oat germplasm. *Euphytica* **2016**, *207*, 273–292. [CrossRef]
109. Leonova, S.; Gnutikov, A.; Loskutov, I.; Blinova, E.; Gustafsson, K.-E.; Olsson, O. Diversity of avenanthramide content in wild and cultivated oats. *Proc. Appl. Bot. Genet. Breed.* **2020**, *181*, 30–47. [CrossRef]
110. Peterson, D.M. Oat antioxidants. *J. Cereal Sci.* **2001**, *33*, 115–129. [CrossRef]
111. Gutierrez-Gonzalez, J.J.; Wise, M.L.; Garvin, D.F. A developmental profile of tocol accumulation in oat seeds. *J. Cereal Sci.* **2013**, *57*, 79–83. [CrossRef]
112. Falk, J.; Krahnstover, A.; van der Kooij, T.A.W.; Schlensog, M.; Krupinska, K. Tocopherol and tocotrienol accumulation during development of caryopses from barley (*Hordeum vulgare* L.). *Phytochemistry* **2004**, *65*, 2977–2985. [CrossRef] [PubMed]
113. Tohno-oka, T.; Kavada, N.; Yoshioka, T. Relationship between grain hardness and endosperm cell wall polysaccharides in barley. *Czech. J. Genet. Plant. Breed.* **2004**, *40*, 116.
114. Piironen, V.; Toivo, J.; Lampi, A.M. Plant sterols in cereals and cereal products. *Cereal Chem.* **2002**, *79*, 148–154. [CrossRef]
115. Ndolo, V.U.; Beta, T. Distribution of carotenoids in endosperm, germ, and aleurone fractions of cereal grain kernels. *Food Chem.* **2013**, *139*, 663–671. [CrossRef] [PubMed]
116. Hussain, A.; Larsson, H.; Kuktaite, R.; Olsson, M.E.; Johansson, E. Carotenoid content in organically produced wheat: Relevance for human nutritional health on consumption. *Int. J. Environ. Res. Public Health.* **2015**, *2*, 14068–14083. [CrossRef] [PubMed]
117. Colasuonno, P.; Marcotuli, I.; Blanco, A.; Maccaferri, M.; Condorelli, G.E.; Tuberosa, R.; Parada, R.; de Camargo, A.C.; Schwember, A.R.; Gadaleta, A. Carotenoid pigment content in durum wheat (*Triticum turgidum* L. var *durum*): An overview of quantitative trait loci and candidate genes. *Front. Plant Sci.* **2019**, *7*, 1347. [CrossRef]
118. Carraro-Lemes, C.F.; Scheffer-Basso, S.M.; Deuner, C.; Berghahn, S. Analysis of genotypic variability in *Avena spp.* regarding allelopathic potentiality. *Planta Daninha* **2019**, *37*, 1–12. [CrossRef]
119. Beleggia, R.; Rau, D.; Laido, G.; Platani, C.; Nigro, F.; Fragasso, M.; de Vita, P.; Scossa, F.; Fernie, A.R.; Nikoloski, Z.; et al. Evolutionary metabolomics reveals domestication-associated changes in tetraploid wheat kernels. *Mol. Biol. Evol.* **2016**, *33*, 1740–1753. [CrossRef]
120. Berni, R.; Cantini, C.; Romi, M.; Hausman, J.-F.; Guerriero, G.; Cai, G. Agrobiotechnology goes wild: Ancient local varieties as sources of bioactives. *Int. J. Mol. Sci.* **2018**, *19*, 2248. [CrossRef]
121. Loskutov, I.G.; Shelenga, T.V.; Konarev, A.V.; Khoreva, V.I.; Shavarda, A.L.; Blinova, E.V.; Gnutikov, A.A. Biochemical aspects of interactions between fungi and plants: A case study of Fusarium in oats. *Agric. Biol.* **2019**, *54*, 575–588. [CrossRef]
122. Matthews, S.B.; Santra, M.; Mensack, M.M.; Wolfe, P.; Byrne, P.F.; Thompson, H.J. Metabolite profiling of a diverse collection of wheat lines using ultraperformance liquid chromatography coupled with time-of-flight mass spectrometry. *PLoS ONE* **2012**, *7*, e44179. [CrossRef] [PubMed]
123. Carreno-Quintero, N.; Bouwmeester, H.J.; Keurentjes, J.J. Genetic analysis of metabolome–phenotype interactions: From model to crop species. *Trends Genet.* **2013**, *29*, 41–50. [CrossRef] [PubMed]
124. Novotelnov, N.V.; Ezhov, I.S. About antibiotic and antioxidant properties of grain yellow pigments. *Proc. Acad. Sci. USSR* **1954**, *99*, 297–300.
125. Khlestkina, E.K. The adaptive role of flavonoids: Emphasis on cereals. *Cereal Res. Commun.* **2013**, *41*, 185–198. [CrossRef]
126. Skadhauge, B.; Thomsen, K.; von Wettstein, D. The role of barley test a layer and its flavonoid content in resistance to fusarium infections. *Hereditas* **1997**, *126*, 147–160. [CrossRef]
127. Atanasova-Penichon, V.; Barreau, C.; Richard-Forget, F. Antioxidant secondary metabolites in cereals: Potential involvement in resistance to fusarium and mycotoxin accumulation. *Front. Microbiol.* **2016**, *7*, 566. [CrossRef]
128. Haikka, H.; Manninen, O.; Hautsalo, J.; Pietilä, L.; Jalli, M.; Veteläinen, M. Genome-wide association study and genomic prediction for *Fusarium graminearum* resistance traits in nordic oat (*Avena sativa* L.). *Agronomy* **2020**, *10*, 174. [CrossRef]
129. Polišenská, I.; Jirsa, O.; Vaculová, K.; Pospíchalová, M.; Wawroszova, S.; Frydrych, J. Fusarium mycotoxins in two hulless oat and barley cultivars used for food purposes. *Foods* **2020**, *9*, 1037. [CrossRef]
130. Das, K.; Roychoudhury, A. Reactive oxygen species (ROS) and response of antioxidants as ROS-scavengers during environmental stress in plants. *Front. Environ.* **2014**, *2*, 53. [CrossRef]

131. Boutigny, A.-L.; Barreau, C.; Atanasova-Penichon, V.; Verdal-Bonnin, M.-N.; Pinson-Gadais, L.; Richard-Forget, F. Ferulic acid, an efficient inhibitor of type B trichothecene biosynthesis and tri gene expression in fusarium liquid cultures. *Mycol. Res.* **2009**, *113*, 746–753. [CrossRef]

132. Gauthier, L.; Bonnin-Verdal, M.-N.; Marchegay, G.; Pinson-Gadais, L.; Ducos, C.; Richard-Forget, F.; Penichon, V.A. Fungal biotransformation of chlorogenic and caffeic acids by *Fusarium graminearum*: New insights in the contribution of phenolic acids to resistance to deoxynivalenol accumulation in cereals. *Int. J. Food Microbiol.* **2016**, *221*, 61–68. [CrossRef] [PubMed]

133. Choo, T.M.; Vigier, B.; Savard, M.E.; Blackwell, B.; Martin, R.; Wang, J.M.; Yang, J.; el-Sayed, M.A.-A. Black barley as a means of mitigating deoxynivalenol contamination. *Crop. Sci.* **2015**, *55*, 1096–1103. [CrossRef]

134. Ferruz, E.; Atanasova-Pénichon, V.; Bonnin-Verdal, M.-N.; Marchegay, G.; Pinson-Gadais, L.; Ducos, C.; Lorán, S.; Ariño, A.; Barreau, C.; Richard, F.-F.; et al. Effects of phenolic acids on the growth and production of T-2 and HT-2 toxins by *Fusarium langsethiae* and *Fsporotrichioides*. *Molecules* **2016**, *21*. [CrossRef] [PubMed]

135. Kumaraswamy, G.K.; Kushalappa, A.C.; Choo, T.M.; Dion, Y.; Rioux, S. Differential metabolic response of barley genotypes, varying in resistance, to trichothecene-producing and -nonproducing(tri5-) isolates of *Fusarium graminearum*. *Plant Pathol.* **2012**, *61*, 509–521. [CrossRef]

136. Søltoft, M.; Jørgensen, L.N.; Svensmark, B.; Fomsgaard, I.S. Benzoxazinoid concentrations show correlation with fusarium head blight resistance in Danish wheat varieties. *Biochem. Syst. Ecol.* **2008**, *36*, 245–259. [CrossRef]

137. Siranidou, E.; Kang, Z.; Buchenauer, H. Studies on symptom development, phenolic compounds and morphological defence responses in wheat cultivars differing in resistance to Fusarium Head Blight. *J. Phytopathol.* **2002**, *150*, 200–208. [CrossRef]

138. Matsukawa, T.; Isobe, T.; Ishihara, A.; Iwamura, H. Occurrence of avenanthramides and hydroxycinnamoylCoA: Hydroxyanthranilate N-hydroxycinnamoyltransferase activity in oat seeds. *Zeitschrift Naturforschung C* **2000**, *55*, 30–36. [CrossRef]

139. Peterson, D.M.; Dimberg, L.H. Avenanthramide concentrations and hydroxycinnamoyl-CoA: Hydroxyanthranilate N-hydroxycinnamoyltransferase activities in developing oats. *J. Cereal Sci.* **2008**, *47*, 101–108. [CrossRef]

140. Bryngelsson, S.; Ishihara, A.; Dimberg, L.H. Levels of avenanthramides and activity of hydroxycinnamoylCoA: Hydroxyanthranilate N-Hydroxycinnamoyl transferase (HHT) in steeped or germinated oat samples. *Cereal Chem.* **2003**, *80*, 356–360. [CrossRef]

141. Dimberg, L.H.; Molteberg, E.L.; Solheim, R.; Frølich, W. Variation in oat groats due to variety, storage and heat treatment I: Phenolic compounds. *J. Cereal Sci.* **1996**, *24*, 263–272. [CrossRef]

142. Gould, K.S. Nature's Swiss army knife: The diverse protective roles of anthocyanins in leaves. *J. Biomed. Biotech.* **2004**, *5*, 314–320. [CrossRef] [PubMed]

143. Dorofeev, V.F.; Yakubtsiner, M.M.; Rudenko, M.I.; Migushova, E.F.; Udachin, R.A.; Merejko, A.F.; Semenova, L.V.; Novikova, M.V.; Gradchaninova, O.D.; Shitova, I.P. *Wheat of the World*; Kolos: Leningrad, Russia, 1976.

144. Shoeva, O.Y.; Gordeeva, E.I.; Arbuzova, V.S.; Khlestkina, E.K. Anthocyanins participate in protection of wheat seedlings from osmotic stress: A case study of near isogenic lines. *Cereal Res. Commun.* **2017**, *45*, 47–56. [CrossRef]

145. Gordeeva, E.I.; Shoeva, O.Y.; Yudina, R.S.; Kukoeva, T.V.; Khlestkina, E.K. Effect of seed pre-sowing gamma-irradiation treatment in bread wheat lines differing by anthocyanin pigmentation. *Cereal Res. Commun.* **2018**, *46*, 41–53. [CrossRef]

146. Shoeva, O.Y.; Khlestkina, E.K. Anthocyanins participate in the protection of wheat seedlings against cadmium stress. *Cereal Res. Commun.* **2018**, *46*, 242–252. [CrossRef]

147. Marbach, I.; Meyer, A.M. Permeability of seed coats to water as related to drying conditions and metabolism of phenolics. *Plant Physiol.* **1974**, *54*, 817–820. [CrossRef] [PubMed]

148. Strumeyer, D.H.; Malin, M.J. Condensed tannins in grain sorghum. Isolation, fractionation, and characterization. *J. Agric. Food Chem.* **1975**, *23*, 909–914. [CrossRef]

149. Gelman, N.S. Wheat grain dehydrogenases: Grain Biochemistry. *Sbornik Acad. Sci. USSR* **1951**, 17–33.

150. Dimberg, L.H.; Gissén, C.; Nilsson, J. Phenolic compounds in oat grains (*Avena sativa* L.) grown in conventional and organic systems. *Ambio* **2005**, *34*, 331–337. [CrossRef]

151. Emmons, C.; Peterson, D.M. Antioxidant activity and phenolic content of oat as affected by cultivar and location. *Crop. Sci.* **2001**, *41*, 1676–1681. [CrossRef]

152. Peterson, D.M.; Wesenberg, D.M.; Burrup, D.E.; Erikson, C.A. Relationships among agronomic traits and grain composition in oat genotypes grown in different environments. *Crop. Sci.* **2005**, *45*, 1249–1255. [CrossRef]

153. Wise, M.L.; Doehlert, D.C.; McMullen, M.S. Association of avenanthramide concentration in oat (*Avena sativa* L.) grain with crown rust incidence and genetic resistance. *Cereal Chem.* **2008**, *85*, 639–641. [CrossRef]

154. Oraby, H.F.; El-Tohamy, M.F.; Kamel, A.M.; Ramadan, M.F. Changes in the concentration of avenanthramides in response to salinity stress in CBF3 transgenic oat. *J. Cereal Sci.* **2017**, *76*, 263–270. [CrossRef]

155. Havrlentová, M.; Deáková, L.; Kraic, J.; Žofajová, A. Can β-D-glucan protect oat seeds against a heat stress? *Nova Biotechnologica Chimica* **2016**, *15*, 107. [CrossRef]

156. Bhatty, R.S.; Macgregor, A.W.; Rossnagel, B.G. Total and acid-soluble β-glucan content in hulless and its relationship to acid-extract viscosity. *Cereal Chem.* **1991**, *68*, 221–227.

157. Bohn, M.; Lüthje, S.; Sperling, P.; Heinz, E.; Dörffling, K. Plasma membrane lipid alterations induced by cold acclimation and abscisic acid treatment of winter wheat seedlings differing in frost resistance. *J. Plant. Physiol.* **2007**, *164*, 146–156. [CrossRef] [PubMed]

158. Valitova, J.; Renkova, A.; Mukhitova, F.; Dmitrieva, S.; Beckett, R.P.; Minibayeva, F.V. Membrane sterols and genes of sterol biosynthesis are involved in the response of *Triticum aestivum* seedlings to cold stress. *Plant. Physiol. Biochem.* **2019**, *142*, 452–459. [CrossRef]
159. Takahashi, D.; Imai, H.; Kawamura, Y.; Uemura, M. Lipid profiles of detergent resistant fractions of the plasma membrane in oat and rye in association with cold acclimation and freezing tolerance. *Cryobiology* **2016**, *72*, 123–134. [CrossRef]

Article

Genetic Diversity and Combining Ability of White Maize Inbred Lines under Different Plant Densities

Mohamed M. Kamara [1], Medhat Rehan [2,3], Khaled M. Ibrahim [4], Abdullah S. Alsohim [3], Mohsen M. Elsharkawy [5], Ahmed M. S. Kheir [6], Emad M. Hafez [1] and Mohamed A. El-Esawi [7,*]

[1]   Department of Agronomy, Faculty of Agriculture, Kafrelsheikh University, Kafr El-Sheikh 33516, Egypt; mohamed.kamara@agr.kfs.edu.eg (M.M.K.); emadhafez2012@agr.kfs.edu.eg (E.M.H.)
[2]   Department of Genetics, Faculty of Agriculture, Kafrelsheikh University, Kafr El-Sheikh 33516, Egypt; medhat.rehan@agr.kfs.edu.eg
[3]   Department of Plant Production and Protection, College of Agriculture and Veterinary Medicine, Qassim University, Burydah 51452, Saudi Arabia; a.alsohim@qu.edu.sa
[4]   Agronomy Department, Faculty of Agriculture, New Valley University, El-Kharga 72511, Egypt; kh_ibrahim75@aun.edu.eg
[5]   Department of Agricultural Botany, Faculty of Agriculture, Kafrelsheikh University, Kafr Elsheikh 33516, Egypt; mohsen.abdelrahman@agr.kfs.edu.eg
[6]   Soils, Water and Environment Research Institute, Agricultural Research Center, Giza 12112, Egypt; ahmed.kheir@arc.sci.eg
[7]   Botany Department, Faculty of Science, Tanta University, Tanta 31527, Egypt
*   Correspondence: mohamed.elesawi@science.tanta.edu.eg

Received: 7 August 2020; Accepted: 31 August 2020; Published: 3 September 2020

Abstract: Knowledge of combining ability and genetic diversity are important prerequisites for the development of outstanding hybrids that are tolerant to high plant density. This work was carried out to assess general combining ability (GCA) and specific combining ability (SCA), identify promising hybrids, estimate genetic diversity among the inbred lines and correlate genetic distance to hybrid performance and SCA across different plant densities. A total of 28 F_1 hybrids obtained by crossing eight adverse inbred lines (four local and four exotic) were evaluated under three plant densities 59,500 (D1), 71,400 (D2) and 83,300 (D3) plants ha^{-1} using spilt plot design with three replications at two locations during 2018 season. Increasing plant density from D1 to D3 significantly decreased leaf angle (LANG), chlorophyll content (CHLC), all ear characteristics and grain yield per plant (GYPP). Contrarily, days to silking (DTS), anthesis–silking interval (ASI), plant height (PLHT), ear height (EHT), and grain yield per hectare (GYPH) were significantly increased. Both additive and non-additive gene actions were involved in the inheritance of all the evaluated traits, but additive gene action was predominant for most traits. Inbred lines L_1, L_2, and L_5 were the best general combiners for increasing grain yield and other desirable traits across research environments. Two hybrids $L_2 \times L_5$ and $L_2 \times L_8$ were found to be good specific combiners for ASI, LANG, GYPP and GYPH. Furthermore, these hybrids are ideal for further testing and promotion for commercialization under high plant density. Genetic distance (GD) among pairs of inbred lines ranged from 0.31 to 0.78, with an average of 0.61. Clustering based on molecular GD has effectively grouped the inbred lines according to their origin. No significant correlation was found between GD and both hybrid performance and SCA for grain yield and other traits and proved to be of no predictive value. Nevertheless, SCA could be used to predict the hybrid performance across all plant densities. Overall, this work presents useful information regarding the inheritance of maize grain yield and other important traits under high plant density.

Keywords: maize; density tolerance; combining ability; gene effects; genetic diversity

1. Introduction

Maize (*Zea mays* L.) is one of the main economic crops that subsidize global food security. It is widely used for food, animal feed, edible oil and fuel worldwide [1]. In Egypt, maize is considered the second most important crop, with the annual production of the grain reaching about 7.30 Mt from approximately 0.94 Mha in 2018 [2]. This production is insufficient to meet the demands of a fast-growing population. The gap between production and consumption is approximately 45% [3]. This gap could be narrowed by further increase in the hybrids yield potential and total yield production from unit area. [4]. Increasing planting density is required to increase grain yield production in maize [5]. The average density of intense maize cultivation in the USA is 97,500 plants ha^{-1} [6]. The recommended planting density in Egypt is 53,533 plants ha^{-1} [7], which is around half the amount used in the USA. The use of lower plant densities decreases light interception, leading to high grain production per plant but low grain production per unit area [8]. The yield production could be maximized by growing maize hybrids that can tolerate high plant density up to 100,000 plants ha^{-1}. However, high plant densities enhance interplant competition for light, nutrients, and water [9]. Additionally, it increases the anthesis–silking interval [10], thereby increasing kernel abortion [11] and reducing single plant yield. Al-Naggar et al. [12] showed that with increased planting density, plant and ear heights increased, whereas chlorophyll content, grains per ear and thousand grain weights decreased. The tolerance of the current Egyptian maize hybrids to high plant densities is low. This probably attributed to their tallness, decumbent leaf, one-eared and large size [7,13]. Conversely, modern maize hybrids in developed countries are characterized by early silking, short anthesis–silking interval and prolificacy, which are essential adaptive traits to high plant density tolerance [10,14–16].

Breeding programs should be directed towards the development of hybrids that are not only high yielding, but also show enhanced adaptability to high plant density tolerance. The successful identification of desirable hybrid combinations depends on the combining ability of the parents and the gene effects involved in the expression of target trait [17]. Furthermore, knowledge of gene action is important to devise an appropriate breeding strategy [18]. General combining ability (GCA) and specific combining ability (SCA) are widely used in selection of good parents and hybrids, respectively [19]. Among different biometrical approaches, the diallel mating design is commonly used by maize breeders to estimate GCA and SCA effects [20–22]. GCA is associated with additive gene effects, whereas SCA is typically associated with non-additive gene effects [23]. Both additive and non-additive gene actions were reported to be important in the inheritance of maize grain yield under high plant density [24]. However, the grain yield and other assessed traits under different plant densities among selected maize inbred lines were mostly controlled by additive gene action [7,25].

The assessment of the diversity and genetic distance in the available maize inbreds is important for a hybrid breeding program, in order to identify inbreds that would produce crosses with good levels of heterosis without testing all hybrids combinations [26,27]. Different types of DNA markers are available to estimate genetic distance. The simple sequence repeat (SSR) markers or microsatellites have been considered as the markers of choice owing to their co-dominant, high polymorphic, multi-allelic nature and high reproducibility [28–30]. However, contradictory results have been reported with respect to the relationship between genetic distance and hybrid performance in maize. Significant correlations were reported between molecular marker-based GD and F_1 hybrid grain yield in maize [31,32]. Whereas, other studies reported no significant correlation [33,34]. The objectives of this study were (i) to estimate GCA of the inbred lines and SCA of the hybrids under different plant densities; (ii) to determine the mode of gene action controlling grain yield and other important agronomic traits; (iii) to identify promising hybrids that yield well at high plant density; and (iv) to assess genetic diversity among the eight inbred lines and correlate genetic distance to hybrid performance and SCA.

2. Results

2.1. Analysis of Variance

The analysis of variance (ANOVA) revealed highly significant mean squares for locations (L), densities (D), hybrids (H) and their interactions (L × D, H × L, H × D and H × D × L) for all the studied characteristics (Table 1). Moreover, general combining ability (GCA) and specific combining ability (SCA) mean squares were highly significant for all the measured traits. The magnitude of GCA mean squares was higher than that of SCA mean squares (the ratio of GCA/SCA was higher than the unity) for all the studied traits, except number of kernels per row (NKPR) trait. Significant mean squares of GCA × L, SCA × L, GCA × D, SCA × D, GCA × L × D, SCA × L × D interactions were detected for all the studied traits, except GCA × L and GCA × L × D for leaf angle (LANG) and chlorophyll content (CHLC), GCA × D for ear diameter (ED) and SCA × D for EHT, LANG and ED were not significant.

Table 1. Analysis of variance for the evaluated crosses under three plant densities combined across two locations for all the studied traits.

SOV	DF	DTS	ASI	PLHT	EHT	LANG	CHLC
Locations (L)	1	1114.26 **	12.87 **	16,592.96 **	6489.21 **	108.64 *	400.29 **
Rep (L)	4	15.14	0.58	325.06	138.65	14.87	10.84
Densities (D)	2	1899.48 **	73.14 **	23,422.30 **	9384.04 **	603.65 **	1585.67 **
L × D	2	308.23 **	22.39 **	9852.25 **	5708.38 **	27.17 **	180.79 **
Error a	8	1.07	0.19	121.77	53.15	3.33	2.34
Hybrids (H)	27	28.04 **	7.89 **	6842.02 **	2056.81 **	425.82 **	119.32 **
GCA	7	57.16 **	12.03 **	11,397.27 **	2447.44 **	836.14 **	162.00 **
SCA	20	17.85 **	6.44 **	5247.69 **	1920.08 **	282.20 **	104.38 **
H × L	27	61.66 **	0.83 **	796.40 **	362.93 **	4.15 **	3.61 **
GCA × L	7	52.48 **	0.88 **	915.53 **	309.35 **	3.16	3.57
SCA × L	20	64.88 **	0.81 **	754.71 **	381.68 **	4.49 **	3.62 *
H × D	54	5.01 **	0.31 **	254.88 **	60.51 **	4.57 **	10.19 **
GCA × D	14	4.54 **	0.30 **	212.94 **	78.99 *	6.63 **	19.13 **
SCA × D	40	5.18 **	0.31 **	269.56 **	54.03	3.85	7.06 **
H × D × L	54	63.25 **	0.88 **	592.24 **	397.54 **	4.11 *	4.30 **
GCA × L × D	14	64.03 **	0.86 **	544.25 **	363.40 **	3.90	2.64
SCA × L × D	40	62.97 **	0.88 **	609.03 **	409.48 **	4.19 *	4.88 **
Error b	324	0.81	0.14	84.07	42.33	2.69	1.93
GCA/SCA		3.20	1.87	2.17	1.27	2.96	1.55

SOV	DF	ED	NRPE	NKPR	TKW	GYPP	GYPH
Locations (L)	1	2.09 *	36.35 **	353.21 **	8232.88 **	12,079.40 **	40.40 **
Rep (L)	4	0. 26	1.35	12.53	225.52	285.31	0.78
Densities (D)	2	21.93 **	88.74 **	2229.25 **	91,176.13 **	50,563.13 **	58.38 **
L × D	2	1.78 **	6.33 **	331.20 **	3151.63 **	14,971.59 **	54.39 **
Error a	8	0.18	0.52	3.09	192.11	56.68	0.30
Hybrids (H)	27	1.01 **	8.75 **	56.88 **	10,944.20 **	9941.33 **	49.21 **
GCA	7	1.07 **	16.50 **	41.59 **	12,835.71 **	13,527.67 **	67.17 **
SCA	20	0.99 **	6.04 **	62.24 **	10,282.17 **	8686.11 **	42.93 **
H × L	27	1.25 **	18.90 **	21.01 **	2126.28 **	1230.60 **	4.24 **
GCA × L	7	1.56 **	20.27 **	17.14 **	2328.04 **	1621.69 **	5.61 **
SCA × L	20	1.14 **	18.42 **	22.36 **	2055.67 **	1093.72 **	3.75 **
H × D	54	0.20 **	0.98 **	8.77 **	360.55 **	187.93 **	0.71 **
GCA × D	14	0.24	1.45 **	10.49 **	560.46 **	166.93 **	0.73 **
SCA × D	40	0.19	0.82 **	8.17 **	290.58 **	195.28 **	0.70 **
H × D × L	54	1.27 **	13.20 **	16.84 **	621.87 **	1517.64 **	5.45 **
GCA × L × D	14	1.45 **	11.20 **	19.16 **	529.08 **	1951.52 **	6.99 **
SCA × L × D	40	1.21 **	13.89 **	16.02 **	654.34 **	1365.78 **	4.91 **
Error b	324	0.15	0.38	2.41	143.75	45.22	0.25
GCA/SCA		1.08	2.73	0.67	1.25	1.56	1.57

* and ** significant at 0.05 and 0.01 levels of probability, respectively. DTS: days to 50% silking, ASI: anthesis–silking interval, PLHT: plant height, EHT: ear height, LANG: leaf angle, CHLC: chlorophyll content, ED: ear diameter, NRPE: number of rows per ear, NKPR: number of kernels per row, TKW: thousand kernel weight, GYPP: grain yield per plant and GYPH: grain yield per hectare.

2.2. Changes in the Studied Traits Due to Increased Plant Density

Across the two locations, the mean of grain yield per plant (GYPP) was significantly decreased as plant density increased from D1 to D2 and D3 by −9.60 and −20.59%, respectively, as compared to D1 (Figure 1A). This reduction was accompanied by reductions in leaf angle (LANG) (−5.97 and −11.23%), chlorophyll content (CHLC) (−5.48 and −12.15%) and all yield attributes; ear diameter (ED) (−7.68 and −14.01%), number of rows per ear (NRPE) (−6.21 and −9.83%), number of kernels per row (NKPR) (−7.38 and −17.77%), and thousand kernel weight (TKW) (−6.39 and −13.13%) at plant density of D2 and D3, respectively, as compared to D1. Conversely, high plant density (D2 and D3) caused a significant increase in grain yield per hectare (GYPH) compared with the low density (D1) by 8.48 and 11.23%, respectively (Figure 1B). Similarly, D2 and D3 caused significant increases in days to 50% silking (DTS) (5.10 and 11.31%), anthesis–silking interval (ASI) (12.87 and 39.88%), plant height (PLHT) (3.78 and 9.75%) and ear height (EHT) (6.64 and 12.86%) as compared with low plant density (D1), respectively.

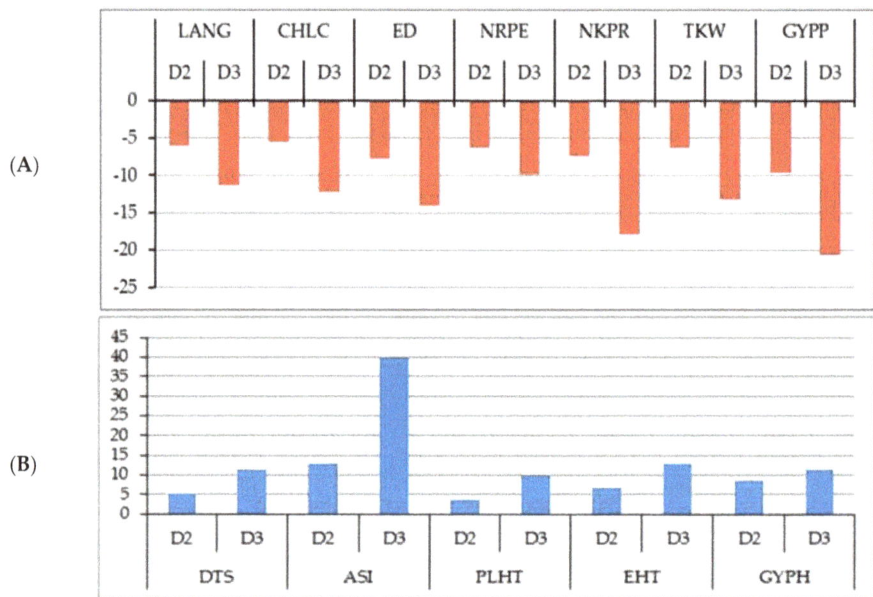

Figure 1. Shows the changes due to increased plant density: (**A**) reduction in leaf angle (LANG), chlorophyll content (CHLC), ear diameter (ED), number of rows per ear (NRPE), number of kernels per row (NKPR), thousand kernel weight (TKW) and grain yield per plant (GYPP); (**B**) increase in days to 50% silking (DTS), anthesis–silking interval (ASI), plant height (PLHT), ear height (EHT) and grain yield per hectare (GYPH) under D2 and D3 in compared with D1.

2.3. Performance of F_1 Hybrids

The mean performances of the 28 F_1 hybrids and the commercial check hybrid SC128 for all the studied characteristics are provided in Supplementary Materials, Table S1. The evaluated hybrids showed a wide variation for all studied traits under all plant densities. The mean values for DTS were 58.22 days in D1, 61.19 days under D2, and 64.80 days in D3 (Table 2). The earliest hybrids were $L_1 \times L_3$ at D1, $L_3 \times L_4$ at D2 and $L_1 \times L_4$ at D3, while the latest hybrids were $L_6 \times L_8$ under D1 and D2 and $L_3 \times L_6$ under D3 (Table 2). A total of 21, 17 and 4 hybrids were significantly earlier than the check hybrid SC128 under D1, D2 and D3, respectively (Supplementary Materials, Table S1). Likewise, the means of ASI were 3.26 days in D1, 3.68 days under D2, and 4.56 days in D3. The longest ASI was shown by the hybrid $L_3 \times L_7$, and the shortest one was shown by $L_2 \times L_5$ under the three plant densities (Table 2). The highest PLHT mean was 263.52 cm in D3, while it was 240.122 cm and 249.20 cm in D1 and D2, respectively. The tallest hybrids were $L_4 \times L_7$ under D1 and D3, and $L_2 \times L_4$ under D2, while the shortest hybrid was $L_2 \times L_6$ under the three plant densities (Table 2). The means of the EHT were 117.86, 125.68 and 133.02 in D1, D2 and D3, respectively. A total of 12, 11 and 14 hybrids were significantly shorter than the check hybrid SC128 under D1, D2 and D3, respectively (Supplementary Materials, Table S1).

Table 2. Minimum, maximum and mean values of all the studied traits under three plant densities across two locations.

Trait	Parameter	D1		D2		D3	
		Value	Hybrid	Value	Hybrid	Value	Hybrid
DTS	Minimum	56.17	$L_1 \times L_3$	58.50	$L_3 \times L_4$	62.52	$L_1 \times L_4$
	Maximum	61.50	$L_6 \times L_8$	65.00	$L_6 \times L_8$	67.60	$L_3 \times L_6$
	Mean	58.22		61.19		64.80	
ASI	Minimum	2.15	$L_2 \times L_5$	2.28	$L_2 \times L_5$	3.12	$L_2 \times L_5$
	Maximum	4.65	$L_3 \times L_7$	5.20	$L_3 \times L_7$	5.65	$L_3 \times L_7$
	Mean	3.26		3.68		4.56	
PLHT (cm)	Minimum	203.17	$L_2 \times L_6$	206.00	$L_2 \times L_6$	213.35	$L_2 \times L_6$
	Maximum	283.00	$L_4 \times L_7$	290.63	$L_2 \times L_4$	304.35	$L_4 \times L_7$
	Mean	240.12		249.20		263.52	
EHT (cm)	Minimum	104.32	$L_3 \times L_6$	109.49	$L_2 \times L_6$	116.42	$L_2 \times L_6$
	Maximum	144.14	$L_6 \times L_7$	151.23	$L_6 \times L_7$	160.29	$L_6 \times L_7$
	Mean	117.86		125.68		133.02	
LANG (°)	Minimum	25.70	$L_4 \times L_5$	24.40	$L_4 \times L_5$	22.00	$L_4 \times L_5$
	Maximum	45.35	$L_3 \times L_7$	42.30	$L_3 \times L_7$	39.05	$L_3 \times L_7$
	Mean	34.03		32.00		30.21	
CHLC (SPAD unit)	Minimum	44.37	$L_7 \times L_8$	41.80	$L_7 \times L_8$	39.39	$L_7 \times L_8$
	Maximum	54.50	$L_2 \times L_8$	53.13	$L_2 \times L_8$	50.20	$L_2 \times L_8$
	Mean	50.34		47.59		44.23	
ED (cm)	Minimum	4.70	$L_1 \times L_7$	4.20	$L_2 \times L_4$	3.60	$L_2 \times L_4$
	Maximum	5.80	$L_1 \times L_8$	5.15	$L_1 \times L_3$	5.00	$L_1 \times L_4$
	Mean	5.16		4.76		4.44	
NRPE	Minimum	13.00	$L_3 \times L_7$	12.30	$L_3 \times L_4$	12.18	$L_1 \times L_3$
	Maximum	16.40	$L_2 \times L_5$	15.20	$L_1 \times L_5$	14.70	$L_1 \times L_5$
	Mean	14.83		13.91		13.37	
NKPR	Minimum	35.20	$L_1 \times L_5$	34.00	$L_1 \times L_5$	30.29	$L_1 \times L_5$
	Maximum	45.10	$L_2 \times L_8$	42.00	$L_2 \times L_8$	37.95	$L_2 \times L_8$
	Mean	40.28		37.31		33.12	
TKW (g)	Minimum	315.00	$L_3 \times L_8$	291.00	$L_5 \times L_6$	276.00	$L_5 \times L_7$
	Maximum	405.00	$L_2 \times L_8$	374.00	$L_1 \times L_4$	353.50	$L_1 \times L_4$
	Mean	356.00		333.24		309.26	
GYPP (g)	Minimum	130.88	$L_3 \times L_7$	122.71	$L_3 \times L_8$	103.75	$L_3 \times L_7$
	Maximum	236.45	$L_2 \times L_8$	215.01	$L_2 \times L_8$	187.44	$L_2 \times L_8$
	Mean	170.11		153.78		135.09	
GYPH ($t\,ha^{-1}$)	Minimum	7.79	$L_3 \times L_7$	8.76	$L_3 \times L_8$	8.78	$L_3 \times L_7$
	Maximum	14.07	$L_2 \times L_8$	15.35	$L_2 \times L_8$	15.61	$L_2 \times L_8$
	Mean	10.12		10.98		11.26	

The hybrid $L_6 \times L_7$ had the highest ear height under the three plant densities, while the hybrids $L_3 \times L_6$ in D1 and $L_2 \times L_6$ under D2 and D3 had the lowest ear heights (Table 2). A total of 13, 20 and 19 hybrids had significantly lower ear placement than the check hybrid SC128 under D1, D2 and

D3, respectively (Supplementary Materials, Table S1). Furthermore, the hybrid $L_4 \times L_5$ displayed the lowest LANG, while $L_3 \times L_7$ gave the highest one under the three plant densities. The means of CHLC were 50.34, 47.59 and 44.23 SPAD units under D1, D2 and D3, respectively. The highest hybrid in CHLC was $L_2 \times L_8$, while the lowest hybrid was $L_7 \times L_8$ across the three plant densities (Table 2). Moreover, the hybrids $L_5 \times L_6$ at D1, $L_3 \times L_4$ at D2 and $L_1 \times L_5$ at D3 significantly surpassed the check hybrid SC128 for this trait (Table S1). The means of ED were 5.16 cm in D1, 4.76 cm under D2, and 4.44 cm in D3. The hybrid $L_1 \times L_7$ at D1 and $L_2 \times L_4$ at D2 and D3 exhibited the lowest ED, while $L_1 \times L_8$, $L_1 \times L_3$ and $L_1 \times L_4$ gave the highest ones under D1, D2 and D3, respectively (Table 2). The mean for the NRPE was 14.83 in D1 and 13.91 in D2, while it was 13.37 in D3. The hybrid $L_2 \times L_5$ under D1 and $L_1 \times L_5$ under D2 and D3 exhibited the highest NRPE, while $L_3 \times L_7$ in D1, $L_3 \times L_4$ under D2 and $L_1 \times L_3$ in D3 had the lowest mean values (Table 2). Additionally, two hybrids under D1, four hybrids at D2 and three hybrids at D3 possessed higher NRPE than the check hybrid SC128 (Supplementary Materials, Table S1). The mean values of the NKPR were 40.28, 37.31 and 33.12 for D1, D2 and D3, respectively. The hybrid $L_2 \times L_8$ had the highest NKPR, but the hybrid $L_1 \times L_5$ displayed the lowest one under the three plant densities. Means of the TKW were 356.0 g, 333.24 g, and 309.26 g in D1, D2, and D3, respectively. The heaviest TKW was assigned for the hybrids $L_2 \times L_8$ under D1 and $L_1 \times L_4$ under D2 and D3, whereas the hybrids $L_3 \times L_8$ in D1, $L_5 \times L_6$ under D2 and $L_5 \times L_7$ under D3 exhibited the lightest TKW (Table 2). Furthermore, four hybrids under D1, five hybrids at D2 and three hybrids at D3 significantly exceeded the check hybrid SC128 for this trait (Supplementary Materials, Table S1). The highest mean of GYPP was 170.11 g in D1, while it was 153.78 and 135.09 g in D2 and D3, respectively. Conversely, the highest mean of GYPH was obtained in D3 (11.26 t ha^{-1}), followed by D2 (10.98 t ha^{-1}) and then by D3 (10.12 t ha^{-1}) (Table 2). The hybrid $L_2 \times L_8$ was the top yielding hybrid and significantly out-yielded the check hybrid SC128 by 9.98, 13.16 and 10.26% under D1, D2 and D3, respectively. Moreover, the hybrid $L_2 \times L_5$ significantly surpassed the check hybrid SC128 by 5.26% only under D2 (Supplementary Materials, Table S1). The optimum plant density for obtaining the highest GYPH was D3 for all hybrids, except the hybrids; $L_2 \times L_7$, $L_3 \times L_4$, $L_3 \times L_7$ and $L_2 \times L_8$, where the optimum density was D2 (Supplementary Materials, Table S1). This indicates that the optimum plant density is genotype dependent and should be identified separately for each hybrid.

2.4. General Combining Ability (GCA) Effects

Estimates of GCA effects are presented in Table 3. High positive values of GCA effects would be of interest for all studied characteristics in question, except DTS, ASI, PLHT, EHT and LANG where high negative values would be desirable from the breeder point of view. Results showed that the highest significant and negative GCA effects under the three plant densities were obtained by the inbred lines L_1 and L_3 for DTS; L_1, L_2 and L_5 for ASI; L_1, L_5, L_6 and L_8 for PLHT; L_3, L_5 and L_8 for EHT and L_1, L_2 and L_4 for LANG. Additionally, the inbred lines L_4 in D1 and D2, as well as L_5 in D3 for DTS; L_4 in D3 and L_8 in D1 and D2 for ASI; L_2 in D3 and L_3 under D1 and D3 for PLHT; and L_5 under D1 and D3 for LANG also expressed significant and negative GCA effects for these traits. In contrast, the inbred lines L_1 in D2 and D3, L_5 under D1 and L_2 under the three plant densities possessed significant and positive GCA effects for CHLC. Regarding ED, the inbred lines L_1 and L_8 in D1 and D3 as well as L_3 in D2 had significant and positive GCA effects.

The highest positive and significant GCA effects for NRPE belonged to L_1 in D2 and D3, L_5 and L_8 in D1 and D3, and L_2 under the three plant densities. Likewise, the inbreds L_3 and L_7 in D1; L_1 and L_6 in D3 and L2 under the three plant densities were determined and considered to be good general combiners for NKPR. The highest positive and significant GCA effects for TKW belonged to L_1 and L_2 under the three plant densities, L_4 under D1 and D2 and L_6 under D3. Furthermore, the inbred lines L_1, L_2 and L_5 under the three plant densities and L_8 under D3 had significant and positive GCA effects for GYPP and GYPH. Based on the summarized results, it can be concluded that parental lines L_1, L_2 and L_5 had the highest GCA effects for grain yield and the majority of studied traits.

Table 3. General combining ability (GCA) effects of the eight parental inbred lines for all the studied traits under three plant densities across two locations.

Inbred Line	DTS			ASI			PLHT			EHT		
	D1	D2	D3	D1	D2	D3	D1	D2	D3	D1	D2	D3
L_1	−0.84 **	−0.94 **	−1.08 **	−0.27 **	−0.26 **	−0.16 **	−7.22 **	−10.48 **	−3.92 **	10.29 **	8.20 **	5.43 **
L_2	−0.21	−0.27	0.07	−0.28 **	−0.31 **	−0.30 **	−0.90	−2.58	−7.66 **	−1.35	−1.28	−1.54
L_3	−0.62 **	−0.81 **	−0.38 **	0.43 **	0.48 **	0.37 **	−3.52 *	−1.12	−4.93 **	−3.74 **	−4.22 **	−3.03 **
L_4	−0.74 **	−0.60 **	−0.20	−0.05	−0.09	−0.23 **	18.43 **	24.20 **	22.27 **	1.48	3.60 **	4.25 **
L_5	0.62 **	0.34 *	−0.56 **	−0.31 **	−0.43 **	−0.30 **	−4.29 **	−3.48 *	−2.93 *	−4.21 **	−6.60 **	−6.32 **
L_6	1.28 **	1.49 **	1.32 **	0.65 **	0.58 **	0.42 **	−8.60 **	−9.06 **	−5.94 **	1.38	1.93	3.43 **
L_7	0.09	0.30 *	0.95 **	0.02	0.14 *	0.13 *	10.17 **	8.45 **	9.75 **	3.13 **	3.12 **	1.33
L_8	0.43 **	0.49 **	−0.12	−0.19 **	−0.11 *	0.07	−4.05 **	−5.94 **	−6.64 **	−6.99 **	−4.75 **	−3.55 **
LSD 0.05	0.28			0.11			2.82			2.00		
LSD 0.01	0.36			0.15			3.70			2.63		

Inbred Line	LANG			CHLC			ED			NRPE		
	D1	D2	D3	D1	D2	D3	D1	D2	D3	D1	D2	D3
L_1	−2.31 **	−2.55 **	−3.35 **	0.07	0.80 **	2.13 **	0.13 *	0.08	0.20 **	−0.14	0.33 **	0.45 **
L_2	−1.81 **	−1.62 **	−0.92 **	1.72 **	2.58 **	3.18 **	0.11	−0.08	−0.19 **	0.39 **	0.43 **	0.50 **
L_3	5.51 **	5.71 **	5.63 **	−1.13 **	−1.32 **	−0.74 **	−0.02	0.13 *	0.06	−0.73 **	−0.55 **	−0.51 **
L_4	−3.77 **	−2.92 **	−2.87 **	−0.20	−0.42	−1.20 **	0.04	−0.03	−0.06	0.13	−0.29 **	−0.19 *
L_5	−0.62 *	−0.13	−0.75 **	1.19 **	−0.38	0.16	−0.09	0.02	−0.02	0.42 **	0.27 **	0.16
L_6	0.84 **	0.55 *	1.34 **	−0.44 *	−0.61 **	−1.81 **	−0.21 **	−0.11	−0.06	0.16	−0.07	−0.08
L_7	1.74 **	1.18 **	1.15 **	−0.97 **	0.00	−1.34 **	−0.13 *	−0.12 *	−0.07	−0.72 **	−0.52 **	−0.37 **
L_8	0.42	−0.21	−0.23	−0.24	−0.66 **	−0.38	0.17 **	0.09	0.13 *	0.48 **	0.40 **	0.04
LSD 0.05	0.50			0.43			0.12			0.19		
LSD 0.01	0.66			0.56			0.15			0.25		

Inbred Line	NKPR			TKW			GYPP			GYPH		
	D1	D2	D3	D1	D2	D3	D1	D2	D3	D1	D2	D3
L_1	−0.55 *	−0.14	0.58 *	12.63 **	16.08 **	17.98 **	12.19 **	9.92 **	8.79 **	0.73 **	0.71 **	0.73 **
L_2	1.47 **	1.06 **	1.15 **	18.79 **	16.50 **	10.23 **	19.20 **	19.23 **	15.67 **	1.14 **	1.37 **	1.30 **
L_3	0.68 **	−0.55 *	−1.17 **	−17.88 **	−10.58 **	−6.60 **	−16.55 **	−16.58 **	−16.95 **	−0.98 **	−1.18 **	−1.39 **
L_4	−0.62 *	0.08	−0.71 **	6.63 **	3.75 *	1.81	−3.93 **	0.90	−0.58	−0.23 **	0.06	−0.05
L_5	−1.43 **	−0.73 **	−0.48 *	0.46	−3.08	−5.10 **	6.92 **	4.29 **	4.19 **	0.41 **	0.31 **	0.34 **
L_6	−0.27	0.46	0.62 *	−5.21 **	−2.58	4.73 *	−13.70 **	−10.97 **	−7.49 **	−0.82 **	−0.78 **	−0.63 **
L_7	0.56 *	−0.01	0.05	−9.04 **	−13.08 **	−13.27 **	−4.05 **	−4.52 **	−6.16 **	−0.24 **	−0.32 **	−0.50 **
L_8	0.15	−0.17	−0.05	−6.38 **	−7.00 **	−9.77 **	−0.08	−2.27 *	2.53 *	0.00	−0.16 *	0.21 **
LSD 0.05	0.48			3.68			2.07			0.15		
LSD 0.01	0.63			4.84			2.72			0.20		

* and ** significant at 0.05 and 0.01 levels of probability, respectively. DTS: days to 50% silking, ASI: anthesis–silking interval, PLHT: plant height, EHT: ear height, LANG: leaf angle, CHLC: chlorophyll content, ED: ear diameter, NRPE: number of rows per ear, NKPR: number of kernels per row, TKW: thousand kernel weight, GYPP: grain yield per plant and GYPH: grain yield per hectare.

2.5. Specific Combining Ability (SCA) Effects

The estimated SCA values under the three plant densities across two locations are presented in Table 4. The hybrids that presented the highest significant and negatives SCA effects (desirable) under the three plant densities were $L_1 \times L_6$, $L_2 \times L_4$, $L_3 \times L_5$, $L_3 \times L_8$, $L_4 \times L_7$ for DTS; $L_1 \times L_7$, $L_2 \times L_5$, $L_2 \times L_7$, $L_2 \times L_8$, $L_3 \times L_4$, $L_3 \times L_6$ and $L_4 \times L_5$ for ASI; $L_1 \times L_4$, $L_2 \times L_6$, $L_2 \times L_7$, $L_2 \times L_8$, $L_3 \times L_4$ and $L_3 \times L_7$ for PLHT; $L_1 \times L_7$, $L_1 \times L_8$, $L_2 \times L_6$ and $L_3 \times L_6$ for EHT and $L_1 \times L_4$, $L_1 \times L_5$, $L_1 \times L_6$, $L_1 \times L_7$, $L_2 \times L_5$, $L_2 \times L_8$, $L_3 \times L_4$, $L_3 \times L_6$, $L_4 \times L_5$, $L_4 \times L_7$ and $L_7 \times L_8$ for LANG. On the contrary, the hybrid combinations; $L_1 \times L_7$, $L_2 \times L_8$, $L_3 \times L_4$ and $L_5 \times L_6$ for CHLC; $L_2 \times L_5$ and $L_2 \times L_7$ for ED; $L_1 \times L_5$, $L_2 \times L_3$, $L_3 \times L_6$ and $L_6 \times L_7$ for NRPE; $L_1 \times L_6$, $L_2 \times L_8$ and $L_6 \times L_7$ for NKPE; $L_1 \times L_4$, $L_1 \times L_6$, $L_2 \times L_5$, $L_2 \times L_8$, $L_3 \times L_5$, $L_4 \times L_5$, $L_6 \times L_7$ and $L_7 \times L_8$ for TKW and $L_1 \times L_3$, $L_1 \times L_6$, $L_2 \times L_5$, $L_2 \times L_8$, $L_3 \times L_4$, $L_3 \times L_6$, $L_4 \times L_5$, $L_6 \times L_7$ and $L_7 \times L_8$ for GYPP and GYPH had the highest significant and positive SCA effects (desirable) under the three plant densities. Moreover, the hybrids $L_1 \times L_5$ in D2 and D3, $L_4 \times L_7$ in D1 and D2 and $L_2 \times L_4$ and $L_5 \times L_7$ under D3 displayed significant and positive SCA effects for GYPP and GYPH. It is notable that the crosses that showed high SCA effects for GYPP and GYPH also showed desirable SCA effects for some other traits, i.e., DTS, LANG, NKPE and TKW for the hybrid $L_1 \times L_6$; ASI, LANG and TKW for the two hybrids $L_2 \times L_5$ and $L_4 \times L_5$; ASI, PLHT, LANG, CHLC, NKPR and TKW for the hybrid $L_2 \times L_8$ and PLHT, NRPE, NKPR and TKW for the hybrid $L_6 \times L_7$.

Table 4. Estimates of specific combining ability (SCA) effects of the 28 F_1 crosses for all the studied traits under the three plant densities across two locations.

Cross	DTS			ASI			PLHT			EHT			LANG			CHLC		
	D1	D2	D3	D1	D2	D3	D1	D2	D3	D1	D2	D3	D1	D2	D3	D1	D2	D3
$L_1 \times L_2$	0.10	0.36	0.28	0.09	0.19	0.53 **	8.70 **	8.46 **	13.08 **	16.34 **	14.76 **	18.45 **	3.69 **	4.89 **	4.69 **	−1.90 **	−2.68 **	−3.97 **
$L_1 \times L_3$	−0.49	−0.42	−0.53	−0.15	0.31 *	0.36 **	2.57	0.49	−7.16 *	7.82 **	11.21 **	9.69 **	3.13 **	3.19 **	3.64 **	0.27	0.95 *	0.44
$L_1 \times L_4$	−0.04	−0.58	−1.03 **	0.34 **	0.32 *	0.42 **	−40.63 **	−36.93 **	−29.11 **	6.74 **	1.02	1.82	−1.29 *	−1.61 **	−1.92 **	−1.66 **	−3.95 **	−4.89 **
$L_1 \times L_5$	0.96 **	0.68 *	1.41 **	0.04	0.08	−0.10	9.59 **	5.85	19.69 **	9.58 **	9.28 **	6.62 **	−2.74 **	−3.20 **	−3.79 **	0.45	1.01 *	2.35 **
$L_1 \times L_6$	−1.05 **	−0.87 **	−1.49 **	−0.10	−0.18	−0.05	6.40 *	7.94 *	3.30	−3.99	−4.10	−6.08 **	−3.21 **	−2.88 **	−1.87 **	0.38	0.84	2.02 **
$L_1 \times L_7$	0.53	1.52 **	0.39	−0.55 **	−0.81 **	−0.89 **	23.63 **	21.17 **	2.15	−20.97 **	−18.24 **	−14.06 **	−2.80 **	−2.98 **	−2.32 **	1.84 **	1.75 **	2.58 **
$L_1 \times L_8$	−0.01	−0.67 *	0.97 **	0.33 **	0.08	−0.27 *	−10.26 **	−6.98 *	−1.95	−15.52 **	−13.93 **	−16.44 **	3.22 **	2.58 **	1.57 **	0.61	2.09 **	1.46 **
$L_2 \times L_3$	1.21 **	2.26 **	1.16 **	0.74 **	0.65 **	0.72 **	22.50 **	34.10 **	35.82 **	5.48 *	3.19	0.98	0.88	0.63	−0.09	−0.65	−1.03 *	−0.57
$L_2 \times L_4$	−0.67 *	−0.84 **	−1.30 **	0.43 **	1.06 **	0.88 **	25.00 **	19.90 **	15.73 **	0.82	0.12	−1.07	1.31 *	0.26	−0.60	−0.91	−0.10	0.39
$L_2 \times L_5$	1.98 **	−0.44	−0.09	−0.53 **	−0.66 **	−0.83 **	−10.73 **	−10.14 **	−5.97	−3.86	−2.44	2.48	−2.74 **	−1.93 **	−1.52 **	−0.97 *	1.04 *	0.04
$L_2 \times L_6$	−0.68 *	0.15	0.83 **	0.01	−0.10	−0.03	−27.25 **	−31.46 **	−36.41 **	−12.74 **	−16.48 **	−18.13 **	−0.71	−2.44 **	−1.03	1.66 **	−1.53 **	−0.52
$L_2 \times L_7$	−1.10 **	−1.15 **	0.50	−0.30 *	−0.61 **	−0.53 **	−6.69 *	−6.98 *	−9.70 **	−4.10	2.13	−0.21	2.50 **	2.93 **	3.40 **	0.04	0.65	1.38 **
$L_2 \times L_8$	−0.84 **	−0.34	−1.38 **	−0.45 **	−0.52 **	−0.75 **	−11.53 **	−13.88 **	−12.56 **	−1.94	−1.28	−2.51	−4.93 **	−4.35 **	−4.86 **	2.73 **	3.64 **	3.25 **
$L_3 \times L_4$	1.19 **	−1.21 **	0.34	−0.89 **	−1.30 **	−1.31 **	−15.53 **	−15.19 **	−12.24 **	3.66	3.71	−1.88	−4.06 **	−3.97 **	−2.44 **	3.44 **	5.64 **	6.31 **
$L_3 \times L_5$	−1.62 **	−0.65 *	−1.04 **	0.61 **	0.77 **	0.62 **	5.89	3.14	0.11	−3.62	1.51	−0.31	1.89 **	1.74 **	1.24 *	−3.44 **	−2.37 **	−3.38 **
$L_3 \times L_6$	−0.27	−1.05 **	1.82 **	−0.87 **	−0.97 **	−0.77 **	5.30	−4.02	0.10	−11.04 **	−10.35 **	−7.78 **	−5.82 **	−3.94 **	−3.81 **	−1.65 **	−3.64 **	−1.92 **
$L_3 \times L_7$	1.41 **	1.89 **	−0.01	0.93 **	0.90 **	0.59 **	−18.37 **	−12.94 **	−13.61 **	−0.76	−5.93 **	−7.95 **	4.18 **	3.53 **	2.14 **	2.44 **	2.35 **	0.92
$L_3 \times L_8$	−1.43 **	−0.80 **	−1.74 **	−0.37 **	−0.36 **	−0.22	−2.36	−5.59	−3.01	−1.55	−3.34	7.25 **	−0.20	−1.18 *	−0.67	−0.42	−1.89 **	−1.81 **
$L_4 \times L_5$	−1.10 **	−0.06	−0.17	−0.47 **	−0.36 **	−0.40 **	−1.56	4.05	9.83 **	−1.54	−3.17	0.54	−3.82 **	−4.42 **	−3.97 **	0.05	−0.77	−1.00 *
$L_4 \times L_6$	−0.16	1.74 **	0.64 *	0.02	0.33 *	0.21	16.55 **	24.75 **	13.42 **	−7.77 **	−5.22 *	−2.11	6.61 **	7.69 **	6.35 **	−2.34 **	−0.54	−1.15 *
$L_4 \times L_7$	−0.67 *	−0.82 **	−0.88 **	0.34 **	0.01	0.25	14.48 **	7.99 *	8.97 **	−1.93	−0.52	0.89	−2.63 **	−3.24 **	−3.67 **	−0.24	−0.28	−0.59
$L_4 \times L_8$	1.44 **	1.79 **	2.39 **	0.23	−0.06	−0.04	1.69	−4.58	−6.59 *	0.03	4.06	1.81	3.88 **	5.28 **	6.25 **	1.66 **	0.01	0.92
$L_5 \times L_6$	0.49	0.05	−1.49 **	0.03	0.02	0.11	−5.53	−7.57 *	−11.28 **	−1.55	−0.52	0.62	0.31	0.40	−0.27	2.56 **	2.02 **	2.80 **
$L_5 \times L_7$	−0.03	1.39 **	0.68 *	−0.27 *	−0.21	−0.12	−4.32	−5.58	−16.94 **	1.50	−0.83	−5.39 *	3.57 **	3.94 **	5.21 **	0.09	−0.99 *	−0.47
$L_5 \times L_8$	−0.67 *	−0.95 **	0.70 *	0.58 **	0.36 **	0.72 **	6.66 *	10.24 **	4.57	−0.51	−3.83	−4.56 *	3.53 **	3.46 **	3.10 **	1.26 **	0.07	−0.33
$L_6 \times L_7$	0.0	−1.91 **	−0.01	0.54 **	0.55 **	0.34 **	−9.99 **	−7.05 *	20.23 **	21.93 **	20.87 **	22.87 **	1.75 **	1.39 *	0.63	0.53	1.64 **	−0.78
$L_6 \times L_8$	1.67 **	1.90 **	−0.29	0.37 **	0.34 **	0.19	14.52 **	17.40 **	10.65 **	15.17 **	15.80 **	10.60 **	1.07	−0.22	0.02	−1.15 *	1.20 *	−0.45
$L_7 \times L_8$	−0.15	−0.91 **	−0.66 *	−0.69 **	0.18	0.36 **	1.26	3.38	8.90 **	4.33	2.52	3.85	−6.57 **	−5.56 **	−5.40 **	−4.70 **	−5.11 **	−3.04 **
LSD 0.05		0.61			0.25			6.23			4.42			1.12			0.95	
LSD 0.01		0.80			0.33			8.19			5.81			1.47			1.24	

Table 4. *Cont.*

Cross	ED			NRPE			NKPR			TKW			GYPP			GYPH		
	D1	D2	D3	D1	D2	D3	D1	D2	D3	D1	D2	D3	D1	D2	D3	D1	D2	D3
$L_1 \times L_2$	−0.13	−0.17	−0.14	−0.05	−0.16	−0.31	−0.74	−0.47	−1.83 **	13.26 **	6.92	3.06	0.63	−1.34	−2.05	0.04	−0.10	−0.16
$L_1 \times L_3$	0.21	0.17	−0.19	−0.69 **	−0.88 **	−1.12 **	−1.34 *	0.43	1.38*	3.93	2.00	−6.11	16.42 **	10.99 **	10.06 **	0.98 **	0.78 **	0.82 **
$L_1 \times L_4$	0.32 *	0.19	0.42 **	0.20	0.76 **	0.50 *	0.96	0.97	0.03	27.43 **	21.67 **	24.48 **	−24.25 **	−22.70 **	−18.21 **	−1.44 **	−1.62 **	−1.51 **
$L_1 \times L_5$	−0.34 **	−0.07	0.11	0.42 *	0.69 **	0.73 **	−3.04 **	−2.34 **	−1.39 **	−33.40 **	−24.50 **	−16.11 **	2.81	10.32 **	12.43 **	0.17	0.74 **	1.04 **
$L_1 \times L_6$	0.03	0.16	0.12	−0.18	−0.36	−0.30	2.61 **	2.77 **	2.93 **	25.26 **	25.00 **	22.06 **	33.52 **	28.56 **	22.11 **	1.99 **	2.04 **	1.85 **
$L_1 \times L_7$	−0.45 **	−0.13	−0.26 *	0.25	0.29	0.15	−0.63	−2.06 **	−1.10 *	−9.90 *	−10.50 *	−6.94	−18.98 **	−21.89 **	−18.42 **	−1.13 **	−1.56 **	−1.55 **
$L_1 \times L_8$	0.35 **	−0.14	−0.06	0.05	−0.33	0.35	2.19 **	0.70	−0.03	−26.57 **	−20.58 **	−20.44 **	−10.14 **	−3.94	−5.92 *	−0.60 **	−0.28	−0.49 **
$L_2 \times L_3$	−0.05	0.08	−0.31*	0.74 **	0.82 **	1.05 **	2.53 **	0.28	−2.66 **	−12.24 **	−7.42	1.14	−15.69 **	−15.14 **	−6.32 **	−0.93 **	−1.08 **	−0.54 **
$L_2 \times L_4$	−0.51 **	−0.46 **	−0.59 **	−1.02 **	−0.04	0.02	−2.07 **	−1.56 **	−0.02	−12.74 **	−9.75 *	−14.77 **	−4.15	2.88	8.81 **	−0.25	0.21	0.74 **
$L_2 \times L_5$	0.42 **	0.29 *	0.57 **	0.79 **	0.39	0.18	0.00	1.13 *	1.03	18.43 **	22.58 **	14.64 **	25.96 **	23.99 **	18.55 **	1.54 **	1.71 **	1.55 **
$L_2 \times L_6$	−0.05	−0.08	0.01	−0.73 **	−0.96 **	−1.18 **	−4.42 **	−3.73 **	−1.99 **	−18.90 **	−14.42 **	−6.69	−33.57 **	−30.75 **	−29.77 **	−2.00 **	−2.20 **	−2.47 **
$L_2 \times L_7$	0.36 **	0.43 **	0.63 **	0.33	−0.31	−0.30	1.45 **	0.44	1.60 **	−25.07 **	−25.92 **	−23.69 **	−21.99 **	−25.20 **	−24.60 **	−1.31 **	−1.80 **	−2.07 **
$L_2 \times L_8$	−0.04	−0.08	−0.17	−0.05	0.27	0.54 *	3.26 **	3.90 **	3.88 **	37.26 **	28.00 **	26.31 **	48.82 **	45.56 **	35.39 **	2.90 **	3.25 **	2.95 **
$L_3 \times L_4$	−0.18	−0.06	0.16	−0.41	−0.76 **	−0.47 *	0.62	−0.34	1.78 **	10.93 **	7.33	−17.94 **	21.41 **	20.99 **	9.83 **	1.27 **	1.50 **	0.80 **
$L_3 \times L_5$	−0.05	0.08	0.12	−0.49 *	−0.12	0.19	2.83 **	2.38 **	0.60	27.10 **	15.17 **	22.98 **	−15.18 **	−17.28 **	−14.84 **	−0.90 **	−1.23 **	−1.25 **
$L_3 \times L_6$	−0.12	−0.09	0.36 **	1.07 **	0.72 **	0.93 **	−0.33	−0.91	−1.05	−2.24	4.67	5.14	16.68 **	10.35 **	14.54 **	0.99 **	0.74 **	1.20 **
$L_3 \times L_7$	−0.01	−0.18	−0.12	−0.36	−0.03	0.11	−2.86 **	−0.35	−0.18	−11.40 **	−10.83 **	−0.86	−17.02 **	1.01	−6.99 **	−1.01 **	0.07	−0.48 **
$L_3 \times L_8$	0.19	0.01	−0.02	0.14	0.25	−0.69 **	−1.44 **	−1.49 **	0.12	−16.07 **	−10.92 **	−4.36	−6.61 **	−10.93 **	−6.29 **	−0.39 *	−0.78 **	−0.54 **
$L_4 \times L_5$	0.19	0.14	−0.06	−0.04	0.38	0.06	−0.57	−0.66	1.09 *	22.60 **	29.83 **	38.56 **	31.54 **	18.23 **	14.79 **	1.88 **	1.30 **	1.24 **
$L_4 \times L_6$	0.22	0.18	0.08	0.21	0.46 *	0.11	0.67	0.15	−2.01 **	3.26	2.33	8.73 *	−9.18 **	−6.02 **	−5.72 *	−0.55 **	−0.43 *	−0.47 **
$L_4 \times L_7$	−0.07	−0.16	−0.11	0.78 **	0.21	0.29	1.84 **	1.72 **	0.26	−23.90 **	−22.17 **	−19.27 **	12.90 **	5.74 *	0.74	0.77 **	0.41 *	0.05
$L_4 \times L_8$	0.03	0.18	0.09	0.28	−1.01 **	−0.52 *	−1.44 **	−0.27	−1.14 *	−27.57 **	−29.25 **	−19.77 **	−28.26 **	−19.12 **	−10.25 **	−1.68 **	−1.37 **	−0.85 **
$L_5 \times L_6$	−0.05	−0.18	−0.36 **	−0.38	−0.91 **	−0.23	1.38 *	0.87	1.30 *	−32.57 **	−35.83 **	−31.86 **	−21.29 **	−10.90 **	−12.19 **	−1.27 **	−0.78 **	−1.01 **
$L_5 \times L_7$	0.06	−0.07	−0.14	−0.90 **	−0.86 **	−0.91 **	−1.65 **	−1.07 *	−2.07 **	−2.74	−5.33	−14.36 **	−4.05	0.14	4.58 *	−0.24	0.01	0.36 *

Table 4. *Cont.*

Cross	ED			NRPE			NKPR			TKW			GYPP			GYPH		
	D1	D2	D3	D1	D2	D3	D1	D2	D3	D1	D2	D3	D1	D2	D3	D1	D2	D3
$L_5 \times L_8$	−0.24	−0.18	−0.24	0.60 **	0.43 *	−0.02	1.06 *	−0.31	−0.57	0.60	−1.92	−13.86 **	−19.79 **	−24.51 **	−23.32 **	−1.18 **	−1.75 **	−1.94 **
$L_6 \times L_7$	0.19	−0.04	−0.31 *	0.46 *	0.69 **	0.49 *	2.79 **	2.35 **	2.28 **	32.93 **	29.17 **	17.81 **	23.51 **	18.01 **	22.66 **	1.40 **	1.29 **	1.87 **
$L_6 \times L_8$	−0.21	0.05	0.09	−0.45 *	0.37	0.18	−2.69 **	−1.50 **	−1.47 **	−7.74	−10.92 **	−15.19 **	−9.67 **	−9.25 **	−11.64 **	−0.58 **	−0.66 **	−0.96 **
$L_7 \times L_8$	−0.09	0.16	0.31 *	−0.57 **	0.02	0.16	−0.93	−1.03	−0.79	40.10 **	45.58 **	47.31 **	25.64 **	22.19 **	22.03 **	1.53 **	1.58 **	1.82 **
LSD 0.05	0.26			0.42			1.06			8.15			4.57			0.34		
LSD 0.01	0.34			0.55			1.39			10.71			6.01			0.45		

* and ** significant at 0.05 and 0.01 levels of probability, respectively. DTS: days to 50% silking, ASI: anthesis–silking interval, PLHT: plant height, EHT: ear height, LANG: leaf angle, CHLC: chlorophyll content, ED: ear diameter, NRPE: number of rows per ear, NKPR: number of kernels per row, TKW: thousand kernel weight, GYPP: grain yield per plant and GYPH: grain yield per hectare.

2.6. SSR Polymorphisms, Genetic Distance (GD) and Cluster Analysis

Out of twenty-two SSR primer pairs analyzed, ten were polymorphic among the eight inbreds studied (Table 5). The primer pairs generated a total of 80 polymorphic fragments (Figure 2). The number of alleles per locus ranged from 2 to 6, with an average number of 2.7 alleles/locus (Table 5). The major allele frequency had an average of 0.59 with a range extended from 0.25 to 0.88. The gene diversity and polymorphic information content (PIC) averaged 0.50 and 0.41, with ranges of 0.22–0.81 and 0.19–0.79, respectively. The umc1033 locus showed the highest gene diversity and PIC (Table 5). Genetic distance estimates based on SSR markers ranged from 0.31 to 0.78 with an average of 0.61 (Table 6). The lowest genetic distance (0.31) was obtained between the inbred lines (L_1 and L_4), whereas the highest genetic distance (0.78) was observed between the inbred lines (L_1 and L_8), (L_2 and L_5), (L_2 and L_6) and (L_2 and L_8). The dendrogram constructed based on GD revealed two main clusters; L_1, L_2, L_3 and L_4 constituted the first group, while L_5, L_6, L_7 and L_8 formed the second one (Figure 3).

Table 5. Number of alleles, major allele frequency, gene diversity and polymorphic information content (PIC) of the ten SSR markers used in this study.

Marker	Ch.	Size Range (bp)	No. of Alleles	Major Allele Frequency	Gene Diversity	PIC
phi308707	1	125–140	2	0.63	0.47	0.36
phi96100	2	150–200	2	0.88	0.22	0.19
phi453121	3	150–200	2	0.50	0.50	0.38
phi072	4	100–150	2	0.75	0.38	0.30
phi024	5	100–200	2	0.50	0.50	0.38
umc1014	6	100–150	3	0.50	0.59	0.51
phi112	7	150–200	3	0.50	0.59	0.51
phi015	8	50–150	3	0.50	0.59	0.51
umc1033	9	50–200	6	0.25	0.81	0.79
phi301654	10	100–150	2	0.88	0.22	0.19
	Mean		2.7	0.59	0.50	0.41

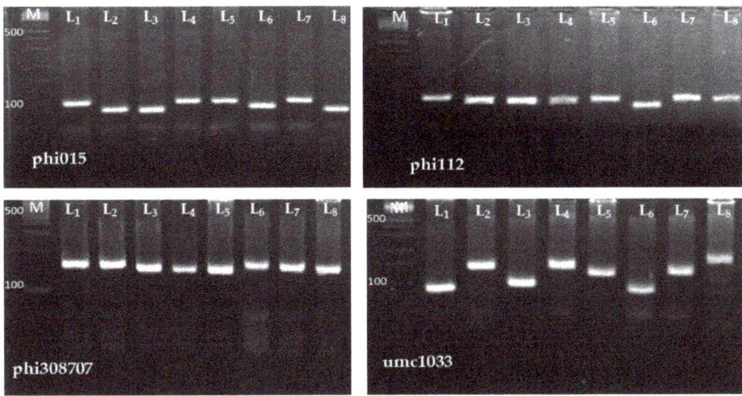

Figure 2. Amplification pattern of representative SSR markers with the eight maize inbred lines (L_1–L_8). M refers to the 100 bp DNA ladder.

Table 6. Genetic distance (GD) matrix among the eight maize inbred lines based on SSR analysis.

Parent	L_1	L_2	L_3	L_4	L_5	L_6	L_7	L_8
L_1	-	0.43	0.53	0.31	0.71	0.71	0.71	0.78
L_2		-	0.43	0.53	0.78	0.78	0.71	0.78
L_3			-	0.43	0.63	0.63	0.63	0.71
L_4				-	0.63	0.71	0.63	0.71
L_5					-	0.63	0.43	0.71
L_6						-	0.43	0.63
L_7							-	0.53
L_8								-

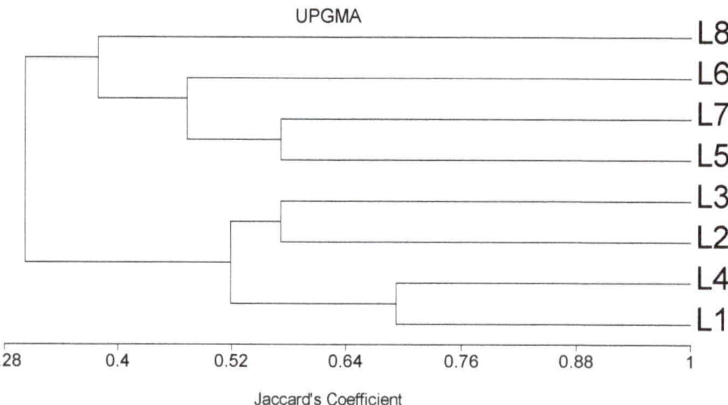

Figure 3. Dendrogram of the eight maize inbred lines constructed from SSR data using (UPGMA) according to Jaccard's coefficients.

2.7. Association between Genetic Distance, F_1 Hybrid Performance and SCA

Correlations between GD estimated for pairs of inbred lines with each of F_1 hybrid performance and SCA were not significant for all measured traits (Table 7, Figure 4A,B). However, significant and positive association was observed between F_1 hybrid performance and SCA for all the studied traits across the three plant densities (Table 7).

Table 7. Correlation coefficients among parental genetic distance (GD), F_1 hybrid performance and SCA for all studied traits across all environments.

Trait	DTS	ASI	PLHT	EHT	LANG	CHLC	ED	NRPE	NKPR	TKW	GYPP	GYPH
r (GD, F_1)	0.20	−0.26	−0.20	−0.60	−0.09	0.30	0.13	0.26	0.04	−0.21	0.05	0.05
r (GD, SCA)	0.01	−0.26	0.00	−0.55	−0.07	0.29	0.11	0.12	−0.25	−0.26	0.04	0.04
r (F_1, SCA)	0.69 **	0.78 **	0.75 **	0.83 **	0.70 **	0.80 **	0.85 **	0.71 **	0.90 **	0.83 **	0.80 **	0.80 **

** significant at 0.01 level of probability.

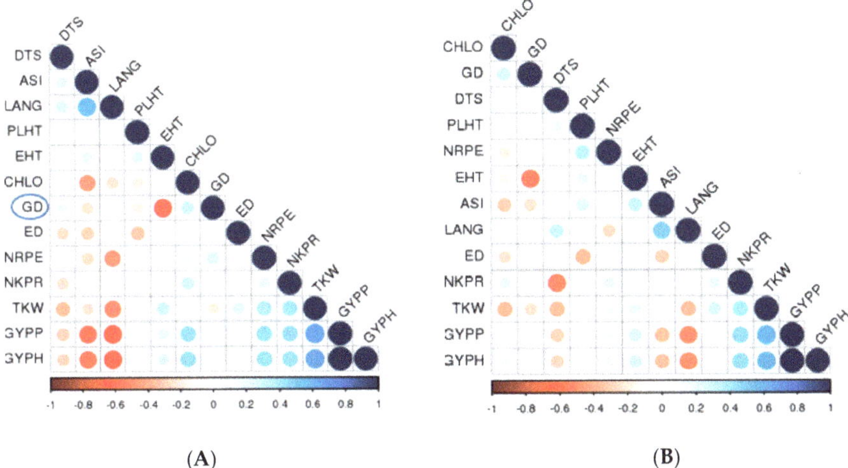

Figure 4. Corrplot depicting correlation coefficient of genetic distance based on molecular data with F$_1$ hybrid performance (**A**) and SCA (**B**) for all studied traits. GD: genetic distance, DTS: days to 50% silking, ASI: anthesis–silking interval, PLHT: plant height, EHT: ear height, LANG: leaf angle, CHLC: chlorophyll content, ED: ear diameter, NRPE: number of rows per ear, NKPR: number of kernels per row, TKW: thousand kernel weight, GYPP: grain yield per plant and GYPH: grain yield per hectare.

3. Discussion

3.1. Analysis of Variance and Hybrid Performance

The significant mean squares of L, D and H observed for all the studied characteristics (Table 2), indicate that the tested locations and densities were dissimilar and there were adequate genetic differences among the hybrids for effective selection of all the studied traits. Significant differences among maize hybrids under different plant densities were also reported [10,35–37]. The presence of significant mean squares for H × D interaction, indicated inconsistent performance of the hybrids across plant densities. In that context, the ranks of maize hybrids differed from one density to another for all measured traits. Therefore, selection of hybrids under various plant densities may be a promising strategy to improve the adaptation of maize hybrids to higher plant density. These results are consistent with the findings of other studies [12,13,36,38].

The significant GCA and SCA effects imply that both additive and non-additive gene effects are involved in governing all traits. The inheritance of a specific trait could be identified based on the ratio of GCA/SCA variances. In the present study, the GCA/SCA ratio was greater than unity for all evaluated characteristics, except NKPR, which indicated the preponderance of additive gene effects in controlling the inheritance of all measured traits, except NKPR which was mainly controlled by non-additive gene action. Therefore, selection breeding methods can be effective for improvement of these traits. This finding is in agreement with that of Mason and Zuber [25] and Al-Naggar et al. [7], who reported that additive genetic effects were important in the inheritance of grain yield and other agronomic traits under different plant densities. However, this result is in contrast to the findings of other studies [36,39], who reported that non-additive gene effects were found to be more important in controlling grain yield inheritance under varying plant densities.

The significant GCA × L and GCA × D interactions mean squares for most traits in the present study indicate that the GCA effects of the inbred lines varied significantly under different environments. This result is in agreement with the findings of several authors [17,26,40,41]. Likewise, the significant SCA × L and SCA × D interactions observed for most traits implied that the performance of the hybrids was not consistent under varying research environments. This suggests the need for extensive evaluation of the hybrids in multiple environments in order to identify high yielding and most stable hybrids tolerant to high plant densities [39].

The highest GYPP of all evaluated hybrids in this study was observed under low density (D1), where competition between plants is minimum [12]. As planting density increases, resources to each plant (water, nutrients and light interception) decrease, increasing plant–plant competition and in turn reducing the assimilated supply to developing cobs and, consequently, resulting in a reduction in grain yield per plant [42–44]. The observed reduction in GYPP due to elevating plant density from D1 to D2 and D3 in this study could be a result of the reduction in all yield attributesED, NRPE, NKPR and TKW. These results are consistent with Tang et al. [45], who stated that increasing plant density in maize leads to a reduction in ear diameter, grains per ear, thousand kernels weight and finally single plant yield. Hashemi et al. [46] also demonstrated that grain yield per plant and all yield components linearly decreased with increasing plant density. Moreover, increasing plant density also reduced LANG and CHLC. The decrease in the leaf angle and chlorophyll content in response to high plant density has also been reported previously in maize [13,47,48].

On the other hand, high plant density (D3) caused significant increases in DTS, ASI, PLHT, EHT and GYPH compared with the low density (D1). Delayed silking and increased ASI period, as symptoms of intense interplant competition for growth resources, can be associated with significant yield reductions [15,49]. Increasing plant density initiated greater stress during pollination that can lead to increasing kernel abortions and decreasing grain fill [8,11]. These two traits (early DTS and short ASI) could be effective indicators for selecting high density tolerance hybrids [50]. The increased values of PLHT and EHT might be related to the stress imposed on maize plants due to competition for light resulting from elevated plant density which potentially increase stem elongation [51,52]. The increase in GYPH with increasing plant density is largely attributed to the higher number of plants per unit area. This suggested that the increase in GYPH due to increased plant density may offset the reduction in GYPP due to competition between plants. These results are in accordance with the results reported in other studies [10,12,53,54].

The two hybrids $L_2 \times L_5$ and $L_2 \times L_8$ had the highest GYPP and GYPH under three plant densities, and could be considered the most highly responsive and tolerant to high plant density. Interestingly, the hybrid $L_2 \times L_8$ significantly outyielded the check hybrid SC128 under all densities; moreover, it had outstanding features, such as short ASI, short plant and ear position, erect leaf under high plant density. Therefore, this hybrid should be tested extensively in multilocation trials and promoted for adoption to high plant density tolerance. Similar to our results, Al-Naggar et al. [12] reported that the selection of hybrids with high grain yield, better plant and ear heights, short ASI, and erect leaf under high plant density stress is important for the development of tolerant hybrids to high plant densities.

3.2. GCA and SCA Estimates

Combining ability analysis helps in the identification of parents with good GCA effects and hybrids with good SCA effects [23]. Selection of parents giving good-performing hybrids is one of the challenges facing breeders. Parents with desirable GCA effect for the target traits can be used to accumulate favorable alleles by recombination and selection [55]. In the current study, high GCA values for the evaluated traits were scattered among the eight inbred lines and changed across plant densities, demonstrating the effects of plant densities on GCA values. Moreover, none of the inbred lines exhibited significant GCA effects for all the measured traits under any of the testing densities. Similar results were reported by other researchers [56,57]. The significant and negative GCA effects were displayed by the inbreds L_1 and L_3 for DTS and L_1, L_2 and L_5 for ASI across the three plant densities, indicating that, these inbreds could be good combiners and possessed favorable alleles towards earliness. Likewise, inbred lines L_5 and L_8 were the best general combiners for reduced plant and ear heights which are important for lodging tolerance especially under high plant density. The inbred line L_2 had the highest positive GCA values for CHLC, NRPE, NKPR and TKW suggesting that this line could be good combiner for improving these traits. Moreover, the best general combiners for

GYPP and GYPH were L_1, L_2, and L_5 under the three plant densities and L_8 under D3. These inbreds could transfer desirable alleles for improved grain yield to their progenies to develop hybrids tolerant to high plant density. The superiority of these inbreds in GCA effects for grain yield was associated with their superiority in GCA effects for some other traits. Interestingly, the inbred line L_1, which had desirable GCA effects for GYPP and GYPH, was also found to be good a general combiner for earliness, short ASI, short PLHT, reduced LANG and increased TKW. Previous findings proved that positive GCA effects for grain yield and negative GCA effects for DTS, PLHT, and LANG traits are a good indicator of high plant density tolerance [13]. Thus, the inbred line L_1 has potential to be used to improve maize grain yield under high plant density.

Estimates of SCA effects provide important information about the non-additive gene effects (dominance and epistatic interaction), which can also be related to hybrid vigor, assisting in the selection of the best hybrid combinations [58]. The highly positive and significant SCA effects for grain yield and its components indicated that the produced hybrids were good specific combiners for developing high-yielding hybrids [1]. In the present study, the most promising specific combiners for grain yield (GYPP or GYPH) and some of its components were $L_1 \times L_3$, $L_1 \times L_6$, $L_2 \times L_5$, $L_2 \times L_8$, $L_4 \times L_5$ and $L_7 \times L_8$ under the three plant densities. These hybrids involved at least one high GCA parent, which could be exploited by conventional breeding procedures. This finding is in line with the result reported in other studies [56,59]. In their studies, high SCA was observed in cross combinations involving one line with high GCA and another with low GCA effects.

Two hybrids, $L_2 \times L_5$ and $L_2 \times L_8$, had desirable significant positive SCA coupled with high mean grain yield under the three plant densities, revealing good correspondence between mean grain yield and SCA effects [1]. Regardless of their significant SCA effects, three crosses $L_3 \times L_4$, $L_3 \times L_6$ and $L_6 \times L_7$, constituted from parents with low × low GCA effects for GYPP and GYPH were not favorable due to insufficient additive variance. This indicates that both GCA and SCA should be taken into consideration in the selection of elite parents for the development of heterotic hybrids [18]. It is notable that none of the hybrids exhibited significant SCA effects for all the traits. However, the hybrids $L_2 \times L_5$, $L_2 \times L_8$ and $L_4 \times L_5$ were found to be good specific combiners for more than one trait, such as ASI, LANG, TKW, GYPP and GYPH. Accordingly, these hybrids would be useful to increase maize grain yield under high plant density for their complementary characteristics, including, short ASI, erect leaf and high grain yield under high plant density. In concordance with the findings reported here, desirable significant SCA under high plant density for ASI, LANG and grain yield has previously been reported by Al-Naggar et al. [13].

3.3. SSR Polymorphisms, Genetic Distance (GD) and Cluster Analysis

The mean number of alleles (2.7) per locus obtained in this study was close to the values reported by other researchers [26,27,34], who detected averages of 2.9, 2.57 and 3.0 alleles per locus, respectively. However, it was lower than the 6.21 alleles/locus reported by Oppong et al. [60] or the 5.7 alleles/locus found by Oyekunle et al. [61] in maize inbred lines using SSR markers. The differences in the means of alleles among different studies could be attributed to the differences in sample size, repeat length and number of the SSR markers involved in the studies [27]. The lower values observed in this study could arise from the small number of lines used for genotyping.

The PIC demonstrates the informativeness of the SSR loci and their potential to detect differences among the inbred lines based on their genetic relationships [62]. Informative markers can be categorized as highly informative (PIC > 0.5), reasonably informative (0.5 < PIC < 0.25) and slightly informative (PIC < 0.25), as reported by Botstein et al. [63]. Accordingly, four markersumc1014, phi112, phi015 and umc1033 with high PIC values, and hence high discriminatory power, were identified. The average gene diversity (0.50) detected among the tested inbred lines in this study indicated high levels of polymorphisms within the inbred lines. This result is in close agreement with the findings reported in other studies [30,64]. The frequency of the most common (major) alleles had an average of 0.59, suggesting that 59.0% of the studied inbreds shared a common major allele at any of the tested loci.

Assessing the genetic diversity is essential for enhancing the yield and conservation strategies of main crops [65–70], such as maize that has high an economic importance [71]. The average genetic diversity existing among all the inbred lines was relatively high (0.61). This indicated that there was considerable genetic diversity among the inbreds based on the microsatellite markers analysis [72]. The largest GD in this study was between the Egyptian (local) and CIMMYT (exotic) inbred lines. The relatively large genetic distance between local and exotic lines, suggesting the opportunity to use these lines for the development of high-yielding and stress-tolerant hybrids. Indeed, the two high-yielding hybrids ($L_2 \times L_5$ and $L_2 \times L_8$) under the three plant densities consisted of local × exotic line combinations. This indicates that novel and complementary alleles existing in the germplasm from the two countries can be exploited for superior maize hybrid development and population improvement [73]. Moreover, it implies the potential benefits of exchanging germplasm between breeding programs for the development of high yielding and density tolerant hybrids.

The dendrogram constructed using the UPGMA clustering grouped the inbred lines into two main clusters, which generally agreed with their origin. One cluster was composed of CIMMYT inbred lines, while the other consisted of local inbreds. This result is consistent with the findings of Mageto et al. [17], who reported that clustering based on GD grouped maize inbred lines according to their origin. Similarly, [34,64] revealed the effectiveness of SSR markers for classifying maize inbreds according to their origin in their studies.

3.4. Association between Genetic Distance, F1 Hybrid Performance and SCA

Our results showed that GD of the parental inbreds was not significantly correlated with the mean of F_1 hybrids for any of the evaluated traits across the tested environments. This implied that the SSR-based GD could not be used to predict the performance of F_1 hybrids in this study. This result is consistent with those reported by [26,33,34,40]. Bernardo [74] attributed this poor correlation to the lack of linkage between genes controlling the trait and markers used to estimate GD, inadequate genome coverage and different levels of dominance among hybrids. Contrary to the current finding, a significant correlation was reported between molecular GD and F_1 hybrid performance [32,75]. There was no significant correlation between GD and SCA for all the traits, suggesting that SSR-based GD might not be effective in predicting SCA effects in the studied materials. Similarly, non-significant association between genetic distances and SCA was reported by [40,76]. However, Betran et al. [75] reported a significant correlation between GD and SCA for maize grain yield. Furthermore, our results showed that SCA effects were significantly correlated with F_1 hybrid performance for all the traits. This indicated that SCA could be used to predict the performance of F_1 hybrids. This result is in agreement with the findings of [17,26].

4. Conclusions

This study revealed a considerable variability among F_1 hybrids for all traits under different plant densities. Additive and non-additive gene effects are involved in the genetic control of all traits, with a predominance of the additive gene action for most traits. Selection of potential hybrids for density tolerance breeding programs should be based on both GCA and SCA effects. The inbred lines L_1 and L_3 were identified as excellent combiners for earliness, L_5 and L_8 for reduced plant and ear heights and L_1, L_2, and L_5 for increased grain yield under varying plant densities. The best hybrids $L_2 \times L_5$ and $L_2 \times L_8$ for grain yield and other multiple traits were identified for further evaluation. The estimated GD based on SSR markers in this study could not be used to predict the hybrids performance and SCA effects. Nevertheless, SCA could be used to predict the hybrids performance across all plant densities. Although SSR determined that GD was not useful in predicting hybrid performance and SCA effects, it was effective in classifying the inbred lines according to their origin, signifying the efficiency of SSR marker for diversity and clustering analyses. The findings of the present study might have important implications for breeding programs designed to improve density tolerance in maize.

5. Materials and Methods

5.1. Plant Materials

Eight white maize (*Zea mays* L.) inbred lines showing clear differences in grain yield and other agronomic characteristics were chosen as parents in this study. Four inbreds (L_1, L_2 L_3 and L_4) were obtained from Maize Research Department, Agricultural Research Center (ARC) in Egypt and the other four (L_5, L_6, L_7 and L_8) were introduced from the International Maize and Wheat Improvement Center (CIMMYT). The parental codes, names and sources of these inbred lines are listed in Table 8.

Table 8. Code, name and source of the parental maize inbred lines.

Parent Code	Name	Source
L1	IL36	ARC-Egypt
L2	IL94	ARC-Egypt
L3	IL53	ARC-Egypt
L4	IL38	ARC-Egypt
L5	CML538	CIMMYT-Mexico
L6	CML203	CIMMYT-Mexico
L7	CML206	CIMMYT-Mexico
L8	CML441	CIMMYT-Mexico

5.2. Production and Evaluation of F_1 Hybrids

In the 2017 season, all possible diallel crosses (excluding reciprocals) were made among the eight inbred lines to obtain seeds of 28 F_1 hybrids. In the 2018 season, the resulting 28 F_1 white hybrids plus the commercial check hybrid SC128 were evaluated under three plant densities, i.e., 59,500 (D1), 71,400 (D2) and 83,300 (D3) plants ha^{-1} at two locations. The two locations were El-Mahmoudia, El-Behira, Egypt ((31°3′ N, 30°48′ E)) in a private farm, and the Experimental Farm, Faculty of Agriculture, Kafrelsheikh University, Egypt ((31°6′ N, 30°56′ E)). A split-plot design in randomized complete blocks (RCB) arrangement with three replications was used in each location. The three plant densities were located at the main plots, while the hybrids were located at the sub plots. Each subplot consisted of one ridge of 6 m long and 0.7 m width. Two seeds were sown in hills at 24, 20 and 17 cm apart, and thereafter (before the 1st irrigation) were thinned to one plant/hill to achieve the three plant densities, i.e., D1, D2 and D3, respectively. Phosphorus at the rate of 476 kg ha^{-1} in the form of calcium super phosphate (15.5% P_2O_5) was added to the soil during seedbed preparation, and potassium sulphate (48% K_2O) at a level of 120 kg ha^{-1} was applied after thinning. Moreover, nitrogen at the rate of 286 kg ha^{-1} was added in two equal doses before the first and second irrigation. All other standard agronomic practices including weed control were followed in each location. Soil analysis was conducted on soil samples collected from 30 cm depth from each location according to Association of Officinal Analytical Chemists (A.O.A.C 2005) [77] (Supplementary Materials, Table S2). Additionally, the meteorological data are presented in the Supplementary Materials, Figure S1.

5.3. Data Collection

Data were collected on days to 50% silking (DTS, days from the planting to 50% extrusion of silks from the plants), anthesis–silking interval (ASI, calculated as the difference between days to 50% silking and days to 50% anthesis), plant height (PLHT, measured in cm as the distance from the soil surface to the top of the first tassel branch) and ear height (EHT, measured in cm as the distance from the soil surface to the base of the topmost ear). Leaf angle (LANG) (°) was measured as the angle between stem and blade of the leaf just above ear leaf. Chlorophyll content (CHLC, SPAD units) was measured by hand-held chlorophyll meter (SPAD-502; Minolta Sensing Co., Ltd., Hangzhou, Japan) from the leaf of the top-most ear. The LANG and CHLC values were recorded on ten guarded plants within each plot, and then the values were averaged per plot. At harvest, ear diameter (ED), number of rows per ear (NKPR), number of kernels per row (NKPR), thousand kernel weight (TKW), grain yield per plant (GYPP, in g plant^{-1}) and grain yield per hectare (GYPH, in ton ha^{-1}) were estimated. Plots were hand-harvested, and the weight of the shelled grain (adjusted to 15.5% grain moisture content) was used to calculate GYPP and GYPH. Grain moisture at harvest was measured using a hand-held moisture meter.

5.4. Molecular Analysis

5.4.1. DNA Isolation

Leaves were sampled from 10 to 15 seedlings of each inbred line after twenty days from planting. Genomic DNA was isolated using CTAB method [78]. DNA quantity as well as quality was assessed using NanoDrop spectrophotometer (ND-1000, USA).

5.4.2. SSR Primers and PCR Amplification

Twenty-two SSR markers were randomly selected from the MaizeGDB database (www.maizegdb.org). The 22 primer pairs were tested to identify the polymorphic ones. Only ten markers were found to be polymorphic and they used for the SSR analysis (Supplementary Materials, Table S3). PCR was performed in a volume of 10 µL reaction mixture containing 1 µL of 20 ng/µL genomic DNA template, 1 unit Taq DNA polymerase (Promega, Madison, WI, USA), 2 mM MgCl2, 0.2 mM dNTPs and 0.5 µM of reverse and forward primer. The PCR reaction was initially started by denaturation at 94 °C for 2 min, followed by 35 cycles consisting of denaturation at 94 °C for 30 sec, 30 sec of annealing at 55 °C, 30 sec of extension at 72 °C and a final extension of 3 min at 72 °C. Amplified products were electrophoresed on 1.5% agarose gel. The gels were stained with ethidium bromide and then distained with tap water and photographed using gel documentation system (UVITEC, Cambridge, UK).

5.5. Statistical Analysis

Analysis of variance (ANOVA) was computed for all data using SAS software (SAS Institute Inc, 2008). Combined analysis of variance of the split-plot design across the two locations was performed if the homogeneity test was non-significant. Least significant difference (LSD) values were calculated to test the significance of differences between means according to Steel et al. [79]. General combining ability (GCA) effects of the parents and specific combining ability (SCA) effects of the hybrids as well as their mean squares were computed according to Griffing's method 4 model I [80], using the DIALLEL-SAS program [81]. The testing of significance of GCA and SCA effects was done at 5% and 1% probability. Pearson's coefficients of correlation (r) were calculated and plotted using the package corrplot [82]. Based on the mean of each trait the reduction or increase due to increased plant density was calculated as follow:

$$\text{Change\%} = 100(\text{D2 or D3} - \text{D1})/\text{D1}$$

5.6. SSR Data Analysis

The amplified bands were scored for each SSR marker based on the presence or absence of bands, generating a binary data matrix of (1) and (0) for each marker. The number of alleles per locus, major allele frequency, gene diversity and polymorphic information content (PIC) were calculated to assess allele diversity of each marker. The value of polymorphic information content (PIC) of each SSR marker was determined as described by Botstein et al. [63] as follows:

$$1 - \sum_{i=1}^{n} P_j^2 - \sum_{i=1}^{n-1} \sum_{j=i+1}^{n} 2P_i^2 P_j^2$$

where P_i and P_j are the frequencies of the *i*th and *j*th allele of a given marker, respectively.

Genetic distances between pairs of inbred lines were calculated according to [83], using the PAST program. The dendrogram tree was generated with the unweighted pair group method using arithmetic averages (UPGMA) by the computational package MVSP version 3.1.

Supplementary Materials: The following are available online at http://www.mdpi.com/2223-7747/9/9/1140/s1. Table S1: Mean performance of the 28 F_1 crosses and the check hybrid SC128 for all the studied traits under the three plant densities across the two locations. Table S2: Physical and chemical soil properties for the two locations during 2018 season. Table S3: List of SSR primers and their sequences used in the present study. Figure S1: Daily maximum temperature (T max), minimum temperature (T min) and solar radiation (SRAD) for the two locations during 2018 season.

Author Contributions: M.M.K., M.R., K.M.I., A.S.A., M.M.E., A.M.S.K., M.A.E.-E., and E.M.H. designed the study, performed the experiments, analyzed the data and wrote the manuscript. All authors have read and agreed to the published version of the manuscript.

Funding: This research received no external funding.

Acknowledgments: Faculty of agriculture, Kafrelsheikh University, Egypt, is thankfully acknowledged for carrying out this work. Tanta University in Egypt is also thankfully acknowledged for the support provided for conducting this work. The Agricultural Research Center (ARC) in Egypt and the International Maize and Wheat Improvement Center (CIMMYT), are thankfully acknowledged for providing us the seeds of the inbred lines used in this study.

Conflicts of Interest: The authors declare no conflict of interest.

References

1. Elmyhun, M.; Liyew, C.; Shita, A.; Andualem, M. Combining ability performance and heterotic grouping of maize (*Zea mays*) inbred lines in testcross formation in Western Amhara, North West Ethiopia. *Cogent Food Agric.* **2020**, *6*, 1727625. [CrossRef]
2. FAOSTAT. Food and Agriculture Organization of the United Nations Statistics Division. Available online: http://faostat.fao.org/site/567/DesktopDefault.aspx (accessed on 23 February 2020).
3. Zohry, A.; Ouda, S.; Noreldin, T. Solutions for maize production consumption gap in Egypt. In Proceedings of the 4th African Regional ICID Conference, Aswan, Egypt, 26–28 April 2016; pp. 24–28.
4. Abdelaal, H.S.A.; McFadden, D.T. Grains production prospects and long run food security in Egypt. *Sustainability* **2019**, *11*, 4457. [CrossRef]
5. Shao, H.; Shi, D.; Shi, W.; Ban, X.; Chen, Y.; Ren, W.; Chen, F.; Mi, G. Nutrient accumulation and remobilization in relation to yield formation at high planting density in maize hybrids with different senescent characters. *Arch. Agron. Soil Sci.* **2020**, 1–17. [CrossRef]
6. Zhao, J.R.; Wang, R.H. Factors promoting the steady increase of American maize production and their enlightenments for China. *J. Maize Sci.* **2009**, *17*, 156–159.
7. Al-Naggar, A.; Atta, M. Elevated plant density effects on performance and genetic parameters controlling maize (*Zea mays* L.) agronomic traits. *J. Adv. Biol. Biotechnol.* **2017**, *12*, 1–20. [CrossRef]
8. Andrade, F.H.; Vega, C.R.; Uhart, S.; Cirilo, A.; Cantarero, M.; Valentinuz, O. Kernel number determination in maize. *Crop. Sci.* **1999**, *39*, 453–459. [CrossRef]
9. Edmeades, G.O.; Bolaños, J.; Elings, A.; Ribaut, J.M.; Banziger, M.; Westgate, M.E.; Westgate, M.; Boote, K. The role and regulation of the anthesis-silking interval in maize. *Physiol. Model. Kernel Set Maize* **2015**, *29*, 43–73. [CrossRef]
10. Sangoi, L.; Gracietti, M.; Rampazzo, C.; Bianchetti, P. Response of Brazilian maize hybrids from different eras to changes in plant density. *Field Crop. Res.* **2002**, *79*, 39–51. [CrossRef]
11. Echarte, L.; Luque, S.L.; Andrade, F.H.; Sadras, V.; Cirilo, A.; Otegui, M.E.; Vega, C.R. Response of maize kernel number to plant density in Argentinean hybrids released between 1965 and 1993. *Field Crop. Res.* **2000**, *68*, 1–8. [CrossRef]
12. Al-Naggar, A.; Shabana, R.A.; Atta, M.M.; Al-Khalil, T.H.; Al-Naggar, A.M.M. Maize response to elevated plant density combined with lowered N-fertilizer rate is genotype-dependent. *Crop. J.* **2015**, *3*, 96–109. [CrossRef]
13. Al-Naggar, A.M.M.; Atta, M.M.; Ahmed, M.; Younis, A. Mean performance, heterobeltiosis and combining ability of corn (*Zea mays* L.) agronomic and yield traits under elevated plant density. *J. Appl. Life Sci. Int.* **2016**, *7*, 1–20. [CrossRef] [PubMed]

14. Hammer, G.L.; Dong, Z.; McLean, G.; Doherty, A.; Messina, C.; Schussler, J.; Zinselmeier, C.; Paszkiewicz, S.; Cooper, M. Can changes in canopy and/or root system architecture explain historical maize yield trends in the U.S. corn belt? *Crop. Sci.* **2009**, *49*, 299–312. [CrossRef]

15. Tokatlidis, I.; Koutroubas, S. A review of maize hybrids' dependence on high plant populations and its implications for crop yield stability. *Field Crop. Res.* **2004**, *88*, 103–114. [CrossRef]

16. Ruffo, M.L.; Gentry, L.; Henninger, A.S.; Seebauer, J.R.; Below, F.E. Evaluating management factor contributions to reduce corn yield gaps. *Agron. J.* **2015**, *107*, 495–505. [CrossRef]

17. Mageto, E.K.; Makumbi, D.; Njoroge, K.; Nyankanga, R. Genetic analysis of early-maturing maize (*Zea mays* L.) inbred lines under stress and nonstress conditions. *J. Crop. Improv.* **2017**, *31*, 560–588. [CrossRef]

18. Kamara, M.M.; El-Degwy, I.S.; Koyama, H. Estimation combining ability of some maize inbred lines using line x tester mating design under two nitrogen levels. *Aust. J. Crop Sci.* **2014**, *8*, 1336.

19. Carena, M.J.; Hallauer, A.R.; Filho, J.M. *Quantitative Genetics in Maize Breeding*; Springer: New York, NY, USA, 2010.

20. Henry, W.B.; Blanco, M.H.; Rowe, D.E.; Windham, G.L.; Murray, S.C.; Williams, W.P. Diallel analysis of diverse maize germplasm lines for agronomic characteristics. *Crop. Sci.* **2014**, *54*, 2547–2556. [CrossRef]

21. Sughroue, J.R.; Hallauer, A.R. Analysis of the diallel mating design for maize inbred lines. *Crop. Sci.* **1997**, *37*, 400–405. [CrossRef]

22. Zhang, Y.; Fan, X.; Yao, W.; Piepho, H.P.; Kang, M.S. Diallel analysis of four maize traits and a modified heterosis hypothesis. *Crop. Sci.* **2016**, *56*, 1115–1126. [CrossRef]

23. Sprague, G.F.; Tatum, L.A. General vs. specific combining ability in single crosses of corn1. *Agron. J.* **1942**, *34*, 923–932. [CrossRef]

24. Al-Naggar, A.; Atta, M.; Shabana, R.; Al-Khalil, T. Heterosis and type of gene action for some adaptive traits to high plant density in maize. *Egypt. J. Plant. Breed.* **2014**, *18*, 189–209. [CrossRef]

25. Mason, L.; Zuber, M.S. Diallel analysis of maize for leaf angle, leaf area, yield, and yield components 1. *Crop. Sci.* **1976**, *16*, 693–696. [CrossRef]

26. Badu-Apraku, B.; Oyekunle, M.; Fakorede, M.A.B.; Vroh, I.; Akinwale, R.O.; Aderounmu, M. Combining ability, heterotic patterns and genetic diversity of extra-early yellow inbreds under contrasting environments. *Euphytica* **2013**, *192*, 413–433. [CrossRef]

27. Akinwale, R.; Badu-Apraku, B.; Fakorede, M.; Vroh-Bi, I. Heterotic grouping of tropical early-maturing maize inbred lines based on combining ability in Striga-infested and Striga-free environments and the use of SSR markers for genotyping. *Field Crop. Res.* **2014**, *156*, 48–62. [CrossRef]

28. Akaogu, I.C.; Badu-Apraku, B.; Adetimirin, V.O.; Vroh-Bi, I.; Oyekunle, M.; Akinwale, R.O. Genetic diversity assessment of extra-early maturing yellow maize inbreds and hybrid performance in Striga-infested and Striga-free environments. *J. Agric. Sci.* **2013**, *151*, 519–537. [CrossRef]

29. Nyaligwa, L.; Hussein, S.; Amelework, B.; Ghebrehiwot, H. Genetic diversity analysis of elite maize inbred lines of diverse sources using SSR markers. *Maydica* **2015**, *60*, M29.

30. Sserumaga, J.P.; Ji, H.; Njoroge, K.; Muthomi, J.; Chemining'wa, G.; Si-myung, L.; Kim, H.; Asea, G.; Makumbi, D. Molecular characterization of tropical maize inbred lines using microsatellite DNA markers. *Maydica* **2014**, *59*, 267–274.

31. Phumichai, C.; Doungchan, W.; Puddhanon, P.; Jampatong, S.; Grudloyma, P.; Kirdsri, C.; Chunwongse, J.; Pulam, T. SSR-based and grain yield-based diversity of hybrid maize in Thailand. *Field Crop. Res.* **2008**, *108*, 157–162. [CrossRef]

32. Singh, P. Genetic distance, heterosis and combing ability studies in maize for predicting F_1 hybrid performance. *Sabrao J. Breed. Genet.* **2015**, *47*, 21–28.

33. Dhliwayo, T.; Pixley, K.; Menkir, A.; Warburton, M. Combining ability, genetic distances, and heterosis among elite CIMMYT and IITA tropical maize inbred lines. *Crop. Sci.* **2009**, *49*, 1201–1210. [CrossRef]

34. Menkir, A.; Melake-Berhan, A.; The, C.; Ingelbrecht, I.; Adepoju, A. Grouping of tropical mid-altitude maize inbred lines on the basis of yield data and molecular markers. *Theor. Appl. Genet.* **2004**, *108*, 1582–1590. [CrossRef] [PubMed]

35. Lashkari, M.; Madani, H.; Ardakani, M.R.; Golzardi, F.; Zargari, K. Effect of plant density on yield and yield components of different corn (*Zea mays* L.) hybrids. *Am-Euras J. Agric. Environ. Sci.* **2011**, *10*, 450–457.

36. Mansfield, B.D.; Mumm, R.H. Survey of plant density tolerance in U.S. maize germplasm. *Crop. Sci.* **2014**, *54*, 157–173. [CrossRef]

37. Trachsel, S.; Vicente, F.M.S.; Suarez, E.A.; Rodriguez, C.S.; Atlin, G.N. Effects of planting density and nitrogen fertilization level on grain yield and harvest index in seven modern tropical maize hybrids (*Zea mays* L.). *J. Agric. Sci.* **2015**, *154*, 689–704. [CrossRef]

38. Kamara, A.Y.; Menkir, A.; Abubakar, A.W.; Tofa, A.I.; Ademulegun, T.D.; Omoigui, L.O.; Kamai, N. Maize hybrids response to high plant density in the Guinea savannah of Nigeria. *J. Crop. Improv.* **2020**, 1–20. [CrossRef]

39. Abd El-Aty, M.; El-Hity, M.; Amer, E.; El-Mouslhy, T. Selection of maize (*Zea mays*) hybrids for plant density tolerance. *Indian J. Agric. Sci.* **2019**, *89*, 951–957.

40. Makumbi, D.; Betrán, J.F.; Bänziger, M.; Ribaut, J.M. Combining ability, heterosis and genetic diversity in tropical maize (*Zea mays* L.) under stress and non-stress conditions. *Euphytica* **2011**, *180*, 143–162. [CrossRef]

41. Badu-Apraku, B.; Oyekunle, M. Genetic analysis of grain yield and other traits of extra-early yellow maize inbreds and hybrid performance under contrasting environments. *Field Crop. Res.* **2012**, *129*, 99–110. [CrossRef]

42. Liu, W.; Tollenaar, M. Response of yield heterosis to increasing plant density in maize. *Crop. Sci.* **2009**, *49*, 1807–1816. [CrossRef]

43. Tollenaar, M.; Wu, J. Yield improvement in temperate maize is attributable to greater stress tolerance. *Crop. Sci.* **1999**, *39*, 1597–1604. [CrossRef]

44. Van Averbeke, W.; Marais, J.N. Maize response to plant population and soil water supply: I. Yield of grain and total above-ground biomass. *S. Afr. J. Plant. Soil* **1992**, *9*, 186–192. [CrossRef]

45. Tang, L.; Ma, W.; Noor, M.A.; Li, L.; Hou, H.; Zhang, X.; Zhao, M. Density resistance evaluation of maize varieties through new "Density–Yield Model" and quantification of varietal response to gradual planting density pressure. *Sci. Rep.* **2018**, *8*, 1–16. [CrossRef] [PubMed]

46. Hashemi, A.M.; Herbert, S.J.; Putnam, D.H. Yield response of corn to crowding stress. *Agron. J.* **2005**, *97*, 839–846. [CrossRef]

47. Ren, B.; Liu, W.; Zhang, J.; Dong, S.; Liu, P.; Zhao, B. Effects of plant density on the photosynthetic and chloroplast characteristics of maize under high-yielding conditions. *Sci. Nat.* **2017**, *104*, 12. [CrossRef]

48. Gou, L.; Xue, J.; Qi, B.; Ma, B.; Zhang, W. Morphological variation of maize cultivars in response to elevated plant densities. *Agron. J.* **2017**, *109*, 1443–1453. [CrossRef]

49. Sangoi, L. Understanding plant density effects on maize growth and development: An important issue to maximize grain yield. *Ciência Rural* **2001**, *31*, 159–168. [CrossRef]

50. Al-Naggar, A.; Shabana, R.; Hassanein, M.S.; Elewa, T.A.; Younis, A.; Metwally, A. Secondary traits and selection environment of plant density tolerance in maize inbreds and testcrosses. *J. Adv. Biol. Biotechnol.* **2017**, *14*, 1–13. [CrossRef]

51. Gyenes-Hegyi, Z.; Pók, I.; Kizmus, L.; Al, E. Plant height and height of the main ear in maize (*Zea mays* L.) at different locations and different plant densities. *Acta Agron. Hung.* **2002**, *50*, 75–84. [CrossRef]

52. Carena, M.; Cross, H. Plant density and maize germplasm improvement in the northern Corn Belt. *Maydica* **2003**, *48*, 105–111.

53. Tollenaar, M.; Lee, E. Yield potential, yield stability and stress tolerance in maize. *Field Crop. Res.* **2002**, *75*, 161–169. [CrossRef]

54. Duvick, D.N. The Contribution of Breeding to Yield Advances in maize (*Zea mays* L.). *Adv. Agron.* **2005**, *86*, 83–145. [CrossRef]

55. Obeng-Bio, E.; Badu-Apraku, B.; Ifie, B.E.; Danquah, A.; Blay, E.T.; Annor, B. Genetic analysis of grain yield and agronomic traits of early provitamin A quality protein maize inbred lines in contrasting environments. *J. Agric. Sci.* **2019**, *157*, 413–433. [CrossRef]

56. Chiuta, N.E.; Charles, M.S. Combining ability of quality protein maize inbred lines for yield and morpho-agronomic traits under optimum as well as combined drought and heat-stressed conditions. *Agronomy* **2020**, *10*, 184. [CrossRef]

57. Gissa, D.W.; Zelleke, H.; Labuschagne, M.T.; Hussien, T.; Singh, H. Heterosis and combining ability for grain yield and its components in selected maize inbred lines. *S. Afr. J. Plant. Soil* **2007**, *24*, 133–137. [CrossRef]

58. Reif, J.C.; Gumpert, F.M.; Fischer, S.; Melchinger, A.E. Impact of interpopulation divergence on additive and dominance variance in hybrid populations. *Genetics* **2007**, *176*, 1931–1934. [CrossRef]

59. Ejigu, Y.G.; Tongoona, P.B.; Ifie, B.E. General and specific combining ability studies of selected tropical white maize inbred lines for yield and yield related traits. *Int. J. Agric. Res. Innov. Technol.* **2017**, *7*, 381–396.

60. Oppong, A.; Bedoya, C.A.; Ewool, M.B.; Asante, M.D.; Thomson, R.N.; Adu-Dapaah, H.; Lamptey, J.N.; Ofori, K.; Offei, S.K.; Warburton, M.L. Bulk genetic characterization of Ghanaian maize landraces using microsatellite markers. *Maydica* **2014**, *59*, 1–8.

61. Oyekunle, M.; Badu-Apraku, B.; Hearne, S.; Franco, J. Genetic diversity of tropical early-maturing maize inbreds and their performance in hybrid combinations under drought and optimum growing conditions. *Field Crop. Res.* **2015**, *170*, 55–65. [CrossRef]

62. Legesse, B.W.; Myburg, A.; Pixley, K.V.; Botha, A. Genetic diversity of African maize inbred lines revealed by SSR markers. *Hereditas* **2007**, *144*, 10–17. [CrossRef]

63. Botstein, D.; White, R.L.; Skolnick, M.; Davis, R.W. Construction of a genetic linkage map in man using restriction fragment length polymorphisms. *Am. J. Hum. Genet.* **1980**, *32*, 314–331.

64. Adu, G.B.; Awuku, F.; Amegbor, I.; Haruna, A.; Manigben, K.; Aboyadana, P. Genetic characterization and population structure of maize populations using SSR markers. *Ann. Agric. Sci.* **2019**, *64*, 47–54. [CrossRef]

65. El-Esawi, M.; Germaine, K.J.; Bourke, P.; Malone, R. AFLP analysis of genetic diversity and phylogenetic relationships of Brassica oleracea in Ireland. *Comptes Rendus Biol.* **2016**, *339*, 163–170. [CrossRef] [PubMed]

66. El-Esawi, M.; Alaraidh, I.A.; Alsahli, A.A.; Ali, H.M.; Alayafi, A.A.; Witczak, J.; Ahmad, M. Genetic variation and alleviation of salinity stress in barley (*Hordeum vulgare* L.). *Molecules* **2018**, *23*, 2488. [CrossRef] [PubMed]

67. El-Esawi, M.; Al-Ghamdi, A.A.; Ali, H.M.; Alayafi, A.A.; Witczak, J.; Ahmad, M. Analysis of genetic variation and enhancement of salt tolerance in French pea (*Pisum sativum* L.). *Int. J. Mol. Sci.* **2018**, *19*, 2433. [CrossRef] [PubMed]

68. El-Esawi, M. Genetic diversity and evolution of Brassica genetic resources: From morphology to novel genomic technologies—A review. *Plant. Genet. Resour.* **2016**, *15*, 388–399. [CrossRef]

69. El-Esawi, M.; Sammour, R. Karyological and phylogenetic studies in the genus *Lactuca* L. (Asteraceae). *Cytologia* **2014**, *79*, 269–275. [CrossRef]

70. El-Esawi, M.; Al-Ghamdi, A.A.; Ali, H.M.; Ahmad, M. Overexpression of AtWRKY30 transcription factor enhances heat and drought stress tolerance in wheat (*Triticum aestivum* L.). *Genes* **2019**, *10*, 163. [CrossRef]

71. Vwioko, E.; Adinkwu, O.; El-Esawi, M.A. Comparative physiological, biochemical, and genetic responses to prolonged waterlogging stress in okra and maize given exogenous ethylene priming. *Front. Physiol.* **2017**, *8*, 632. [CrossRef]

72. Xia, X.; Reif, J.; Hoisington, D.; Melchinger, A.; Frisch, M.; Warburton, M. Genetic diversity among CIMMYT maize inbred lines investigated with SSR markers. *Crop. Sci.* **2004**, *44*, 2230–2237. [CrossRef]

73. Adebayo, M.A.; Menkir, A.; Gedil, M.; Blay, E.; Gracen, V.; Danquah, E.; Funmilayo, L. Diversity assessment of drought tolerant exotic and adapted maize (*Zea mays* L.) inbred lines with microsatellite markers. *J. Crop. Sci. Biotechnol.* **2015**, *18*, 147–154. [CrossRef]

74. Bernardo, R. Relationship between single-cross performance and molecular marker heterozygosity. *Theor. Appl. Genet.* **1992**, *83*, 628–634. [CrossRef] [PubMed]

75. Betran, F.; Beck, D.; Bänziger, M.; Edmeades, G. Genetic analysis of inbred and hybrid grain yield under stress and nonstress environments in tropical maize. *Crop Sci.* **2003**, *43*, 807–817. [CrossRef]

76. Parentoni, S.; Magalhães, J.; Pacheco, C.; Santos, M.; Abadie, T.; Gama, E.; Guimarães, P.; Meirelles, W.; Lopes, M.; Vasconcelos, M.; et al. Heterotic groups based on yield-specific combining ability data and phylogenetic relationship determined by RAPD markers for 28 tropical maize open pollinated varieties. *Euphytica* **2001**, *121*, 197–208. [CrossRef]

77. AOAC, C.A. *Official Methods of Analysis of the Association of Analytical Chemists International*; Official Methods: Gaithersburg, MD, USA, 2005.

78. Doyle, J. DNA Protocols for plants. In *Molecular Techniques in Taxonomy*; Springer Science and Business Media LLC: New York, NY, USA, 1991; pp. 283–293.

79. David, F.N.; Steel, R.G.D.; Torrie, J.H. *Principles and Procedures of Statistics, A Biometrical Approach*, 3rd ed.; McGraw Hill Inc. Book Co.: New York, NY, USA, 1997; pp. 352–358.

80. Griffing, B. Concept of general and specific combining ability in relation to diallel crossing systems. *Aust. J. Biol. Sci.* **1956**, *9*, 463–493. [CrossRef]

81. Zhang, Y.; Kang, M.S.; Lamkey, K.R. DIALLEL-SAS05: A comprehensive program for griffing's and gardner-eberhart analyses. *Agron. J.* **2005**, *97*, 1097–1106. [CrossRef]

82. Wei, T.; Simko, V.; Levy, M.; Xie, Y.; Jin, Y.; Zemla, J. Corrplot: Visualization of a Correlation Matrix. Available online: https://github.com/taiyun/corrplot (accessed on 15 December 2019).
83. Jaccard, P. Nouvelles recherches sur la distribution florale. *Bull. Soc. Vaud. Sci. Nat.* **1908**, *44*, 223–270.

Article

Rice Breeding in Russia Using Genetic Markers

Elena Dubina [1], **Pavel Kostylev** [2,*], **Margarita Ruban** [1], **Sergey Lesnyak** [1], **Elena Krasnova** [2] **and Kirill Azarin** [3]

[1] Federal Scientific Rice Centre, Belozerny, 3, 350921 Krasnodar, Russia; lenakrug1@rambler.ru (E.D.); arrri_kub@mail.ru (M.R.); Sergei_l.a@mail.ru (S.L.)

[2] Agrarian Research Center "Donskoy", Nauchny Gorodok, 3, 347740 Zernograd, Russia; krasnovaelena67@mail.ru

[3] Department of Genetics, Southern Federal University, 344006 Rostov-on-Don, Russia; azarinkv@sfedu.ru

* Correspondence: p-kostylev@mail.ru

Received: 17 October 2020; Accepted: 12 November 2020; Published: 15 November 2020

Abstract: The article concentrates on studying tolerance to soil salinization, water flooding, and blast in Russian and Asian rice varieties, as well as hybrids of the second and third generations from their crossing in order to obtain sustainable paddy crops based on domestic varieties using DNA markers. Samples IR 52713-2B-8-2B-1-2, IR 74099-3R-3-3, and NSIC Rc 106 were used as donors of the *SalTol* tolerance gene. Varieties with the *Sub1A* locus were used as donors of the flood resistance gene: Br-11, CR-1009, Inbara-3, TDK-1, and Khan Dan. The lines C101-A-51 (Pi-2), C101-Lac (Pi-1, Pi-33), IR-58 (Pi-ta), and Moroberekan (Pi-b) were used to transfer blast resistance genes. Hybridization of the stress-sensitive domestic varieties Novator, Flagman, Virazh, and Boyarin with donor lines of the genes of interest was carried out. As a result of the studies carried out using molecular marking based on PCR in combination with traditional breeding, early-maturing rice lines with genes for resistance to salinity (*SalTol*) and flooding (*Sub1A*), suitable for cultivation in southern Russia, were obtained. Introgression and pyramiding of the blast resistance genes *Pi-1, Pi-2, Pi-33, Pi-ta*, and *Pi-b* into the genotypes of domestic rice varieties were carried out. DNA marker analysis revealed disease-resistant rice samples carrying 5 target genes in a homozygous state. The created rice varieties that carry the genes for blast resistance (Pentagen, Magnate, Pirouette, Argamac, Kapitan, and Lenaris) were submitted for state variety testing. The introduction of such varieties into production will allow us to avoid epiphytotic development of the disease, preserving the biological productivity of rice and obtaining environmentally friendly agricultural products.

Keywords: rice; salinity; submergence tolerance; blast; SSR markers; PCR analysis

---

1. Introduction

Rice (*Oryza sativa* L.) is the most important food crop for more than half of the world's population (China, Japan, India, Bangladesh, etc.). Biotic and abiotic stressors are the main obstacles to increasing global crop production and expanding rice production. It was found that only about 10% of the world's agricultural land is located in areas that do not suffer from stress factors [1].

Decreased rice yields in adverse climatic conditions threaten global food security. Genetic loci that ensure productivity in difficult conditions exist in the germplasm of cultivated plants, their wild relatives, and species adapted to extreme conditions [2].

One-fifth of the world's irrigated land (North Africa, Central and South-East Asia, etc.) is adversely affected by high soil salinity [3]. About 45 million hectares in the world are subject to soil salinization [4]. In the Russian Federation, rice is grown on an area of about 200 thousand hectares, of which about 80 thousand hectares are saline [5]. The decline in productivity on saline soils can be overcome by developing rice varieties tolerant to salinity and introducing them into agricultural production. Several non-allelic genes provide tolerance to salinity during ontogenesis [6]. The main locus of salt tolerance is *SalTol*, which was first identified in some rice varieties [7,8]. This locus is mapped on chromosome 1 and its main function is to control the balance of Na+/K+ ions in rice plants [9].

One of the serious abiotic stress factors for rice, which inhibits plant growth and affects crop yield, is prolonged submersion of plants under water, which often happens to large areas of land in the rice-growing regions of South-East Asia [5]. Rice dies if total flooding lasts more than two weeks. A negative effect on the growth and development of rice plants at this time is exerted by a lack of oxygen (O_2) and limited diffusion of carbon dioxide (CO_2). Lack of light due to turbid flood water in the rice paddies during this period limits the ability of plants to photosynthesize and can even lead to their death [4–6].

Scientists in Asia have found the *Sub1A* gene, which regulates the response of plant cells to ethylene and gibberellin, leading to restriction of carbohydrate intake and dormancy of shoots under water, which contributes to tolerance to immersion [10,11]. In Russia, this gene can be used to develop varieties resistant to a large layer of water during the germination phase, which will become an effective way to protect rice from weeds without herbicides. Most weeds die under water without oxygen, and rice can survive. To develop such varieties, it is necessary to combine in one genotype genes with increased energy of initial growth, the ability to anaerobic germination, resistance to prolonged flooding and lodging.

In all countries of the world, including Russia, blast is among the most dangerous fungal diseases of rice and causes large yield losses in the years of epiphytoty. The most effective way to protect rice without fungicides is to grow blast-resistant varieties. More than 50 genes of resistance to this pathogen are known: *Pi-1, Pi-2, Pi-33, Pi-b, Pi-ta, Pi-z*, etc. [12]. Combining several effective resistance genes with their contribution on the genetic basis of the best varieties is an effective breeding strategy for resistance to variable fungal pathogens. Lines with a combination of 3–5 resistance genes show an increase and broadening of the spectrum of blast resistance in comparison with lines with separate genes. A number of successful breeding programs have already been carried out abroad to develop blast-resistant rice varieties by the gene pyramiding method using marker breeding [13].

Resistance to various biotic and abiotic factors is one of the traits that are difficult to assess when the assessment of the breeding material is possible only in the presence of an appropriate stress factor. At present, during the breeding of agricultural plants for resistance, the splitting population obtained from the crossing of resistance sources with genotypes that have productivity is tested against a natural background, or artificial infection is carried out under controlled conditions. This procedure, although it gives excellent results, is quite lengthy and costly. In addition, there are always susceptible plants that have escaped damage [14].

The use of DNA markers allows us to speed up the assessment and conduct selection without phenotypic assessment, at an early stage, regardless of the external conditions. In recent years, great progress has been made in the development of molecular marking technologies and their application to control complex agronomic traits using marker breeding [15]. The technology of molecular marking of resistance loci makes it possible to quickly select plant forms with target genes without using provocative backgrounds [16]. The identification of molecular markers linked to genes of resistance to these factors facilitates breeding work. The use of DNA markers brings the breeding of agricultural plants to a qualitatively new level, making it possible to evaluate genotypes directly and not through phenotypic manifestations, which, ultimately, is realized in the accelerated development of varieties with a complex of valuable traits [14]. Therefore, it is relevant to develop new rice varieties by marking [17].

The purpose of the study was the development of initial rice material using DNA markers for breeding highly productive varieties resistant to biotic and abiotic environmental stress factors: soil salinity, prolonged flooding, and blast.

2. Materials and Methods

We used samples from the collection of the Institute of Agricultural Genetics (Vietnam) as donors of the transferred salt tolerance gene: IR 52713-2B-8-2B-1-2, IR 74099-3R-3-3, and NSIC Rc 106, which were crossed with the early–maturing Krasnodar variety Novator. These varieties carry the *SalTol* locus, which has been mapped near the centromeric region of the first chromosome. RM493 and RM7075 [18] were used as flanking SSR-markers of this locus, with the greatest difference in the length of microsatellite repeats between the parental forms.

Varieties with the *Sub1A* locus were used as donors of the flooding resistance gene: BR-11, CR-1009, Inbara-3, TDK-1, and Khan Dan. The early-ripening variety Novator and rice lines with the introgressed genes for blast resistance *Pi-2* and *Pi-33* were also taken as recipients. The *Sub1A* locus is mapped to an interval of 0.06 morganides in chromosome 9 [11]. We used microsatellite markers for the *Sub1A* gene, CR25K and SSR1A. The *Sub1A* gene was identified by molecular marking based on PCR using specific primers.

When transferring blast resistance genes, lines C101-A-51 (*Pi-2*), C101-Lac (*Pi-1, Pi-33*), IR-58 (*Pi-ta*), and Moroberekan (*Pi-b*) were used. To identify the Pi-1 gene, we used primer pairs of the flanking microsatellite SSR markers RM224 and RM144; for the Pi-2 gene, we used RM527 and SSR140; for the Pi-33 gene, RM310 and RM72; for the *Pi-b* and *Pi-ta* genes, intragenic markers developed in the laboratory of biotechnology, Federal Scientific Rice Centre. They are localized on chromosomes 11, 6, 8, 2, and 12, respectively (Table 1) [19,20].

Table 1. Nucleotide sequences of codominant markers for identification of the allelic status of genes Pi-1, Pi-2, Pi-a, and Pi-b.

Resistance Gene	Chromosomal Localization of Gene	Marker		Sequence
Pi-2	6	Rm 527	F	GGC TCG ATC TAG AAA ATC CG
			R	TTG CAC AGG TTG CGA TAG AG
		SSR140	F	AAG GTG TGA AAC AAG CTA GCA
			R	TTC TAG GGG AGG GGT GTA GAA
Pi-33	8	Rm 72	F	CCG GCG ATA AAA CAA TGA G
			R	GCA TCG GTC CTA ACT AAG GG
		Rm310	F	CCA AAA CAT TTA AAA TAT CATG
			R	GCT TGT TGG TCA TTA CCA TTC
Sub1A	9	Sub1A203	F	GAT GT GT GGAGGAGAAGT GA
			R	GGTAGAT GCCGAGAAGT GTA
		Rm 7481	F	CGACCCAATATCTTTCTGCC
			R	ATTGGTCGTGCTCAACAAG
SalTol	1	Rm 493	F	GTACGTAAACGCGGAAGGTGACG
			R	CGACGTACGAGATGCCGATCC
		Rm 7075	F	TATGGACTGGAGCAAACCTC
			R	GGCACAGCACCAATGTCTC

The early-ripening released rice varieties Boyarin, Flagman, and Virage served as the paternal form. During plant hybridization, pneumocastration of maternal forms and pollination by the Twell method were used [21]. Hybrid plants were grown on checks of Federal State Unitary Enterprise "Proletarskoe" (Rostov region) and the Federal State Unitary Elite Seed-growing Enterprise "Krasnoe" of the Federal Scientific Rice Centre, Krasnodar region. From the selected rice leaves, genomic DNA was isolated under laboratory conditions at the Federal Scientific Rice Centre, the Academy of Biology and Biotechnology of the Southern Federal University, and the All-Russian Research Institute of Agricultural Biotechnology. PCR products were separated by electrophoresis in 2.5% agarose and 8% acrylamide gels. The experimental data were statistically processed using Microsoft Excel and Statistica 6 software.

The account of the degree of damage to plants (in percentages) was carried out on the 14th day after inoculation, in accordance with the express method for assessing rice varietal resistance to blast. The assessment was carried out by taking two indicators into account: the type of reaction (in points) and the degree of damage (in percentages), using the ten-point scale of the International Rice Research Institute [12]:

- resistant: 0–1 point—no damage, small brown spots, covering less than 25% of the total leaf surface;
- medium resistant: 2–5 points—typical elliptical blast spots, 1–2 cm long, covering 26–50% of the total leaf surface;
- susceptible: 6–10 points—typical blast spots of elliptical shape, 1–2 cm long, covering 51% or more of the total leaf surface.

The intensity of disease development (IDD, %) was calculated by the formula (Equation (1)):

$$\mathrm{IDD} = \sum (a \times b)/n \times 9 \tag{1}$$

where IDD is the intensity of disease development (%), $\sum (a \times b)$ is the sum of the products of the number of infected plants multiplied by the corresponding damage point, and n is the number of recorded plants (pcs).

Depending on the damage points, all varieties wee conventionally divided into 4 groups: resistant, intermediate, susceptible, and strongly susceptible.

3. Results and Discussion

The development of blast-resistant varieties and their rapid introduction into production is the most promising solution in the fight against this disease. However, the development of resistant varieties is one of the most difficult areas of breeding. The causative agents of the disease have a great potential for variability, which, combined with its colossal reproduction capabilities, provides the pathogen with the highest adaptive capabilities. Combining several effective genes of resistance on a genetic basis of the best varieties widely used in production is an effective breeding strategy for long-term resistance to variable fungal pathogens.

Based on the use of DNA marker breeding (marker-assisted selection (MAS)—breeding with use of DNA markers towards genes of interest), we introduced 5 blast resistance genes into domestic rice varieties adapted to the agro-climatic conditions of rice cultivation in southern Russia.

A series of crosses made it possible to obtain rice lines based on the varieties Boyarin, Flagman and Virage with the introgressed and pyramided blast resistance genes *Pi-1*, *Pi-2*, *Pi-33*, *Pi-ta*, and *Pi-b* in a homozygous state. During all cycles of backcrossing, the transfer of the dominant alleles of each such gene in the offspring was controlled by closely linked molecular markers. Plants with no resistance alleles in the genotype were discarded.

At the first stage of work in 2005 at Agrarian research center "Donskoy", 6 hybrids were obtained from crossing the varieties Boyarin and Virage with three donors of blast resistance carrying the *Pi-1*, *Pi-2*, and *Pi-33* genes. After analysis at the Federal Scientific Rice Centre, homozygous forms were identified for the dominant alleles.

At the second stage of work (2008), after crossing the *Pi-1 + 33* × Boyarin and *Pi-2* × Boyarin hybrids between themselves, it was possible to obtain forms with three pyramided genes simultaneously: *Pi-1*, *Pi-2*, and *Pi-33* in a homozygous state.

At the third stage of work (2010), they were hybridized with varieties—donors of the *Pi-ta* and *Pi-b* genes—for combining 5 genes. There were two types of crosses: ((*Pi-1 + 2 + 33*) × *Pi-ta*) × *Pi-b* and *Pi-1 + 2 + 33* × (*Pi-ta* × *Pi-b*).

Leaves were selected from the best F_2 hybrid plants for DNA analysis at All-Russian Research Institute of Agricultural Biotechnology and the Federal Scientific Rice Centre using one marker for each gene. Based on the analysis results, it was possible to isolate two rice samples that were homozygous for all five dominant alleles. Reanalysis of the leaves of these samples confirmed last year's results, i.e., homozygosity for the dominant alleles of all five loci.

Figures 1 and 2 show the panicles of two lines, 1225/13 and 1396/13, which show the presence of dominant alleles at five loci in the homozygous state: *Pi-1*, *Pi-2*, *Pi-33*, *Pi-b*, and *Pi-ta*.

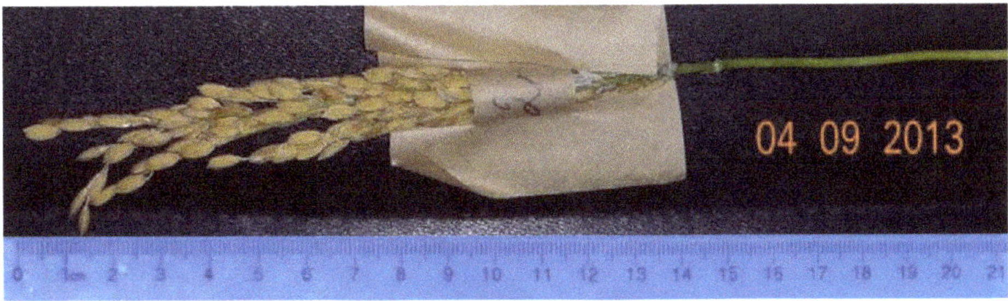

Figure 1. Panicle of the early-ripening line 1225/13.

Figure 2. Panicle of the mid-ripening line 1396/13.

Line 1225/13 is early maturing, matures in 110 days, and dwarfish (80 cm), with a small panicle (15 cm) (Figure 1).

The second line 1396/13 is mid-ripening, the period to maturity is 120 days, and it is taller (100 cm), with a large long panicle (22 cm) (Figure 2).

Against the infectious background in the Federal Scientific Rice Centre, the index of disease development (IDD) in this line was only 1.4%, while the variety Novator was damaged by 67.7%. The results of the analysis made it possible to send these lines to the breeding nursery in 2014–2015 for testing for yield and blast resistance. The variety **Pentagen** (1396/13), carrying 5 genes for blast resistance, is widely used in hybridization with high-yielding Russian varieties.

In the process of work at the Federal Scientific Rice Centre in 2007–2008, crosses were carried out and F_1 hybrids were obtained from the combination (Flagman × C101-Lac) × (Flagman × C-101-A-51), which have the blast resistance genes *Pi-33* and *Pi-2* in their genotypes, respectively. The resulting F_1 generation was used in backcrosses with the recipient parental forms. It should be noted that the F_1 plants had a high degree of sterility (up to 95%). After the first series of backcrosses in 2008, the BC_1 and BC_2 generations were obtained in artificial climate chambers. In BC_1 populations, fertility increased and averaged about 50%. Starting from the first backcrossing, marker control was carried out for the presence of transferred donor alleles in the hybrid offspring. In 2009, plants of the BC_3 and BC_4 generation were obtained. Among these plants, we selected the forms with the shortest growing season and the highest panicle fertility. From the BC_4F_1 stage (the first self-pollination of rice plants, which makes it possible to transfer the donor allele to a homozygous state), individual selection was carried out. Segregation for Pi-2, Pi-33, and Sub1A genes fit into the Mendelian framework: in the second generation as a result of DNA analysis of the obtained plants, the ratio was 1:2:1 by genotype and 3:1 by phenotype.

Plants were selected that were closest in morphotype to the recipient parental form and had donor genes for resistance to the pathogen *Pyricularia oryzae* Cav. in their genotype in a homozygous state [22].

Figure 3 shows the results of PCR analysis for identification of the *Pi-33* blast resistance gene in the BC_4F_3 hybrid material.

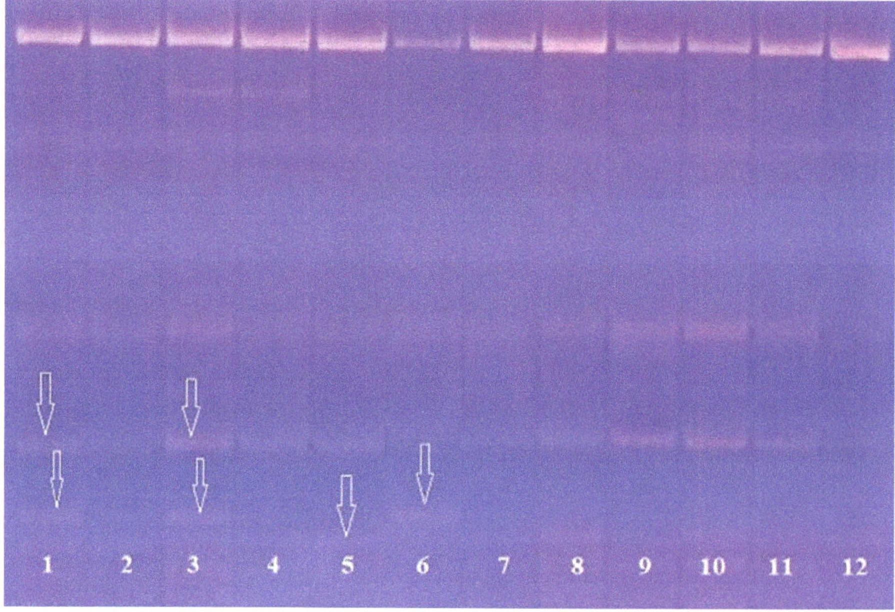

Figure 3. Electrophoregram of genomic DNA amplification products at the loci RM310 and RM72: 1–4, 7–12, analyzed hybrid BC_4F_3 plants; 5, donor line of the Pi-33 gene C101-Lac; Flagman, maternal form.

The figure shows that plants Number 2, 4, and 7–12 are homozygotes for the dominant allele; plants Number 1 and 3 are heterozygous. The size of the PCR product in varieties with the dominant allele of the Pi-33 gene, which determines the resistance, is 198 bp; in varieties with a recessive allele, it is 152 bp.

In 2015–2016, the resulting rice lines with introgressed blast resistance genes *Pi-2* and *Pi-33* were crossed with the variety Khan Dan (Vietnamese breeding): the donor of the *Sub1A* gene. This work was performed for obtaining breeding material with combined genes for disease resistance and tolerance to prolonged immersion of plants under water. In 2017–2020, F_4 and BC_2F_3 generations were obtained using climate chambers at the Federal Scientific Rice Centre (All-Russian Rice Research Institute, Krasnodar, Russia).

To increase economic efficiency and reduce labor costs, multi-primer systems have been developed to identify two genes (*Pi* and *Sub1A*) in a hybrid material simultaneously.

At the first stage, when we selected DNA markers for reliable interpretation of PCR products and identification of non-specifically amplified fragments, the following parameters were taken into account: the annealing temperature of specific pairs introduced into the reaction mixture, the difference in the size of PCR products synthesized during amplification with specific primer pairs (at least 100 base pairs), and the self-complementarity of the primer sequences.

The results of testing the combination of primer pairs flanking the marker regions of the *Pi-2 + Sub1A* genes are shown in Figure 4.

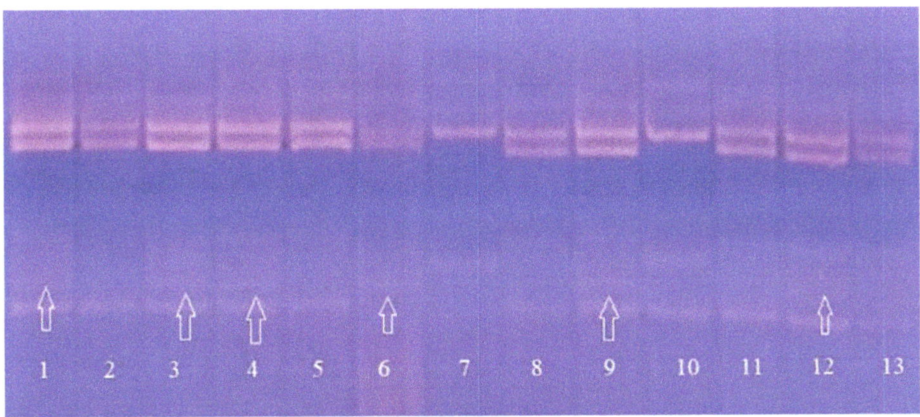

Figure 4. Electrophoregram of multiplex PCR of genomic DNA amplification products at the loci RM527 and SSR140 for the Pi-2 gene and at the Sub1A203 locus for the *Sub1A* gene: 1–5, 9–13, analyzed hybrid plants of the BC$_2$F$_3$ generation; 6, Khan Dan, donor of the *Sub1A* gene; 7, Flagman, maternal form; 8, C101Lac-A-51, donor line of the *Pi-2* gene.

The electrophoregram shows that when PCR with such a combination of molecular markers is carried out, the target products specific for DNA markers of the desired genes are reliably amplified. Samples Number 3 and 12 have dominant alleles of the genes *Pi-2* and *Sub1A* in a homozygous state in their genotype; Samples 1, 4, 5, and 9 are homozygous for the *Sub1A* gene and have the *Pi-2* gene in the genotype in a heterozygous state; Sample 10 is a recessive homozygote for two target genes and was rejected. The size of the PCR product in varieties with the dominant allele of the *Pi-2* gene, which determines the resistance, is 233 bp. The size of the PCR product in varieties with the dominant allele of the *Sub1A* gene, which determines the resistance, is 118 bp. Clear identification on the electrophoregram makes it possible to reliably identify the presence of dominant alleles of the target genes.

The introduction of such varieties into production will allow us to avoid epiphytic development of the disease, preserving the biological rice productivity, and obtaining environmentally friendly agricultural products.

Magnat is the first cultivar in Russia created at the Agrarian Research Center Donskoy together with the Federal Scientific Rice Centre by the method of marker selection from a hybrid population (C101A-51 × Boyarin) × (C101 LAC × Boyarin) with transfer of blast resistance genes. Sample C101 LAC is a donor of the genes *Pi-1* and *Pi-33*, and C101A-51 is a *Pi-2*. The growing season is 125 ± 1 days and the plant height is 96 ± 2 cm. The panicle is erect and compact, 17.5 ± 0.5 cm long, and bears 185 ± 5 spikelets. The grain is oval, 8.3 ± 0.2 mm long, 3.1 ± 0.1 mm wide, and 2.2 ± 0.1 mm thick and weighs 24.0 ± 2.0 mg. The yield of the Magnat variety was 8.25 t/ha, which is 1.1 t/ha higher than that of the Boyarin standard.

The rice variety **Pirouette** was bred at the Agrarian Research Center Donskoy, together with the Federal Scientific Rice Centre, by the method of stepwise hybridization and marker breeding from a hybrid population (C101-A-51 (*Pi-2*) × Boyarin) × (C101-Lac (*Pi-1 + 33*) × Virazh). It contains three blast resistance genes: *Pi-1*, *Pi-2*, and *Pi-33*. The variety is mid-ripening, the growing season from flooding to full ripeness is 124 ± 1 days. The average yield of the variety Pirouette was 9.57 t/ha, which is

1.13 t/ha higher than that of the standard variety Yuzhanin. Plant height is 88 ± 2 cm; the panicle is erect, compact, and 17.5 ± 0.5 cm long and carries 165 ± 5 spikelets. The spikelets are oval, 8.9 ± 0.2 mm long, and 3.7 ± 0.1 mm wide. The weight of 1000 grains is 31.6 ± 2.0 g. The variety is resistant to lodging and shedding, is cold-tolerant, and germinates well from under a layer of water. It has been ncluded in the Register of Breeding Achievements of the Russian Federation for the North Caucasus region since 2020.

The rice variety **Kapitan** was bred at the Agrarian Research Center Donskoy in cooperation with the Federal Scientific Rice Centre by the method of triple backcrossing and marker breeding from the Flagman × IR-36 hybrid population. The variety is mid-ripening and the growing season from the flooding to full ripeness is 120 ± 1 days. On average, over the years of competitive testing, the yield of the variety Kapitan was 8.13 t/ha, which is 0.64 t/ha higher than that of the variety Yuzhanin. A higher yield of this variety is formed due to more grain in the panicle and an increased weight of the caryopsis. The average height of plants is 90 ± 2 cm; the panicle is semi-inclined, compact, and 18.5 ± 0.5 cm long; and the average number of spikelets is 140 ± 10 pieces (Figure 5). The grains are oval, 9.5 ± 0.2 mm long, and 3.6 ± 0.1 mm wide. The average weight of 1000 grains is 35.0 ± 2.1 g. The variety carries the *Pi-ta* gene and is resistant to blast, lodging, and shedding. The variety has been under state testing since 2019.

Figure 5. Rice panicles of the variety Kapitan.

The rice variety **Argamak** was bred at the Agrarian Research Center Donskoy by individual multiple selection of plants with the largest panicles from a hybrid population Il. 14 (*Pi-1, Pi-2, Pi-33*) × Kuboyar. The variety belongs to the mid-ripening group, and the growing season from flooding to full ripeness is 119 days. On average, over the years of competitive testing (2017–2019), the yield of the variety was 8.79 t/ha, which is 1.59 t/ha higher than that of the variety Yuzhanin. The maximum yield was formed in 2019: 10.1 t/ha, 2.55 more than the standard. The high yield of this variety was formed due to the greater grain content of the panicle than that of the standard and the increased density of the stem. Plant height is 93 ± 2 cm on average; the panicle is erect, compact, and 16 ± 0.5 cm long; the average number of spikelets is 142 ± 6 pieces. The grains are oval,

8.4 ± 0.2 mm long, 3.3 ± 0.1 mm wide. Weight of 1000 grains—31.1 ± 1.9 g. The variety is resistant to blast, lodging, and shedding. It has been tested at state varietal testing since 2020.

The rice variety **Lenaris** (Federal Scientific Rice Centre) had shown high adaptability, non-lodging, and the possibility for straight combine harvesting. Its yield was 10.6 t/ha. Plants had high spikelet fertility and short stems (77 ± 5 cm) and were resistant to the Krasnodar population of *P. oryzae* as well. Their panicle is slightly drooping and compact; its length is 18 ± 1.0 cm. The mass of 1000 grains is about 30.4 ± 1.8 g.

In 2013–2014, the Agrarian Research Center Donskoy conducted crosses and obtained F_1–F_2 hybrids of the variety Novator with Asian donor rice varieties carrying the *SalTol* and *Sub1A* genes. The hybrids of the second generation varied significantly in terms of quantitative traits: growing season (from early ripening to non-flowering), plant height (75–122 cm), panicle length (14–25 cm), number of filled grains (80–206 pcs), number of spikelets (99–300 pcs), panicle density (4.4–16.6 pcs/cm), 1000-grain weight (26.3–34.9 g), grain weight per panicle (2.1–5.5 g), etc.

Hybridization of the salt-sensitive domestic variety Novator with the lines IR52713-2B-8-2B-1-2, IR74099-3R-3-3, and IR61920-3B-22-2-1 (NSIC Rc 106)—*SalTol* locus donors—was carried out. The first generation of hybrids was used to generate an F_2 hybrid population. From the populations of plants of the second generation, 90 early-ripening samples with well-ripened grains (30 in each combination of crossing) were selected, which were analyzed by PCR for the presence of introduced *SalTol* alleles. As an example, Figure 2 shows the data of electrophoretic analysis of PCR products with the Rm493 marker. The donor allele of the parental line NSIC Rc 106, designated as 2.2, was found in a homozygous state in Sample 282. The rest of the plants, whose amplification spectra are presented in this electrophoregram, carried the alleles of the donor and the variety Novator; that is, they were heterozygous for the *SalTol* locus (Figure 2). Similar results were obtained during DNA analysis of the studied rice samples with the RM7075 marker (Figures 6 and 7).

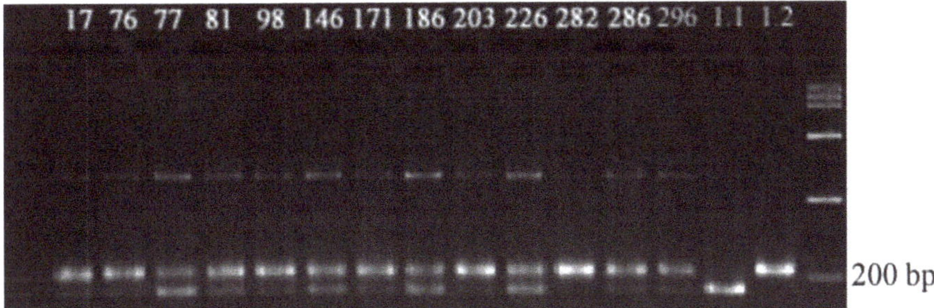

Figure 6. Electrophoregram of genomic DNA amplification products with RM 493: 1.1, Novator; 1.2, NSIC Rc 106; 17–296, hybrid plants NSIC Rc 106 × Novator; DNA marker (100–1500 bp).

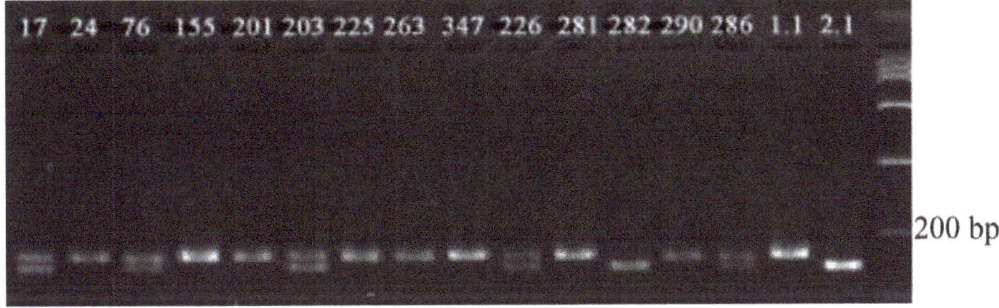

Figure 7. Electrophoregram of genomic DNA amplification products with RM 7075: 1.1, Novator; 2.1, NSIC Rc 106; 17–286, hybrid plants NSIC Rc 106 × Novator; DNA marker (100–1500 bp).

In general, according to the results of DNA analysis of F_2 hybrids, 17 plants homozygous for the dominant allele of the *SalTol* locus were identified, 29 samples carried *SalTol* in a heterozygous state, and 44 plants showed only recessive alleles inherited from the variety Novator.

Segregation for *SalTol* genes did not fit into the Mendelian framework, since the sample was unrepresentative due to selection. Plants with recessive alleles of the gene prevailed, and the number of salt-tolerant dominant homozygotes was less than the expected number. This is due to the linkage of *SalTol* genes with genes unfavorable for plants in our conditions: photosensitivity, late maturity, spikelet shedding, and spinosity.

Testing plants under salinity in the early stages of development is a quick, common method based on simple criteria. It was shown that at the initial vegetation stage, the length of the root and shoots and seed germination are potential indicators of resistance to the effects of increased salt concentrations [18,19]. Evaluation of the potential salt tolerance of the studied rice hybrids and their parental forms revealed significant variations in salinity tolerance depending on the genotype. The greatest decrease in seed germination—52%—was found in the salt-sensitive variety Novator. The line NSIC Rc 106 and second-generation plants, which were homo- and heterozygous for the *SalTol* locus, showed the highest resistance by seed germination (germination decrease of less than 5%). The donor lines IR52713-2B-8-2B-1-2, and IR74099-3R-3-3 and hybrid combinations obtained on their basis with the *SalTol* gene in a homozygous state also showed high resistance for this trait.

The least suppression of growth indices, as well as in the case of seed germination, was recorded in the lines NSIC Rc 106, IR52713-2B-8-2B-1-2, IR74099-3R-3-3, and *SalTol* homozygous plants from the F_2 generation; the greatest decrease in the length of shoots and roots under salt stress was shown in the variety Novator and in hybrid plants that did not inherit the *SalTol* locus according to molecular analysis data. Thus, DNA analysis made it possible to simplify the breeding scheme and obtain salt-tolerant F_2 hybrids carrying the *SalTol* locus in a homozygous state. These results indicate that the developed codominant markers of the *SalTol* locus RM 493 and RM 7075 are an effective tool for marker-assisted selection of salt-tolerant forms based on domestic rice genotypes.

Rice samples with *SalTol* genes in 2018–2020 were studied in a control nursery and in competitive variety testing; productive forms were identified.

At the same time, in 2013, hybrids were obtained by crossing the variety Novator with donors of the *Sub1A* gene. The Asian varieties turned out to be late-ripening and photosensitive and did not flower under our conditions. Hybridization was carried out only with the help of artificial climate chambers. The first generation in 2013 was characterized by a high degree of sterility (90–95%) and brown color of the flowering scales during maturation, which indicates significant genetic differences between the parental forms. In the second generation in 2014, a very large spectrum of splitting was observed in terms of the growing season, plant height, panicle length and shape, number of spikelets, and spinosity (Table 2).

Table 2. Variations in the quantitative traits in F_2 hybrids from crossing submergence-resistant samples with the variety Novator, 2014.

Trait	Crossing Combination			
	BR 11 × Novator	**CR-1009 × Novator**	**Inbara 3 × Novator**	**TDK-1 × Novator**
Plant height, cm	71–129 (97.5)	57–131 (89.4)	60–149 (100.2)	45–138 (99.9)
Panicle length, cm	11.5–27 (18.4)	10–26 (17.7)	9.5–32 (19.1)	9–27 (18.9)
Number of grains, pcs	10–220 (77.1)	2–201 (50.3)	4–343 (60.2)	4–180 (55.3)
Number of spikelets, pcs	57–322 (174.6)	38–273 (133.8)	18–411 (137.0)	17–261 (122.2)
Spikelet length, mm	6.1–10.1 (8.1)	6.8–9.8 (8.0)	6.1–11.9 (9.1)	7.2–11.3 (9.3)
Spikelet width, mm	2.3–3.8 (3.1)	2.5–4.0 (3.1)	2.1–3.8 (2.9)	2.3–3.9 (3.0)
Mass of 1000 grains, g	11–38 (25.4)	10–35 (23.2)	12–37 (25.9)	13–39 (25.8)
Mass of grain from the panicle, g	0.72–5.54 (1.98)	0.03–5.90 (1.22)	0.06–5.42 (1.54)	0.07–6.09 (1.79)

Note: The average value is indicated in brackets.

This wide range of variability is not observed in other crops. This is due to the genetic and ecological-geographical remoteness of the crossed forms. In each combination, about 400 plants were selected for morphometric and genetic analysis. Among the F_2 hybrids, we managed to select the best plants according to many traits, combining early maturity, optimal plant height, good grain size in panicles, non-shattering, and fertility of spikelets (Table 3).

Table 3. Selected F_2 hybrid plants from crossing submergence-resistant samples with Novator, 2014.

Hybrid	Duration, Days	Plant Height, cm	Panicle Length, cm	Number of Grains in Panicle, pcs	Mass of 1000 Grains, g
Novator, st	112	97.5	16.5	110	31.8
176(BxN) *	120	108.0	21.5	146	27.0
334 (BxN)	118	96.7	17.3	145	21.6
34 (CxN)	121	81.2	16.5	109	25.3
390 (CxN)	122	82.0	17.5	122	27.7
273 (IxN)	120	95.4	15.1	151	24.4
507 (IxN)	119	97.2	19.0	138	29.3
81 (TxN)	123	96.5	17.2	159	32.0
393 (TxN)	121	97.3	14.9	152	25.2

Note *: (BxN), BR 11 × Novator; (CxN), CR-1009 × Novator; (IxN, Inbara 3 × Novator; (TxN), TDK-1 × Novator.

PCR analysis of leaves was carried out in 20 plants of each of the four hybrids, as a result of which, forms with the *Sub1A* flood resistance gene were isolated. The electrophoretic analysis of PCR products with the RM 7481 marker is shown in Figure 8. The donor allele of the parental line CR-1009 was homozygous in Samples 2, 3, 5, 9, 13 and 17. Plants 2, 4, 6–8, 10, 11, 16, 18, and 19 were heterozygous at the *Sub1A* locus; that is, they carried both the alleles of the donor and the recessive alleles inherited from the variety Novator. Thus, according to the results of PCR analysis with the RM 7481 marker, 14 homozygotes of F_2 plants at the *Sub1A* locus were identified, 40 samples carried *Sub1A* in a heterozygous state, and 22 plants inherited only the recessive allele from the variety Novator.

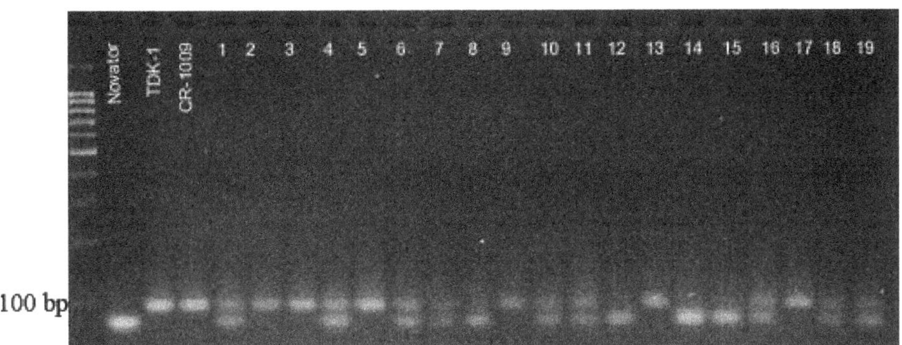

Figure 8. Electrophoregram of the amplification products of rice genomic DNA with the primer RM 4781. 1–19, F_2 (Novator × CR-1009); TDK-1 and CR-1009, donor of the *Sub1A* gene. Molecular weight marker, 1 kb.

Of the analyzed BR-11 × Novator hybrid plants, the Sub1A gene (in homo- and heterozygous state) was present in nine, i.e., in a ratio of 9:11, although with monohybrid segregation, it should have been 15:5. In the hybrid combination CR-1009 × Novator, F_2 segregated in a ratio of 18:2, i.e., almost all of the selected plants had the *Sub1A* gene. In the hybrids Inbara-3 × Novator and TDK-1 × Novator, segregation took place in a ratio of 14:6 or approximately 3:1, i.e., close to Mendelian.

The deviations in segregation of the two combinations can be explained by the influence of selection and gene linkage. A total of 55 plants with the target gene in the homo- and heterozygous state were isolated from 80 plants of four hybrids. The selected samples with the *Sub1A* gene in 2015 were reproduced in the Federal State Unitary Enterprise "Proletarskoye"of the Rostov Region, where the best F_3 plants were selected from them for DNA analysis.

In F_3 plants, significant morphological and biological segregation continued. Significant variation was noted for the growing season, plant height, size of panicles and caryopses, fertility, grain shedding, etc. The best forms were selected from them and leaves were taken for DNA analysis. At the next stage of work, in 2016–2020, constant lines carrying the *Sub1A* gene in a homozygous state were selected and tested for yield and resistance to prolonged water flooding. As a result, rice varieties for herbicide-free technologies will be developed, vigorously overcoming a deep layer of water in the germination phase with minimal seed loss.

4. Conclusions

1. As a result of the studies carried out using molecular marking based on PCR in combination with traditional breeding, early-maturing rice lines with genes for resistance to salinization (*SalTol*) and to flooding (*Sub1A*), which are suitable for cultivation in the south of Russia, were isolated.

2. Rice lines have been developed, the genotype of which contains five effective blast resistance genes (*Pi-1*, *Pi-2*, *Pi-33*, *Pi-ta*, and *Pi-b*). The introduction of such varieties into production will allow us to avoid epiphytotic development of the disease, preserving the biological productivity of rice and obtaining environmentally friendly agricultural products.

3. Samples of the F_4 and BC_2F_3 generations were obtained with combined blast resistance (*Pi*) and prolonged flooding tolerance (*Sub1A*) genes as a factor in the control of weeds in the homo- and heterozygous state, which was confirmed by the data of their DNA analysis. The testing of the obtained rice breeding resources for submergence tolerance under laboratory conditions made it possible to select tolerant rice forms that will be studied in the breeding process for a complex of agronomically valuable traits. Their use will reduce the use of chemical plant protection products against diseases and weeds, thereby increasing the ecological status of the rice-growing industry.

The research was carried out with the financial support of the Kuban Science Foundation in the framework of the scientific project № 20.1/1.

Author Contributions: Conceptualization, P.K.; methodology, E.D. and K.A.; validation, P.K.; formal analysis, E.D.; investigation, E.D., P.K., M.R., S.L. and E.K.; resources, E.D.; writing—original draft preparation, E.D., P.K. and K.A.; project administration, E.D.; funding acquisition, E.D. All authors have read and agreed to the published version of the manuscript.

Funding: This research was funded by Federal Scientific Rice Centre.

Conflicts of Interest: The authors declare no conflict of interest.

References

1. Kumar, V.; Bhagwat, S.G. Microsatellite (SSR) Based Assessment of Genetic Diversity among the Semi-dwarf Mutants of Elite Rice Variety WL112. *Int. J. Plant Breed. Genet.* **2012**, *6*, 195–205. [CrossRef]
2. Singh, S.; Mackill, D.; Ismail, A. Responses of Sub1 rice introgression lines to submergence in the field: Yield and grain quality. *Field Crops Res.* **2009**, *113*, 23. [CrossRef]
3. Negrao, S.; Courtois, B.; Ahmadi, N.; Abreu, I.; Saibo, N.; Oliveira, M.M. Recent updates on salinity stress in rice: From physiological to molecular responses. *Crit. Rev. Plant Sci.* **2011**, *30*, 329–377. [CrossRef]
4. Food and Agriculture Organization. Report of Salt Affected Agriculture. Available online: http://www.fao.org/ag/agl/agll/spush (accessed on 16 October 2010).
5. Ladatko, N.A. The growth of rice plants under conditions of chloride salinity. *Rice Breed.* **2006**, *9*, 37–41.
6. Mekawy, A.M.; Assaha, D.V.; Yahagi, H.; Tada, Y.; Ueda, A.; Saneoka, H. Growth, physiological adaptation, and gene expression analysis of two Egyptian rice cultivars under salt stress. *Plant Physiol. Biochem.* **2011**, *187*, 17–25. [CrossRef] [PubMed]

7. Ren, Z.H.; Gao, J.P.; Li, L.G.; Cai, X.L.; Huang, W.; Chao, D.Y.; Zhu, M.Z.; Wang, Z.Y.; Luan, S.; Lin, H.X. A rice quantitative trait locus for salt tolerance encodes a so-dium transporter. *Nat. Genet.* **2005**, *37*, 1141–1146. [CrossRef] [PubMed]

8. Takehisa, H.; Shimodate, T.; Fukuta, Y.; Ueda, T.; Yano, M.; Yamaya, T.; Kameya, T.; Sato, T. Identification of quantitative trait loci for plant growth of rice in paddy field flooded with salt water. *Field Crops Res.* **2004**, *89*, 85–95. [CrossRef]

9. Platten, J.D.; Egdane, J.A.; Ismail, A.M. Salinity tolerance, Na$^+$ exclusion and allele mining of HKT1; 5 in *Oryza sativa* and *O. glaberrima*: Many sources, many genes, one mechanism? *BMC Plant Biol.* **2013**, *13*, 2–16. [CrossRef] [PubMed]

10. Xu, K.; Mackill, D.J. A major locus for submergence tolerance mapped on rice chromosome 9. *Mol. Breed.* **1996**, *2*, 219–224. [CrossRef]

11. Xu, K.; Xu, X.; Fukao, T.; Canlas, P.; Maghirang-Rodriguez, R.; Heuer, S.; Ismail, A.M.; Bailey-Serres, J.; Ronald, P.C.; Mackill, D.J. Sub1A is an ethylene-response-factor like gene that confers submergence tolerance to rice. *Nature* **2006**, *442*, 705–708. [CrossRef] [PubMed]

12. Kovalenko, E.D.; Gorbunov, Y.B.; Kovalyova, A.A.; Dudenko, V.P.; Kolomiec, T.M. *Guidelines for Assessing the Resistance of Rice Varieties to the Blast Causative Agent*; VASKHNIL: Moscow, Russia, 1988; 30p.

13. Choi, H.C.; Kim, Y.G.; Hong, H.C.; Hwang, H.G.; Ahn, S.N.; Han, S.S.; Ryu, J.D.; Moon, H.P. Development of blast-resistant rice multi-line cultivars and their stability to blast resistance and yield performance. *Korean J. Breed.* **2006**, *38*, 83–89.

14. Mukhina, Z.M. *Using DNA Markers to Study the Genetic Diversity of Plant Resources: Monographic*; Education-South: Krasnodar, Russian, 2008; p. 84.

15. Singh, D.; Kumar, A.; Kumar, A.S.; Chauhan, P.; Kumar, V.; Kumar, N.; Singh, A.; Mahajan, N.; Sirohi, P.; Chand, S.; et al. Marker assisted selection and crop management for salt tolerance: A review. *Afr. J. Biotechnol.* **2011**, *10*, 14694–14698. [CrossRef]

16. Sabouri, H.; Rezai, A.M.; Moumeni, A.; Kavousi, A.; Katouzi, M.; Sabouri, A. QTLs mapping of physiological traits related to salt tolerance in young rice seedlings. *Biol. Plant* **2009**, *53*, 657–662. [CrossRef]

17. Kostylev, P.I.; Redkin, A.A.; Krasnova, E.V.; Dubina, E.V.; Ilnitskaya, E.T.; Yesaulova, L.V.; Mukhina, Z.M.; Kharitonov, E.M. Creating rice varieties, resistant to pirikulariosis by means of DNA markers. *Vestn. Russ. Acad. Agric. Sci.* **2014**, *1*, 26–28.

18. Linh, L.H.; Xuan, T.D.; Ham, L.H.; Ismail, A.M.; Khanh, T.D. Molecular Breeding to Improve Salt Tolerance of Rice (*Oryza sativa* L.) in the Red River Delta of Vietnam. *Int. J. Plant Genom.* **2012**, *2012*. [CrossRef] [PubMed]

19. Fukuta, Y.; Xu, D.; Kobayashi, N. Genetic characterization of universal differential varieties for blast resistance developed under the IRRI-Japan Collaborative Reseaarch Project using DNA markers in rice (*Oryza sativa* L.). *JIRCAS Work. Rep.* **2009**, *63*, 35–68.

20. Kostylev, P.I. *Methods of Breeding, Seed Farming Techniques and High-Quality Rice: Monographic*; Rostov-on-Don: Zernograd, Russia, 2011; 267p.

21. Dubina, E.; Kostylev, P.; Garkusha, S.; Ruban, M.; Pischenko, D. Marker assisted rice breeding for resistance to biotic and abiotic stressors. In Proceedings of the BIO Web of Conference, XI International Scientific and Practical Conference "Biological Plant Protection is the Basis of Agroecosystems Stabilization", Krasnodar, Russia, 21–24 September 2020. [CrossRef]

22. Dubina, E.; Mukhina, Z.; Kharitonov, E.; Shilovskiy, V.; Kharchenko, E.; Esaulova, L.; Korkina, N.; Maximenko, E.; Nikitina, I. Creation of blast disease-resistant rice varieties with modern DNA-markers. *Russ. J. Gen.* **2015**, *8*, 752–756. [CrossRef]

Publisher's Note: MDPI stays neutral with regard to jurisdictional claims in published maps and institutional affiliations.

Article

Nanopore RNA Sequencing Revealed Long Non-Coding and LTR Retrotransposon-Related RNAs Expressed at Early Stages of Triticale SEED Development

Ilya Kirov [1,2,*], Maxim Dudnikov [1,2], Pavel Merkulov [1], Andrey Shingaliev [1], Murad Omarov [1,3], Elizaveta Kolganova [1], Alexandra Sigaeva [1], Gennady Karlov [1] and Alexander Soloviev [1]

[1] Laboratory of Marker-Assisted and Genomic Selection of Plants, All-Russia Research Institute of Agricultural Biotechnology, Timiryazevskaya str. 42, 127550 Moscow, Russia; max.dudnikov.07@gmail.com (M.D.); paulmerkulov97@gmail.com (P.M.); kronstein491@yandex.ru (A.S.); muradok98@gmail.com (M.O.); liza.colg@gmail.com (E.K.); alexandra9954@gmail.com (A.S.); karlovg@gmail.com (G.K.); A.Soloviev70@gmail.com (A.S.)

[2] Kurchatov Genomics Center of ARRIAB, All-Russia Research Institute of Agricultural Biotechnology, Timiryazevskaya Street, 42, 127550 Moscow, Russia

[3] Faculty of Computer Science, National Research University Higher School of Economics, Pokrovsky Boulvar, 11, 109028 Moscow, Russia

* Correspondence: kirovez@gmail.com

Received: 24 November 2020; Accepted: 15 December 2020; Published: 17 December 2020

Abstract: The intergenic space of plant genomes encodes many functionally important yet unexplored RNAs. The genomic loci encoding these RNAs are often considered "junk", DNA as they are frequently associated with repeat-rich regions of the genome. The latter makes the annotations of these loci and the assembly of the corresponding transcripts using short RNAseq reads particularly challenging. Here, using long-read Nanopore direct RNA sequencing, we aimed to identify these "junk" RNA molecules, including long non-coding RNAs (lncRNAs) and transposon-derived transcripts expressed during early stages (10 days post anthesis) of seed development of triticale (AABBRR, $2n = 6x = 42$), an interspecific hybrid between wheat and rye. Altogether, we found 796 lncRNAs and 20 LTR retrotransposon-related transcripts (RTE-RNAs) expressed at this stage, with most of them being previously unannotated and located in the intergenic as well as intronic regions. Sequence analysis of the lncRNAs provide evidence for the frequent exonization of Class I (retrotransposons) and class II (DNA transposons) transposon sequences and suggest direct influence of "junk" DNA on the structure and origin of lncRNAs. We show that the expression patterns of lncRNAs and RTE-related transcripts have high stage specificity. In turn, almost half of the lncRNAs located in Genomes A and B have the highest expression levels at 10–30 days post anthesis in wheat. Detailed analysis of the protein-coding potential of the RTE-RNAs showed that 75% of them carry open reading frames (ORFs) for a diverse set of GAG proteins, the main component of virus-like particles of LTR retrotransposons. We further experimentally demonstrated that some RTE-RNAs originate from autonomous LTR retrotransposons with ongoing transposition activity during early stages of triticale seed development. Overall, our results provide a framework for further exploration of the newly discovered lncRNAs and RTE-RNAs in functional and genome-wide association studies in triticale and wheat. Our study also demonstrates that Nanopore direct RNA sequencing is an indispensable tool for the elucidation of lncRNA and retrotransposon transcripts.

Keywords: long non-coding RNAs; seed development; Nanopore sequencing; retrotransposons; triticale

---

1. Introduction

Long non-coding RNAs (lncRNAs) are a diverse set of RNAs longer than 200 bp with no or very little coding potential. Traditionally, lncRNAs are considered to be protein non-coding, although many of them carry small open reading frames and encode functional peptides in different plants [1]. A broad range of functionality has been described for lncRNAs in plants, including the miRNA sponge, protein scaffolding, and post-transcriptional regulation of target genes via antisense pairing followed by mRNA decoy [2]. Based on their genomic localization regarding other genes, lncRNAs are broadly grouped into different classes: (1) lincRNAs or long non-coding intergenic RNAs; (2) NAT lncRNAs or natural anti-sense lncRNAs; (3) intronic lncRNAs, located in the introns; and (4) sense lncRNAs [3]. LncRNAs have been identified in many plant species and their expression in response to various abiotic and biotic stresses has been investigated, although our knowledge about lncRNA functions is still very limited [4–9]. However, the catalogue of lncRNAs for some plants with complex polyploidy and repeat-rich genomes, including wheat (*Triticum aestivum* L.), remains mainly underexplored [10].

Intergenic space and introns are frequently sources of lncRNA origin, although these genomic regions are enriched by insertions or remnants of transposable elements (TEs) [11]. Corroborating this, lncRNA sequences are more often associated with TEs than protein-coding genes and the bulk of them possess TE-related sequences [11–16]. For example, up to 75% of human lncRNAs have at least one exon with sequences originating from TEs [12,17]. A similar trend was demonstrated in some plant species, including maize, where 65% of lncRNAs had similarities to TEs [18]. More intriguingly, the TE-derived sequences may trigger the origin of new lncRNAs, providing a positive feedback loop with the evolution of lncRNAs [11,19]. TE-derived lncRNAs can have important and conserved biological functions [20]. For example, the rice lncRNA MIKKI is derived from LTR retrotransposons and has been shown to sequestrate miR171 and prevent degradation of its targets, mRNAs of SCARECROW-Like (SCL) transcription factors, in roots [20]. It is important to note that TEs can become transcriptionally and transpositionally active under certain circumstances, including stressful conditions, and in some developmental stages (e.g., microsporogenesis) and tissues (e.g., developing endosperm) [21–23]. However, individual TE transcripts and their coding potential have only been studied episodically in plants. The association of lncRNAs with repeat sequences like TEs makes the annotation of many lncRNAs challenging because of the ambiguity in the mapping of repeat-derived short RNAseq reads. RNAseq reads with multiple mapping positions in the genome sequence are frequently discarded from further analysis. Although some tools have been developed so far to overcome these obstacles, most of the lncRNA identification pipelines still ignore ambiguously mapped reads [24,25]. Thus, long-read sequencing technologies provide a great opportunity for transcriptome exploration, including the identification of transcribed repeats (e.g., transposable elements) or repeat-related transcripts [26]. Moreover, Panda and Slotkin [26] showed that by using Nanopore long cDNA reads, it is even possible to trace the expression of individual TEs from multicopy families in *Arabidopsis* and maize. The application of long-read technologies to the exploration of the lncRNA repertoire in plants has been demonstrated for several plant species, including *Oryza sativa* L. ssp. *japonica* [27], *Populus simonii* [28], *P. qiongdaoensis* [29], poplar "Nanlin 895" [30], *Trifolium pratense* [31], *Cardamine violifolia* [32], and *Vigna angularis* [33]. These studies demonstrated that long-read sequencing can be used to obtain a comprehensive catalogue of lncRNAs.

Grain development is one of the most important and practically relevant biological processes. It involves massive biochemical, physiological, and transcriptomic changes [34–36]. Wheat grain development is divided into five stages: (i) undifferentiated embryo and cellularization of the endosperm (0–7 days post anthesis (dpa)); (ii) embryo differentiation the embryo with formation of the main cell types (transfer cells, aleurone, starchy endosperm and the surrounding cells) (7–14 dpa); (iii) root and leaf primordia differentiation, full kernel development and the milk-ripe stage (14–21 dpa); (iv) further growth and differentiation of primary and lateral roots, and the dough stage of endosperm (21–31 dpa); (v) fully differentiated embryo (31–50 dpa) [34]. Studies of the transcriptome during wheat seed development have been extensively elucidated using RNAseq sequencing [35,37–43]. Global transcriptome analysis of developing seeds has shown that the expression of protein-coding genes is highly dynamic. Recently, Madhavan et al. (2020) used publicly available Illumina RNAseq reads and identified 443 lncRNAs expressed during the grain filling stage (14 and 30 dpa) [44]. It is currently unknown which types of lncRNA are expressed during other stages of wheat seed development.

Here, we used Nanopore long-read sequencing to discover lncRNAs, TE transcripts, and TE-related lncRNAs that are specifically expressed during the cell proliferation stage of seed development (10 dpa) in triticale (× *Triticosecale* Wittmack, AABBRR genome, $2n = 6x = 42$) an interspecific hybrid between wheat and rye (*Secale cereale* L.). We identified 796 lncRNAs and 20 LTR retrotransposon-derived transcripts, with most of them being previously unannotated. The majority of the detected retrotransposon RNAs had a single intron, carried open reading frames (ORFs) encoding for a diverged set of GAG proteins, and were encoded by potentially autonomous and non-autonomous retrotransposons. The lncRNAs were also expressed during wheat seed development and had high stage specificity. Moreover, we found that lncRNA loci were biased toward frequent TE sequence exonization and were mainly located in the intergenic regions of A and B genomes of triticale. Our work explored the lncRNA landscape during the early stage of wheat and triticale seed development and provides a unique dataset for further functional studies of lncRNA and TEs, and their implications for seed development. Finally, the identified lncRNAs can be further incorporated into genome-wide association studies for marker-assisted improvement of the bread quality of modern triticale genotypes.

2. Material and Methods

2.1. Plant Material and DNA Isolation

For this study, the spring triticale line "L8665" obtained from the Department of Genetics, Russian State Agrarian University, was used. For DNA isolation, seeds of this line were germinated in the dark (room temperature) during 5–7 days and genomic DNA was isolated by the cetyltrimethylammonium bromide (CTAB) protocol [45].

2.2. Sample Collection and RNA Isolation

Plants of the spring triticale line "L8665" were grown in a greenhouse under natural light conditions. Developing seeds at 10 days post anthesis and flag leaves were dissected and placed into liquid nitrogen. RNA was isolated by the ExtractRNA kit (Evrogen, Moscow, Russia) following the manufacturer's instructions. The RNA concentration and integrity were estimated by Nanodrop (Nanodrop Technologies, Wilmington, CA, USA) and gel electrophoresis using an 1.2% agarose gel with ethidium bromide staining.

2.3. RT-PCR

For RT-PCR, RNA was treated by DNAse I (Qiagen, Hilden, Germany, Q-79254) following the manufacturer's instruction and used for cDNA synthesis. cDNA was synthesized using the MMLV RT kit (Evrogen, Moscow, Russia). Primers used for RT-PCR and the expected fragment lengths are listed in Table 1. The CDC (Cell division control protein, AAA-superfamily of ATPases; Ta54227) gene was used as a reference because of its high expression stability as shown by a previous study [46].

Table 1. Primers used for RT-PCR.

Target	Primer Sequences	Amplicon Size
Lnc001	lncTR001/F: AGGTTGCAAGTCTCTTGCTCTTGA lncTR001/R: TCATGCCCGCTAAGAATTACAGTGT	RNA/DNA = 500 bp/~1100 bp
Lnc002	lncTR002/F: TGGGTTGTGACTTGTGATACGCA lncTR002/R: CGGTTAGGGCTGGGCTGAATG	RNA/DNA = 300 bp/300 bp
Lnc003	lncTR003/F: ACAGTATGAAGCTAGCCGGCTTG lncTR003/R: TATCCTGTCGTCCTCTCGTCTCG	RNA/DNA = 303 bp/303 bp
CDC (the cell division control protein), Ta54227 [47]	CDC/F: GCCTGGTAGTCGCAGGAGAT CDC/R: ATGTCTGGCCTGTTGGTAGC	RNA/DNA = 76 bp/76 bp
gRNA TaeST2.19707.1	gRNATae_19/F: ATTACACCCCCAAACCGCCAAAT gRNATae_19/R:TGGGGAATTTTCCACACCCACTT	RNA/DNA = 490 bp/490 bp
shGAG TaeST2.19707.1	shGAGTae_19/F: TTGATTGCCGCCTGGTTATCACA shGAGTae_19/R: AGTGGGAATCGGAGGAACTGGAA	RNA/DNA = 560 bp/3200 bp
gRNA TaeST2.45518.1	gRNATae_19/F: ATTACACCCCCAAACCGCCAAAT gRNATae_19/R: TGGGGAATTTTCCACACCCACTT	RNA/DNA = 417 bp/417 bp
shGAG TaeST2.45518.1	shGAGTae_45/F: GCTTACTCTTGTCTACTCCACGCA shGAGTae_45/R: GGACTGGAGAAGCGAATGCATCT	RNA/DNA = 500 bp/897 bp

The PCR conditions were 94 °C for 1 min; 35 cycles of 94 °C for 1 min, 58 °C for 1 min, and 72 °C for 1 min; and a final elongation of 72 °C for 3 min.

2.4. Nanopore Direct RNA Sequencing and Transcript Assembly

Poly-A+ mRNA was purified from 100 µg of total RNA by the Dynabeads mRNA DIRECT Kit (ThermoFisher Scientific, Waltham, MA, USA) following the manufacturer's instructions. Final poly-A+ RNA concentration was measured using a Quantus Fluorometer (Promega Corporation, Madison, WI, USA) and checked by gel electrophoresis. For Nanopore sequencing, a library was prepared from 1 µg Poly-A+ using the nanopore Direct RNA Sequencing Kit SQK-RNA002 (Oxford Nanopore Technologies, Oxford, UK). Direct RNA Sequencing (DRS) was carried out using MinION and flow cell FLO-MIN106. Basecalling was performed by Guppy (Version 4.0.11).

For transcript assembly, sequences of the A and B genomes and unanchored scaffolds (Un) of *Triticum aestivum* ([48]) downloaded from the EnsemblPlants server (https://plants.ensembl.org/Triticum_aestivum/Info/Index) were combined with *Secale cereale* genome sequences [49] into a single fasta file. Reads were mapped to the obtained fasta file by minimap2 software [50] with the '-ax splice' argument. The obtained sam file was converted to a bam file by samtools [51] (samtools view -Sb) followed by sorting of the bam file by bamtools [52]. Transcript assembly was performed by StringTie2 [53] with the following arguments: –L –j 2 –f 0.05. The obtained gtf file was converted to gff format by the gffread tool [54]. The sorted bam and gff files were used for read mapping visualization by the locally installed JBrowse [55].

To obtain a high-confidence set of transcripts, we extracted transcript sequences using gtf from StringTie2 assembly by the gffread tool [54]. The reads were then mapped back to the transcripts using minimap2 software (settings: -ax map-ont), and the bam file with only primary alignments and mapping quality > 30 was obtained using samtools with –F 256 –q 30 –b parameters. This bam file was used to count the number of reads per transcripts and transcripts with > 5 DRS reads were selected.

2.5. Long Non-Coding RNA Prediction

To identify lncRNAs, all high-confidence transcripts with a length of >200 bp were selected using biopython [56]. Transcripts with open reading frame (ORF) lengths of >300 bp predicted by getorf were filtered out. Protein coding potential was calculated by three tools: LncFinder [57], PLEK [58], and CNCI [59]. LncFinder [57] was run in Rstudio Version 1.2.1335 (http://www.rstudio.com/) with R version 3.6.0. The following parameters were utilized: parallel.cores = 20, SS.features = TRUE, format = "DNA", frequencies.file = "wheat", svm.model = "wheat". PLEK [58] and CNCI [59] were run with the default settings. Only transcripts identified by all three tools as non-coding and without similarity to Pfam domains were classified as lncRNAs.

To identify lncRNAs with exons overlapping the annotated TEs, we intersected CLARITE TE annotation (https://urgi.versailles.inra.fr/download/iwgsc/IWGSC_RefSeq_Annotations/v1.0/) with the genomic coordinates of the assembled exons.

2.6. Retrotransposon-Related Transcript Annotation

LTR retrotransposons (RTEs) were predicted in the genome using LTRharvest 1.5.10 with the default parameters [60] and LTRdigest 1.5.10 [61] with the following parameters: -aaout yes -pptlen 10 30 -pbsoffset 0 3 -pdomevalcutoff 0.001. Hidden markov model (HMM) profiles of RTE domains were downloaded from the GyDB database [62]. The gff3 file from the LTRdigest analysis was treated by a custom Python script (https://github.com/Kirovez/LTR-RTE-analysis/blob/master/LtrDiParser_v2.2.py) to extract sequences of LTR retrotransposons possessing similarity to any RTE domains including GAG, reverse transcriptase, RNAse H, aspartic protease, and integrase. To identify retrotransposon-related transcripts, the TEsorter tool (parameters: -eval 0.00001 -db gydb) was run with the set of all high-confidence transcripts. The transcripts with a similarity to RTE domains were manually checked in the locally installed JBrowse [55]. TEsorter data were also used for RTE classification. We also ran TEsorter with confident transcripts and the RExDB [63] database (-db rexdb) to identify the transcripts of Class II transposons but no transcripts were detected.

2.7. GAG Protein Analysis

To find GAG proteins, ORFs were predicted for RTE transcripts and Blastp with corresponding proteins was run followed. GAG proteins were aligned by MAFFT [64] with the standard parameters and a phylogenetic tree was constructed using iTOL [65]. The multiple alignment visualization was carried out in Jalview version 2.11.1.3 [66]. RNA-binding motifs (CX2CX4HX4C, where X is any amino acid) were identified by a custom Python script (https://github.com/Kirovez/LTR-RTE-analysis/blob/master/RBM_GAG_screen.py).

2.8. Gene Ontology Enrichment

Gene ontology (GO) enrichment analyses was performed using ShinyGO v0.61 [67] (http://bioinformatics.sdstate.edu/go/) with an false discovery rate (FDR) <0.01.

2.9. Expression Analysis

For RNA-Seq analysis of lncRNA and RTE transcript expression in different organs and development stages, publicly available data were used (Table 2). Reads were mapped on the de novo assembled transcriptome by Hisat2 [68] with the default options. The obtained files with alignments were used to calculate RPM (read per million reads) values for every transcript. For this purpose, we used Salmon v0.8.1 [69] and the quant command with the default parameters.

Table 2. Publicly available RNAseq data used in this study.

SRA Accession	Number of Reads	Development Stage/Organs	Reference
ERR392055	26,791,465	10 dpa/seed	[39]
ERR392076	29,714,230	20 dpa/seed	[39]
ERR392069	31,433,795	30 dpa/seed	[39]
SRR10522394	39,611,224	Leaves	[70]
SRR1175868	64,825,850	Pistils	[71]

2.10. Extrachromosomal Circular DNA Isolation

Extrachromosomal circular DNA (eccDNA) was isolated and amplified according to the protocol of Lanciano et al. [72] with several modifications. Briefly, 5 µg of genomic DNA was treated by Plasmid-Safe ATP-Dependent DNAse (Epicenter, Madison, WI, USA) for 48 h according to the manufacturer's instructions. DNA precipitation was carried out by 0.1 volume 3 M sodium acetate and 2.5 volume absolute ethanol, followed by overnight incubation at −20 °C. After centrifugation, the eccDNA pellet was obtained and exposed to the rolling circle amplification (RCA) reaction by the Illustra TempliPhi 100 Amplification Kit (GE Healthcare, Chicago, IL, USA) for 65 h at 28 °C. Detection of the eccDNA of LTR retrotransposon TaeST2.45518.1 (named '*MIG*', location in wheat genome: 7B: 312,336,869 … 312,341,902 (5 kb)) was performed by inverse PCR with specific primers: Forward: CACACCACTAGCAACCTCCA ; Reverse: TGCTTGTGACAAGATGGGCA. The PCR conditions were 94 °C for 1 min; 35 cycles of 94 °C for 1 min, 58 °C for 1 min, 72 °C for 1 min; and final elongation at 72 °C for 3 min.

2.11. Statistics and Visualization

Statistical analysis was done in Rstudio Version 1.2.1335 (http://www.rstudio.com/) with R version 3.6.0. Bar plots, density plots, and box plots were drawn by ggplot2 [73]. Heatmaps were constructed by the ComplexHeatmap [74] R package.

3. Results

3.1. Direct Oxford Nanopore RNA Sequencing

Total RNA was isolated from whole spikes of hexaploid triticale (AABBRR) cv. L8665 at 10 days post anthesis (dpa). This RNA was used for direct RNA sequencing by MinION (Oxford Nanopore). In total, 1,100,000 direct RNA sequencing (DRS) reads with N50~1.1 kb were obtained (Figure 1). To assemble the transcripts, reads were mapped to the genome sequence created artificially by combining the A and B genome sequences of the wheat chromosome-level assembly and rye draft genome contigs. Overall, 82,785 transcripts from 74,904 loci were assembled with 47,378 and 26,169 loci located in genomes A/B (AB lncRNAs) and R. A total of 36,490 transcripts had >5 mapped DRS reads, representing a set of transcripts with high confidence that was used for further analysis.

To estimate the triticale seed development phase used for Nanopore sequencing, we determined the expression values (reads per million mapped reads (RPM)) of the key genes involved in starch biosynthesis (expression started at Phase 1), the genes of storage proteins (high molecular weight glutenins and gliadins), and those of the *wbm* protein (Table 3), which are expressed during the grain filling stage. We identified the expression of starch biosynthesis genes, while no genes of storage proteins or the *wbm* protein were expressed. This suggests that the seeds used for Nanopore sequencing were in the early stages of seed development (before 14 dpa).

Thus, using direct Nanopore RNA sequencing, we assembled a high-confidence transcriptome of triticale seed at the early development stage and detected the expression of key genes known to be involved in the biological process (starch biosynthesis) occurring at this stage.

3.2. Long Non-Coding RNA Prediction

The assembled high-confidence set of transcripts was used for long non-coding RNA (lncRNA) prediction. The following criteria were applied to distinguish lncRNAs from protein-coding transcripts: (1) transcript length of >200 bp; (2) transcripts with an ORF length of <300 bp; (3) transcripts classified as non-coding by three tools including LncFinder [57], PLEK [58], and CNCI [59]; and (4) transcripts with no similarity to any Pfam domains. Using these criteria, we identified 796 triticale lncRNAs (Supplementary Files S1 and S2) encoded by 780 loci in Genomes A (167 lncRNAs), B (212 lncRNAs), and R (410 lncRNAs) and in unanchored wheat scaffolds (seven lncRNAs) (Figure 1A). Most of the lncRNAs had lengths of <1000 bp (Figure 1B). LncRNAs (386 transcripts) encoded by the loci of

Genomes A and B or unanchored wheat scaffolds (AB lncRNAs) were used for further analysis because of the significantly better annotation of these genomes compared with the R genome. Intersection of the genome position of the AB lncRNA loci with lncRNA and mRNA loci previously annotated in the A or B wheat genomes showed that 281 (73%) of the triticale AB lncRNAs are located in the previously unannotated genomic regions (Figure 1C). This number is significantly higher (Fisher's Exact Test, *p*-value < 2.2×10^{-16}) than that for non-lncRNA AB transcripts (10%, 2523), pointing to the underexplored nature of lncRNA loci. Moreover, only 13% (106) of the AB lncRNAs in our dataset were previously known.

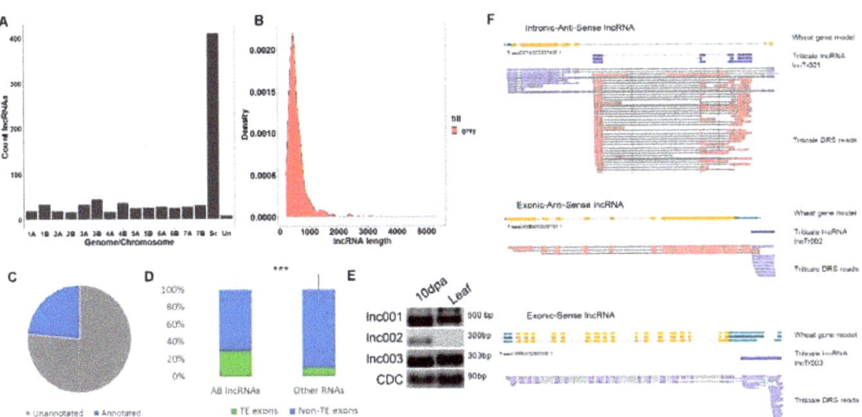

Figure 1. Identification and classification of triticale long non-coding RNAs (lncRNAs). (**A**) Bar plot showing the number of triticale lncRNA loci located on different chromosomes of Genomes A, B, and R and unanchored wheat contigs (Un). Sc indicates loci mapped to *Secale cereale* contigs. (**B**) Density plot of the lncRNA length distribution. (**C**) Pie graph showing the portion of the AB lncRNAs located in the unannotated genomic regions. (**D**) The portion of genes encoding AB lncRNAs and other RNAs (not classified as lncRNAs) that have exons with transposable elements. Three stars (***) indicate significant differences based on Fisher's Exact Test for Count Data, p-value < 0.001. (**E**) RT-PCR with specific primers on three genic lncRNAs of distinct types (**F**), and RNA isolated from developing grain (10 days post anthesis (dpa)) and flag leaves. CDC: reference gene (the cell division control protein). (**F**) Types of genic lncRNAs and the corresponding wheat genes.

Table 3. Expression of the key genes of the starch biosynthesis pathway and wheat storage proteins.

Wheat Gene ID	Gene Expression, +/−	Reads per Million (RPM)	Gene Annotation	Genomic Coordinates
TraesCS4A02G418200	+	76	GBSS/Starch synthase, chloroplastic/amyloplastic	4A:688,097,145–688,100,962
TraesCS4B02G029700	+	7	(BGC1) Flo6/5′-AMP-activated protein kinase subunit beta-2 (PTST)	4B:21,937,120–21,944,075
TraesCS4A02G284000	+	3		4A:590,660,989–590,667,561
TraesCS7B02G139700	+	6	ISA	7B:175,999,323–176,007,332
TraesCS7A02G251400	+	15		7A:235,460,629–235,468,417
TraesCS6A02G048900	−	0		6A:24,921,651–24,922,607
TraesCS6A02G049200	−	0	α/β-gliadins	6A:25,203,493–25,204,413
TraesCS6A02G049100	−	0		6A:25,107,550–25,108,401
TraesCS6A02G049600	−	0		6A:25,472,841–25,473,704
TraesCS1A02G007400	−	0	γ-gliadin-A3	1A:4,033,339–4,034,196
TraesCS7A02G531903	−	0	wbm	7A:710,471,331–710,471,679
TraesCS1A02G317311	−	0	HMW Glu-1Ax	1A:508,723,999–508,726,319
TraesCS1B02G329711	−	0	HMW Glu-1Bx	1B:555,765,127–555,766,152

LncRNAs are frequently associated with transposons. We found that 111 (29%) AB lncRNAs had exons that had transposon sequences with a length of >50 bp (Figure 1D). This was significantly more than that expected by chance (10%, Fisher's exact test, p-value $< 2.2 \times 10^{-16}$). Of the TE-related lncRNAs, 61, 47, and 3 lncRNAs had exons with similarity to Class I, Class II, and unclassified TEs, respectively.

The AB lncRNAs were classified regarding the position of the annotated wheat protein-coding genes, resulting in 17 genic lncRNAs. Gene ontology analysis revealed that the genes overlapping with genic lncRNAs were significantly enriched (FDR < 0.01) in several Gene Ontology categories including "vesicle-mediated transport", "lipid modification", "ATPase activity", and "hydrolase activity". The genic lncRNAs were of different types, with Intronic-antisense (two lncRNAs, Figure 1F (top)), exonic-anti-sense (five lncRNAs, Figure 1F (middle)), and exonic-sense (nine lncRNAs, Figure 1F (bottom)) transcripts being the most common types. In addition, exonic (sense and anti-sense) lncRNAs were found. The expression levels of three types of genic lncRNA (depicted in Figure 1F) belonged to the distinct types estimated in developing grain (10 dpa) and flag leaves (Figure 1E). The RT-PCR results suggested that two lncRNAs (lnc001 and lnc003) were expressed in both samples, while the expression of one type of lncRNA (lnc002) only occurred in developing grain (10 dpa).

Altogether, we identified hundreds of previously unknown genic and intergenic lncRNAs of triticale and showed that they frequently possess the remnants of Class I and Class II transposable elements.

3.3. AB lncRNAs Are Prone to Tissue-Specific Expression

To find lncRNAs with possible specific roles during seed development, we used wheat RNAseq data to estimate lncRNA expression in several seed developmental stages, leaf tissues, and pistils of wheat.

A non-zero expression value in at least one condition was obtained for 351 AB lncRNAs. To estimate any biases in AB lncRNA expression compared with all high-confidence transcripts assembled from DRS reads, we calculated the tissue-specificity index (TSI). The results showed that the TSI was significantly (according to the Wilcoxon rank sum test with continuity correction, p-value $< 2.2 \times 10^{-16}$) higher for lncRNAs, suggesting high tissue specificity of lncRNA expression (Figure 2A). We further found that 46% of the lncRNAs had their highest level of expression at 10, 20 or 30 dpa, with 107 lncRNAs having their maximum expression level at 10 dpa; this is in accordance with the type of our triticale sample (10 dpa) used for RNA isolation (Figure 2B). Moreover, we identified 95 AB lncRNAs for which >90% of the sum of RPKM (reads per kilobases per million reads) values across all the samples accounted for the 10, 20, and 30 DPA stages (Figure 2C).

Thus, the expression pattern of the identified lncRNAs was found to be tissue-specific, with almost half of the lncRNAs demonstrating the maximum expression level during seed development.

3.4. Retrotransposon-Related Transcripts Encoding GAG Proteins Are Expressed during Early Seed Development in Triticale

Early embryonic and endosperm development may be accompanied by the activation of transposable element (TE) activity [22]. Therefore, we analyzed the assembled transcripts for the presence of open reading frames encoding TE-related proteins. We focused on the transcripts of Genomes A/B because of the high quality of wheat genome assembly and annotation compared with the rye genome. No transcripts corresponding to DNA transposons were detected. However, we found 20 transcripts (RTE-RNAs) carrying a single ORF with similarities to distinct proteins of LTR retrotransposons. Surprisingly, we found no transcripts encoding for the full set of RTE proteins (GAG and POL). To check whether any RTE-RNAs were encoded by LTR retrotransposons with detectable LTR sequences (RTEs), we predicted RTEs in Genomes A and B. The results showed that 5 and 10 RTE-RNAs (15) were transcribed from full-length (potentially autonomous) or non-complete (one or more RTE domains were not detected while both LTRs were present) RTEs, respectively. Thus, almost 25% of the RTE-RNAs were found to be transcribed from potentially autonomous RTEs.

For five RTE-RNAs, no associated RTEs were predicted, but 75% (15) of the RTE-RNAs were found to carry an ORF encoding a single GAG protein (GAG-RNAs). Of those, eight (53%) and five (33%) GAG-RNAs were encoded by full-length (Figure 3A) and non-complete copies of LTR retrotransposons, respectively. For two (14%) GAG-RNAs, no corresponding RTEs were identified (Table 4). It should be noted that four GAG-RNA genes were found to be located in the introns of three annotated protein-coding genes in the sense or anti-sense orientation, including TraesCS2B02G261900 (sense and anti-sense), TraesCS5A02G298800 (anti-sense), and TraesCS1B02G222500 (sense). Two GAG-RNA genes (TaeST2.11597.1 and TaeST2.11598.1) were found to be located in the introns of the same gene (TraesCS2B02G261900).

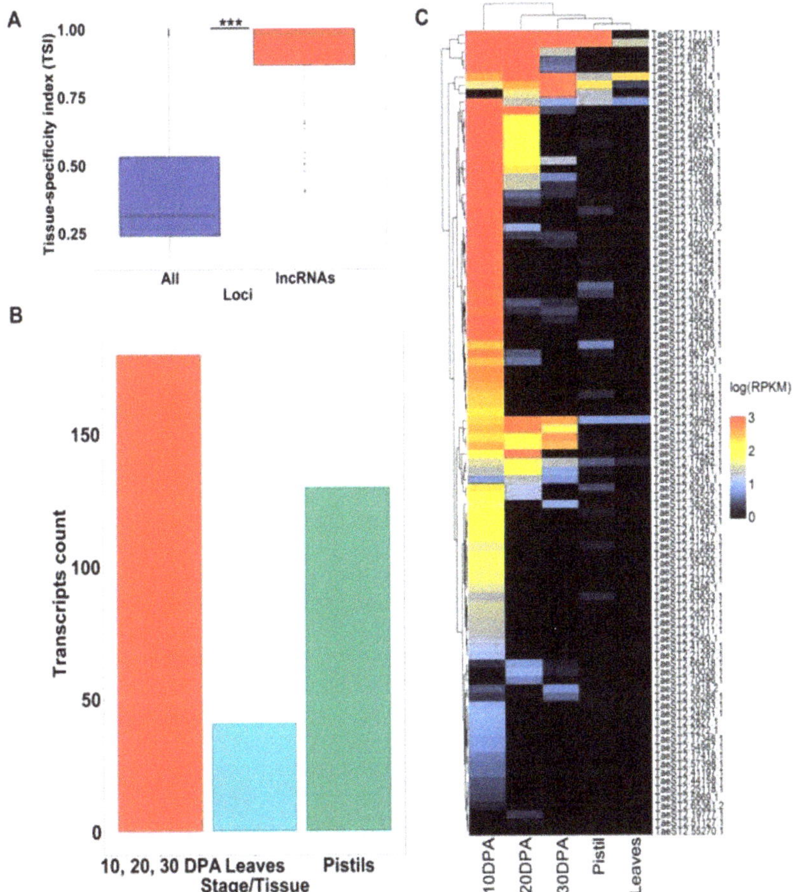

Figure 2. Expression patterns of triticale AB lncRNAs based on wheat RNAseq data. (**A**) Boxplot of the tissue-specificity index (TSI) for lncRNAs and all AB high-confidence transcripts assembled from direct RNA sequencing (DRS) reads (TSI values close to 1 represent high tissue specificity). Stars indicate significant differences estimated by the Wilcoxon rank sum test with continuity correction (*p*-value < 2.2×10^{-16}). (**B**) Bar plot showing the number of AB lncRNAs with the maximum expression (reads per kilobases per million reads, RPKM) level at a specific stage of wheat development (10, 20, and 30 days post anthesis (dpa)) or tissue (leaves and pistils). (**C**) Heatmap of expression values (log(RPKM)) of AB lncRNAs with >90% of RPKM values based on wheat RNAseq data (10, 20, and 30 dpa; pistils and leaves).

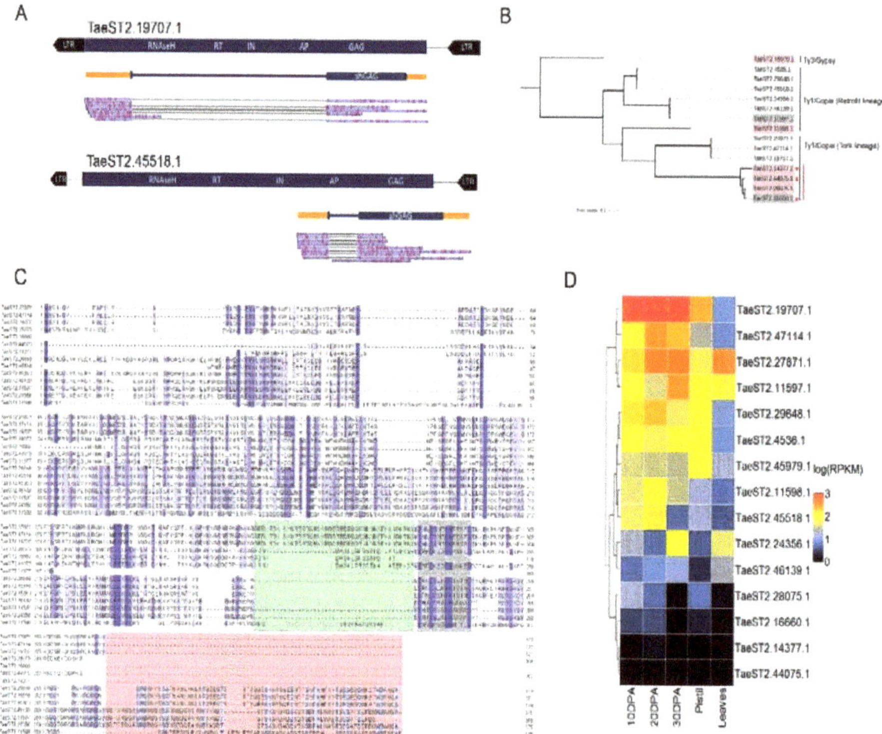

Figure 3. Identification, phylogenetic analysis, and expression pattern of RTE-related transcripts. (**A**) Full-length LTR retrotransposons of Tork (TaeST2.19707.1) and Retrofit (TaeST2.45518.1) lineages expressing a short isoform encoding the GAG protein (shGAG). The dark blue rectangles show ORFs encoding all retrotransposon proteins (genome scheme) and the GAG protein (isoform scheme). Orange color indicates untranslated regions (UTRs). The Nanopore direct RNA read alignment on the LTR retrotransposon genome sequence is also shown. (**B**) Neighbor-joining phylogenetic tree built from 15 GAG proteins. Red and gray highlight the GAG proteins expressed from RTEs without one or more encoded proteins and from loci without predicted RTEs, respectively. Red stars indicate GAG proteins with <300 aa. The vertical red line indicates a group of GAGs encoded by truncated retrotransposons possessing only GAG ORFs. (**C**) Multiple alignment of Ty1/Copia GAG proteins. Green and red highlight variable regions between the Tork and Retrofit regions. Gray shows the RNA-binding motif ($CX_2CX_4HX_4C$, where X is any amino acid). (**D**) Heatmap of the log(RPKM) expression of isoforms encoding GAG proteins in wheat leaves and pistils, and during seed development.

Table 4. Number of retrotransposon-related transcripts (RTE-RNAs) in different groups based on completeness of associated LTR retrotransposons (RTEs) and the type of predicted RTE proteins encoded by open reading frames (ORFs) (reverse transcriptase, RT; aspartic protease, AP; RNAse H).

	RTE Group			
Encoded proteins	Full-length RTEs (all domains are detectable)	Non-complete RTEs (one or more canonical protein domains are not detectable)	No associated RTEs identified	Total number of RTE-RNAs
GAG	8	5	2	15
Other RTE proteins (AP, RT, RNAse H)	1	1	3	5
Total number of RTE-RNAs	**9**	**6**	**5**	**20**

We further focused on GAG-encoding RTE-RNAs (GAG-RNAs) as the most represented group. Oxford Nanopore RNA sequencing provides a unique opportunity to analyze the exon–intron structure of GAG-RNA-encoded loci and predict the deduced full GAG protein sequences. Classification of the deduced GAG proteins showed that 85% (14) and 15% (1) of them belong to Ty1/Copia and Ty3/Gypsy elements, respectively. Based on the information from the transcript assembly, we grouped all GAG-RNAs into three categories based on the number of introns they possessed: (1) a single intron, (2) two introns, and (3) no introns. The vast majority (86%, 13) of GAG-RNAs were found to carry a single intron (Figure 3A), while one GAG-RNA from Ty3/Gypsy had no introns, showing that the exon–intron structure may differ between the two LTR retrotransposon superfamilies. To understand the functional role of splicing in the generation of GAG-RNA transcripts, we also predicted ORFs for unspliced RNA variants corresponding to the regions between two LTRs. We observed that unspliced transcripts expressed from the whole RTEs had significantly longer ORFs, resulting in GAG proteins being fused with other RTE proteins. Additionally, for two GAG-RNA encoding RTEs (TaeST2.19707.1 and TaeST2.45518.1, Figure 3A), only one very long ORF (>4000 bp) was predicted, while other RTEs were found to have two or three ORFs encoding distinct proteins. Thus, we showed that the splicing of the GAG-RNA isoform is critical to ensure the production of the entire GAG protein.

We next performed a comparison and phylogenetic analysis of the amino acid sequences of the 15 GAG proteins. The multiple alignment revealed significant differences between one Ty3/Gypsy GAG (TaeST2.45979, GAG length 514 aa) and the Ty1/Copia GAG proteins. We compared the Ty1/Copia GAG sequences and found that Ty1/Copia GAGs originated from two RTE lineages, Tork (seven sequences) and Retrofit (seven sequences). Phylogenetic tree analysis revealed three and two groups of highly similar (up to 99%) GAG proteins in the Retrofit and Tork lineages, respectively (Figure 3B). In addition, we detected pronounced sequence divergence between the GAG proteins of Tork and Retrofit lineages. The GAG protein of the Retrofit lineage has a ~50 aa-specific C-end which is not found in the Tork GAG sequences (Figure 3C). In turn, Tork GAGs have specific amino acid sequences before the RNA-binding motif (RBM) site. These differences are also reflected in the phylogenetic tree's topology, where the branches corresponding to the Tork and Retrofit Ty1/Copia lineages are readily distinguishable (Figure 3B).

In addition, we noticed that the divergence of groups in a single lineage correlated well with the completeness of the RTEs expressing corresponding GAG-RNAs; the most diverged group of GAG was the one with the Tork lineage and encoded by truncated retrotransposons (Figure 3B) that had a single ORF-encoding GAG.

We then analyzed the GAG protein sequences in more detail. In particular, we estimated the presence of the RNA-binding motif (RBM)($CX_2CX_4HX_4C$, where X is any amino acids), a specific part of GAG proteins, which is responsible for GAG–RNA interactions. We found that RBM could be identified in all except three GAG proteins, including one Ty3/Gypsy (TaeST2.45979.1) and two

Ty1/Copia elements (TaeST2.14377.1, TaeST2.16660.1) of the Tork lineage. The Ty1/Copia GAG-RNAs without the CX2CX3GHX4X motif (TaeST2_14377 and TAeST_16660) are truncated GAG proteins with smaller protein lengths (175 aa and 182 aa vs >300 aa for the full-length Ty1/GAG protein) and probably very limited functionality. Altogether, the results of the phylogenetic analysis suggest that a divergent set of GAG proteins, including truncated GAGs with no RNA-binding capacity, is expressed during triticale seed development.

We then estimated the expression patterns of the GAG-RNAs in several seed developmental stages and in leaf tissues and pistils of wheat. Nine of the 15 GAG-RNA loci had a specific expression pattern with maximum expression levels at early developmental stages or in pistils. For two GAG-RNA loci (TaeST2.14377.1 and TaeST2.44075.1), the expression level was too low, indicating a possible triticale-specific expression pattern (Figure 3D). Thus, the expression data showed that most of the identified triticale GAG-RNAs were also expressed during wheat seed development, and some RTEs expressed genomic RNA (gRNA) as well as a short isoform (shGAG) carrying ORF for GAG protein.

Overall, our results suggested that tens of transcripts encoded by full-length and truncated LTR retrotransposon copies are expressed at early stages of triticale seed development. Three-fourths of these RNAs carry ORFs for encoding a set of GAG proteins of variable length and phylogenetic diversity.

3.5. A Full-Length Ty1/Copia LTR Retrotransposon Is Active in Triticale Seeds

To transpose in the genome, RTEs need to express the full-length genomic RNA (gRNA). Although we did not detect gRNA for the RTEs expressing TaeST2.19707.1 (RTE3B, location in wheat genome: 3B:555,156,557 … 555,163,131 (6.58 kb)) and TaeST2.45518.1 (named '*MIG*', location in wheat genome: 7B:312,336,869 … 312,341,902 (5 kb)) shGAG RNAs by Nanopore sequencing (Figure 3A), we performed RT-PCR with primer pairs designed to detect (a) shGAG isoforms and (b) gRNA isoforms. The RT-PCR analysis resulted in detection of the expression levels of shGAG and gRNAs in developing triticale seeds (10 dpa) and flag leaves, although the gRNA expression level was lower (Figure 4A). Next, we assessed whether the RTEs were capable of transposing. To answer this question, we determined the generation of the extrachromosomal circular DNA (eccDNA) by these RTEs using inverted PCR. EccDNAs are byproducts of RTE activity in plants [71]. We first determined that inverted PCR with the designed primers did not produce PCR products with genomic DNA. Unfortunately, the primers on RTE3B produced PCR fragments with genomic DNA; therefore, the activity of RTE3B could not be assessed by inverted PCR. We continued eccDNA detection only for *MIG* (Supplementary File S3). For this, we enriched the eccDNA fraction by exonuclease treatment of total genomic DNA using an enzyme that specifically cut linear DNA while leaving circular DNA molecules intact. The product was then amplified by rolling circle amplification, and inverted PCR was carried out. We enriched eccDNA in genomic DNA isolated from developing grain at 10 dpa, as well as glume and lemma tissue. The specific products were detected only for eccDNA isolated from developing grain, and no products were obtained with eccDNA of glume and lemma tissues. Thus, our results showed that RTE *MIG* expresses both shGAG RNA and gRNA isoforms and has transposition activity.

Here, we provide experimental evidence suggesting that some detected RTE-RNAs originate from autonomous LTR retrotransposons with ongoing transposition activity in triticale at early stages of seed development.

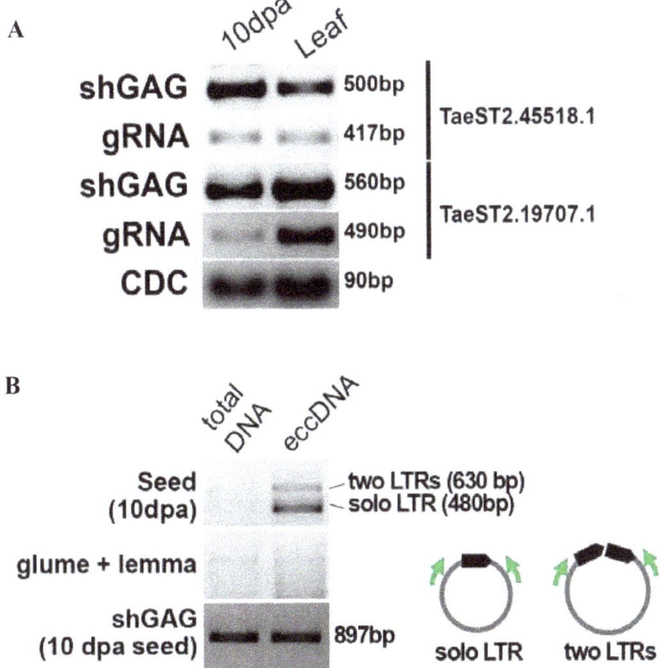

Figure 4. Expression and extrachromosomal circular DNA (eccDNA) formation. (**A**) RT-PCR detection of the shGAG (a short isoform carrying ORFs for GAG protein) and gRNA isoforms of RTE3B and *MIG* full-length RTEs. CDC: reference gene (the cell division control protein). (**B**) Inverted PCR with genomic DNA and eccDNA-enriched fractions obtained from developing seeds (10 dpa) and glume and lemma triticale tissues. The positions of the primers on eccDNA are shown in the small representation in the right. PCR with shGAG primers with total and eccDNA-enriched DNA was used as a control.

4. Discussion

4.1. A Set of Intergenic lncRNAs Detected by Nanopore Sequencing Is Expressed during the Early Stage of Triticale Seed Development

Wheat seed development is a dynamic multistage process that involves significant changes in the transcriptome landscape. Here, we uncovered the lncRNA transcriptome of triticale seed at 10 days post anthesis (dpa), corresponding to the second stage of grain development, known as embryo differentiation (7–14 dpa) [34]. This stage is characterized by the formation of Type A starch granules and the expression of genes involved in starch biogenesis [36]. In turn, the genes encoding for the main storage proteins (the high molecular weight (HMW) glutenins and gliadin [36]) and the genes involved in storage protein biogenesis (e.g., wbm [75]) are expressed during the mid-development stage (14–21 dpa). In agreement with this, we detected the expression levels of the starch metabolism genes (Table 3), while expression of the storage protein genes (HMW glutenins and gliadins) and the wbm gene (found recently in the analyzed line (L8665) [76]) was not detected by Nanopore read analysis. These results prove that the analyzed triticale transcriptome corresponds to the early stage of grain development.

Our analysis showed that hundreds (798) of lncRNAs were expressed during this stage. Surprisingly, we found that 87% of the A and B lncRNAs were expressed from as yet unannotated regions of Genomes A and B. LncRNAs are often underrepresented in plant genome annotation. For example, 8009 lncRNAs were previously identified in the intergenic regions of barley [77], and 1760 unannotated lncRNAs were identified in foxtail millet [78]. The annotation of the lncRNAs expressed from the intergenic space can be challenging because of the biological properties of lncRNAs, including their high tissue- and stage-specificity. Indeed, we identified a high tissue-specificity index for triticale lncRNAs, suggesting that a large number of lncRNAs are expressed during a narrow time window. Additionally, lncRNAs often possess exons containing transposon-similar multicopy sequences that make the transcript assembly difficult because of unambiguity in short RNAseq read mapping. Here, we found that almost 30% of the triticale lncRNAs possessed exons with similarity to TEs. This is in accordance with previous reports on rice [79], where 73% lncRNAs were found to overlap with different TEs. Furthermore, 9.18% of sunflower lncRNAs have TE-related exons [80]. In these terms, RNAseq-based lncRNA identification can underestimate the number of lncRNAs in a cell or lead to transcript misassembly, because short reads from the repeat regions are often mismapped or discarded from the analysis [25]. However, notably, the study of plant lncRNAs and transposon-derived transcripts has mostly been limited by short read RNAseq data. Here, to escape lncRNA identification biases because of the short length of the RNAseq reads, we applied Nanopore direct RNA sequencing. This approach allowed us to precisely determine the transcribed regions in a complex wheat genome. In addition, Nanopore direct RNA sequencing is strand-specific and can be used to identify natural anti-sense lncRNAs. The application of third-generation sequencing could help to illuminate "the dark side" of developing seed transcriptomes involving lncRNAs and transposon-derived loci, thereby overcoming the obstacles in lncRNA discovery by short RNAseq reads. It will be especially useful for crop plants where lncRNA loci can be included into genome-wide association studies (GWAS) to determine which of them can influence key agronomical traits. Notably, lncRNA loci have not been well used for GWAS analysis so far [81]. We believe that the triticale lncRNAs identified in this study could act as valuable new targets for marker-assisted selection.

4.2. Transcripts Encoding Diverse GAG Proteins Are Expressed during the Early Stage of Seed Development

Seed development is accompanied by epigenetic relaxation, which may trigger retrotransposon (RTE) activity [21–23]. Here, we identified 20 transcripts with similarities to RTE proteins. Interestingly, we found that 75% of the RTE-related transcripts carry ORFs encoding GAG proteins, the main component of the RTE virus-like particle. It is known that during their lifecycle, RTEs have to produce significantly more GAG proteins that other RTE proteins. To ensure that there are excess GAG proteins compared with other RTE proteins, an *Arabidopsis* RTE called *EVD* encodes a special isoform (shGAG) encoding the GAG protein [82]. Because almost no systematic studies of RTE transcripts have been carried out at the single isoform level in crop plants, it was not previously clear whether shGAG transcript production is a common pattern for plant species. In this study, we detected isoform production for several full-length triticale RTEs and showed their transcription in wheat, implying that shGAG is a very common pattern for plants. Moreover, we showed that one of the full-length RTEs, TaeST2.45518.1, is capable of producing extrachromosomal circular DNAs (eccDNA), which are a byproduct of RTE activity and have been used to isolate transpositionally active RTEs [74]. Whether this RTE is active in the triticale embryo and can produce copies that are transmitted to the next generation or whether it is active in other tissues of developing grains (e.g., endosperm) warrants further investigation.

Based on the knowledge of splicing patterns of GAG-encoding isoforms, we were able to predict the GAG protein sequence and analyze it in more detail. Our results point to the existence of a divergent group of GAG proteins expressing during triticale and wheat seed development. These GAG proteins and the RTEs encoding them have several distinct features: (1) the RTEs encoding these GAG proteins are non-autonomous elements and possess no similarity to POL proteins; (2) the lengths of most of these GAG proteins (169–270 aa) are less than that of the conventional GAG protein (>300 aa); and (3) half of these GAG proteins lack the RNA-binding motif and cannot interact with RTE RNAs. The elements encoding GAG-RNA consist of two LTRs and the internal part is similar to the GAG protein. This structure makes these elements very similar to the previously described

TR-GAG elements (terminal repeat with GAG domain) ([83]) found in many angiosperm species. However, the TR-GAG elements described in the current paper are classified as Ty1/Copia, while GAG proteins of previously identified TR-GAG elements are similar to both Ty1/Copia and the Ty3/Gypsy superfamily. While further functional and evolutionary studies are required, we suggest that these GAG loci are intermediate products of RTE diversification or "domestication". The "domestication" of GAG proteins has been documented in animals and insects [84–87]. Because of the short lengths of these GAG proteins, it can be also suggested that they may be involved in control of the activity of functional RTEs via incorporation into their virus-like particles. The mechanism of copy number control of RTEs via virus-like particles (VLP) misassembly, which is caused by a truncated GAG form, known as dominant-negative factor, has been well described for yeast [88,89]. No such examples have been described in plants. In the future, it will be intriguing to check whether TR-GAG proteins are capable of interacting with normal GAG proteins of full-length RTEs, which, as we showed here, are expressed in the same developmental stage.

Previously, Nanopore long-read sequencing of transcriptomes was used to annotate expressed transposons in *Arabidopsis* mutants lacking key systems of TE suppression, resulting in elevated expression of transposons [26]. The authors detected the expression of almost 1300 TEs. However, elucidation of transposon expression a "wild-type" genetic background would provide a unique opportunity to trace natural evolutionary forces shaping plant retrotranscriptome and to more deeply understand the features of host–transposon interactions. Recent studies on maize and sunflower [26,90] and our current results point to the great advantage of Nanopore RNA sequencing to decipher RTE expression in crop plants, even those with complex genomes such as triticale. Together with the growing number of publicly available long-read RNA sequencing datasets, this opens a broad avenue for studies of transposon expression in plants on the isoform-based level.

5. Conclusions

Here, using Nanopore direct RNA sequencing, we identified hundreds of previously unknown lncRNAs and LTR retrotransposon-derived transcripts that are expressed in the early stages of triticale and wheat seed development. We showed that triticale lncRNAs often possess similar sequences to transposons and their expression has high stage and tissue specificity, with half of the lncRNAs having the highest expression level at 10–30 days post anthesis in wheat. In addition, we found that most of the detected retrotransposon-related RNAs have a single intron, carry ORFs encoding for a divergent set of GAG proteins, and are encoded by potentially autonomous and non-autonomous retrotransposons. Of these, we identified one Ty1/Copia LTR retrotransposon that produces extrachromosomal circular DNA, and we suggest that it has transposition activity in developing triticale seeds. Finally, this study identified a unique set of lncRNAs and LTR retrotransposons expressed in the early stages of seed development, which we believe will be useful for further exploration of their functional potential and the association with phenotypic variation in triticale and wheat.

Supplementary Materials: The following are available online at http://www.mdpi.com/2223-7747/9/12/1794/s1, Supplementary File S1: Genomic location and overlapped genes for 976 Triticale lncRNAs, Supplementary file S2. Fasta sequences of 796 lncRNAs, Supplementary file S3. Sequence of RTE7B LTR retrotransposon that is expressed and produce eccDNA in triticale developing seed at 10 dpa.

Author Contributions: Conceptualization, I.K. and A.S. (Alexander Soloviev); methodology, I.K.; formal analysis, I.K., M.D., P.M., A.S. (Andrey Shingaliev), M.O., E.K., and A.S. (Alexandra Sigaeva); writing—original draft preparation, I.K.; writing—review and editing, I.K., A.S., and G.K.; funding acquisition, A.S. and G.K. All authors have read and agreed to the published version of the manuscript.

Funding: This research was funded by the Ministry of Education and Science of Russian Federation (Goszadanie No 0431-2019-0005).

Conflicts of Interest: The authors declare no conflict of interest.

Data Availability Statement: Nanopore data produced for this study are available in Sequence Read Archive (SRA) NCBI under Bioproject Accession PRJNA683988.

References

1. Fesenko, I.; Kirov, I.; Kniazev, A.; Khazigaleeva, R.; Lazarev, V.; Kharlampieva, D.; Grafskaia, E.; Zgoda, V.; Butenko, I.; Arapidi, G.; et al. Distinct types of short open reading frames are translated in plant cells. *Genome Res.* **2019**, *29*, 1464–1477. [CrossRef]

2. Rai, M.I.; Alam, M.; Lightfoot, D.A.; Gurha, P.; Afzal, A.J. Classification and experimental identification of plant long non-coding RNAs. *Genomics* **2019**, *111*, 997–1005. [CrossRef]

3. Ma, L.; Bajic, V.B.; Zhang, Z. On the classification of long non-coding RNAs. *Rna Biol.* **2013**, *10*, 925–933. [CrossRef]

4. Szcześniak, M.W.; Rosikiewicz, W.; Makałowska, I. CANTATAdb: A Collection of Plant Long Non-Coding RNAs. *Plant Cell Physiol.* **2015**, *57*, e8. [CrossRef]

5. Paytuví Gallart, A.; Hermoso Pulido, A.; Anzar Martínez de Lagrán, I.; Sanseverino, W.; Aiese Cigliano, R. GREENC: A Wiki-based database of plant lncRNAs. *Nucleic Acids Res.* **2016**, *44*, D1161–D1166. [CrossRef]

6. Xuan, H.; Zhang, L.; Liu, X.; Han, G.; Li, J.; Li, X.; Liu, A.; Liao, M.; Zhang, S.J.G. PLNlncRbase: A resource for experimentally identified lncRNAs in plants. *Gene* **2015**, *573*, 328–332. [CrossRef]

7. Jin, J.; Lu, P.; Xu, Y.; Li, Z.; Yu, S.; Liu, J.; Wang, H.; Chua, N.-H.; Cao, P. PLncDB V2.0: A comprehensive encyclopedia of plant long noncoding RNAs. *Nucleic Acids Res.* **2020**. [CrossRef]

8. Jha, U.C.; Nayyar, H.; Jha, R.; Khurshid, M.; Zhou, M.; Mantri, N.; Siddique, K.H.M. Long non-coding RNAs: Emerging players regulating plant abiotic stress response and adaptation. *BMC Plant Biol.* **2020**, *20*, 466. [CrossRef]

9. Lucero, L.; Ferrero, L.; Fonouni-Farde, C.; Ariel, F. Functional classification of plant long noncoding RNAs: A transcript is known by the company it keeps. *New Phytol.* **2020**. [CrossRef]

10. Budak, H.; Kaya, S.B.; Cagirici, H.B. Long Non-coding RNA in Plants in the Era of Reference Sequences. *Front. Plant Sci.* **2020**, *11*. [CrossRef]

11. Palazzo, A.F.; Koonin, E.V. Functional Long Non-coding RNAs Evolve from Junk Transcripts. *Cell* **2020**. [CrossRef]

12. Kapusta, A.; Kronenberg, Z.; Lynch, V.J.; Zhuo, X.; Ramsay, L.; Bourque, G.; Yandell, M.; Feschotte, C. Transposable elements are major contributors to the origin, diversification, and regulation of vertebrate long noncoding RNAs. *PLoS Genet.* **2013**, *9*, e1003470. [CrossRef]

13. Liu, J.; Jung, C.; Xu, J.; Wang, H.; Deng, S.; Bernad, L.; Arenas-Huertero, C.; Chua, N.H. Genome-wide analysis uncovers regulation of long intergenic noncoding RNAs in Arabidopsis. *Plant Cell* **2012**, *24*, 4333–4345. [CrossRef]

14. Kelley, D.; Rinn, J. Transposable elements reveal a stem cell-specific class of long noncoding RNAs. *Genome Biol.* **2012**, *13*, R107. [CrossRef]

15. Zhao, T.; Tao, X.; Feng, S.; Wang, L.; Hong, H.; Ma, W.; Shang, G.; Guo, S.; He, Y.; Zhou, B.; et al. LncRNAs in polyploid cotton interspecific hybrids are derived from transposon neofunctionalization. *Genome Biol.* **2018**, *19*, 195. [CrossRef]

16. De Quattro, C.; Pe, M.E.; Bertolini, E. Long noncoding RNAs in the model species *Brachypodium distachyon*. *Sci. Rep.* **2017**, *7*, 11252. [CrossRef]

17. Ramsay, L.; Marchetto, M.C.; Caron, M.; Chen, S.H.; Busche, S.; Kwan, T.; Pastinen, T.; Gage, F.H.; Bourque, G. Conserved expression of transposon-derived non-coding transcripts in primate stem cells. *BMC Genom.* **2017**, *18*, 214. [CrossRef]

18. Lv, Y.; Hu, F.; Zhou, Y.; Wu, F.; Gaut, B.S. Maize transposable elements contribute to long non-coding RNAs that are regulatory hubs for abiotic stress response. *BMC Genom.* **2019**, *20*, 864. [CrossRef]

19. Xue, L.; Wu, H.; Chen, Y.; Li, X.; Hou, J.; Lu, J.; Wei, S.; Dai, X.; Olson, M.S.; Liu, J.; et al. Evidences for a role of two Y-specific genes in sex determination in Populus deltoides. *Nat. Commun.* **2020**, *11*, 5893. [CrossRef]

20. Cho, J.; Paszkowski, J. Regulation of rice root development by a retrotransposon acting as a microRNA sponge. *eLife* **2017**, *6*, e30038. [CrossRef]

21. Warman, C.; Panda, K.; Vejlupkova, Z.; Hokin, S.; Unger-Wallace, E.; Cole, R.A.; Chettoor, A.M.; Jiang, D.; Vollbrecht, E.; Evans, M.M.S.; et al. High expression in maize pollen correlates with genetic contributions to pollen fitness as well as with coordinated transcription from neighboring transposable elements. *PLoS Genet.* **2020**, *16*, e1008462. [CrossRef]

22. Anderson, S.N.; Stitzer, M.C.; Zhou, P.; Ross-Ibarra, J.; Hirsch, C.D.; Springer, N.M. Dynamic Patterns of Transcript Abundance of Transposable Element Families in Maize. *G3 Genes|Genomes|Genet.* **2019**, *9*, 3673. [CrossRef]

23. Slotkin, R.K.; Vaughn, M.; Borges, F.; Tanurdžić, M.; Becker, J.D.; Feijó, J.A.; Martienssen, R.A. Epigenetic Reprogramming and Small RNA Silencing of Transposable Elements in Pollen. *Cell* **2009**, *136*, 461–472. [CrossRef]

24. Slotkin, R.K. The case for not masking away repetitive DNA. *Mob. DNA* **2018**, *9*, 15. [CrossRef]

25. Lanciano, S.; Cristofari, G. Measuring and interpreting transposable element expression. *Nat. Rev. Genet.* **2020**, *21*, 721–736. [CrossRef]

26. Panda, K.; Slotkin, R.K. Long-Read cDNA Sequencing Enables a "Gene-Like" Transcript Annotation of Transposable Elements. *Plant Cell* **2020**, *32*, 2687–2698. [CrossRef]

27. Zhang, G.; Sun, M.; Wang, J.; Lei, M.; Li, C.; Zhao, D.; Huang, J.; Li, W.; Li, S.; Li, J.; et al. PacBio full-length cDNA sequencing integrated with RNA-seq reads drastically improves the discovery of splicing transcripts in rice. *Plant J.* **2019**, *97*, 296–305. [CrossRef]

28. Xu, J.; Du, R.; Meng, X.; Zhao, W.; Kong, L.; Chen, J. Third-Generation Sequencing Indicated that LncRNA Could Regulate eIF2D to Enhance Protein Translation under Heat Stress in *Populus simonii*. *Plant Mol. Biol. Rep.* **2020**. [CrossRef]

29. Xu, J.; Zheng, Y.; Pu, S.; Zhang, X.; Li, Z.; Chen, J. Third-generation sequencing found LncRNA associated with heat shock protein response to heat stress in Populus qiongdaoensis seedlings. *BMC Genom.* **2020**, *21*, 572. [CrossRef]

30. Chao, Q.; Gao, Z.-F.; Zhang, D.; Zhao, B.-G.; Dong, F.-Q.; Fu, C.-X.; Liu, L.-J.; Wang, B.-C. The developmental dynamics of the Populus stem transcriptome. *Plant Biotechnol. J.* **2019**, *17*, 206–219. [CrossRef]

31. Chao, Y.; Yuan, J.; Li, S.; Jia, S.; Han, L.; Xu, L. Analysis of transcripts and splice isoforms in red clover (*Trifolium pratense* L.) by single-molecule long-read sequencing. *BMC Plant Biol.* **2018**, *18*, 300. [CrossRef]

32. Rao, S.; Yu, T.; Cong, X.; Xu, F.; Lai, X.; Zhang, W.; Liao, Y.; Cheng, S. Integration analysis of PacBio SMRT- and Illumina RNA-seq reveals candidate genes and pathway involved in selenium metabolism in hyperaccumulator Cardamine violifolia. *BMC Plant Biol.* **2020**, *20*, 492. [CrossRef]

33. Zhu, Z.; Chen, H.; Xie, K.; Liu, C.; Li, L.; Liu, L.; Han, X.; Jiao, C.; Wan, Z.; Sha, A. Characterization of Drought-Responsive Transcriptome During Seed Germination in Adzuki Bean (*Vigna angularis* L.) by PacBio SMRT and Illumina Sequencing. *Front. Genet.* **2020**, *11*, 996. [CrossRef]

34. Rogers, S.O.; Quatrano, R.S. Morphological Staging of Wheat Caryopsis Development. *Am. J. Bot.* **1983**, *70*, 308–311. [CrossRef]

35. Rangan, P.; Furtado, A.; Henry, R.J. The transcriptome of the developing grain: A resource for understanding seed development and the molecular control of the functional and nutritional properties of wheat. *BMC Genom.* **2017**, *18*, 766. [CrossRef]

36. Wan, Y.; Poole, R.L.; Huttly, A.K.; Toscano-Underwood, C.; Feeney, K.; Welham, S.; Gooding, M.J.; Mills, C.; Edwards, K.J.; Shewry, P.R.; et al. Transcriptome analysis of grain development in hexaploid wheat. *BMC Genom.* **2008**, *9*, 121. [CrossRef]

37. Pellny, T.K.; Lovegrove, A.; Freeman, J.; Tosi, P.; Love, C.G.; Knox, J.P.; Shewry, P.R.; Mitchell, R.A.C. Cell Walls of Developing Wheat Starchy Endosperm: Comparison of Composition and RNA-Seq Transcriptome. *Plant Physiol.* **2012**, *158*, 612–627. [CrossRef]

38. Wei, L.; Wu, Z.; Zhang, Y.; Guo, D.; Xu, Y.; Chen, W.; Zhou, H.; You, M.; Li, B. Transcriptome analysis of wheat grain using RNA-Seq. *Front. Agric. Sci. Eng.* **2014**, *1*, 214–222. [CrossRef]

39. Pfeifer, M.; Kugler, K.G.; Sandve, S.R.; Zhan, B.; Rudi, H.; Hvidsten, T.R.; Mayer, K.F.X.; Olsen, O.-A. Genome interplay in the grain transcriptome of hexaploid bread wheat. *Science* **2014**, *345*, 1250091. [CrossRef]

40. Yu, Y.; Zhu, D.; Ma, C.; Cao, H.; Wang, Y.; Xu, Y.; Zhang, W.; Yan, Y. Transcriptome analysis reveals key differentially expressed genes involved in wheat grain development. *Crop J.* **2016**, *4*, 92–106. [CrossRef]

41. Gillies, S.A.; Futardo, A.; Henry, R.J. Gene expression in the developing aleurone and starchy endosperm of wheat. *Plant Biotechnol. J.* **2012**, *10*, 668–679. [CrossRef]

42. Yamasaki, Y.; Gao, F.; Jordan, M.C.; Ayele, B.T. Seed maturation associated transcriptional programs and regulatory networks underlying genotypic difference in seed dormancy and size/weight in wheat (*Triticum aestivum* L.). *BMC Plant Biol.* **2017**, *17*, 154. [CrossRef]

43. Guan, Y.; Li, G.; Chu, Z.; Ru, Z.; Jiang, X.; Wen, Z.; Zhang, G.; Wang, Y.; Zhang, Y.; Wei, W. Transcriptome analysis reveals important candidate genes involved in grain-size formation at the stage of grain enlargement in common wheat cultivar "Bainong 4199". *PLoS ONE* **2019**, *14*, e0214149. [CrossRef]

44. Madhawan, A.; Sharma, A.; Bhandawat, A.; Rahim, M.S.; Kumar, P.; Mishra, A.; Parveen, A.; Sharma, H.; Verma, S.K.; Roy, J. Identification and characterization of long non-coding RNAs regulating resistant starch biosynthesis in bread wheat (*Triticum aestivum* L.). *Genomics* **2020**, *112*, 3065–3074. [CrossRef]

45. Rogers, S.O.; Bendich, A.J. Extraction of DNA from milligram amounts of fresh, herbarium and mummified plant tissues. *Plant Mol. Biol.* **1985**, *5*, 69–76. [CrossRef]

46. Paolacci, A.R.; Tanzarella, O.A.; Porceddu, E.; Ciaffi, M. Identification and validation of reference genes for quantitative RT-PCR normalization in wheat. *BMC Mol. Biol.* **2009**, *10*, 11. [CrossRef]

47. Ryan, P.R.; Dong, D.; Teuber, F.; Wendler, N.; Mühling, K.H.; Liu, J.; Xu, M.; Salvador Moreno, N.; You, J.; Maurer, H.-P.; et al. Assessing How the Aluminum-Resistance Traits in Wheat and Rye Transfer to Hexaploid and Octoploid Triticale. *Front. Plant Sci.* **2018**, *9*. [CrossRef]

48. International Wheat Genome Sequencing, C. A chromosome-based draft sequence of the hexaploid bread wheat (*Triticum aestivum*) genome. *Science* **2014**, *345*, 1251788. [CrossRef]

49. Bauer, E.; Schmutzer, T.; Barilar, I.; Mascher, M.; Gundlach, H.; Martis, M.M.; Twardziok, S.O.; Hackauf, B.; Gordillo, A.; Wilde, P.; et al. Towards a whole-genome sequence for rye (*Secale cereale* L.). *Plant J. Cell Mol. Biol.* **2017**, *89*, 853–869. [CrossRef]

50. Li, H. Minimap2: Pairwise alignment for nucleotide sequences. *Bioinformatics* **2018**, *34*, 3094–3100. [CrossRef]

51. Li, H.; Handsaker, B.; Wysoker, A.; Fennell, T.; Ruan, J.; Homer, N.; Marth, G.; Abecasis, G.; Durbin, R.; Genome Project Data Processing, S. The Sequence Alignment/Map format and SAMtools. *Bioinformatics* **2009**, *25*, 2078–2079. [CrossRef]

52. Barnett, D.W.; Garrison, E.K.; Quinlan, A.R.; Stromberg, M.P.; Marth, G.T. BamTools: A C++ API and toolkit for analyzing and managing BAM files. *Bioinformatics* **2011**, *27*, 1691–1692. [CrossRef]

53. Kovaka, S.; Zimin, A.V.; Pertea, G.M.; Razaghi, R.; Salzberg, S.L.; Pertea, M. Transcriptome assembly from long-read RNA-seq alignments with StringTie2. *Genome Biol.* **2019**, *20*, 278. [CrossRef]

54. Pertea, G.; Pertea, M. GFF Utilities: GffRead and GffCompare. *F1000Research* **2020**, *9*. [CrossRef]

55. Buels, R.; Yao, E.; Diesh, C.M.; Hayes, R.D.; Munoz-Torres, M.; Helt, G.; Goodstein, D.M.; Elsik, C.G.; Lewis, S.E.; Stein, L.; et al. JBrowse: A dynamic web platform for genome visualization and analysis. *Genome Biol.* **2016**, *17*, 66. [CrossRef]

56. Cock, P.J.; Antao, T.; Chang, J.T.; Chapman, B.A.; Cox, C.J.; Dalke, A.; Friedberg, I.; Hamelryck, T.; Kauff, F.; Wilczynski, B.; et al. Biopython: Freely available Python tools for computational molecular biology and bioinformatics. *Bioinformatics* **2009**, *25*, 1422–1423. [CrossRef]

57. Han, S.; Liang, Y.; Ma, Q.; Xu, Y.; Zhang, Y.; Du, W.; Wang, C.; Li, Y. LncFinder: An integrated platform for long non-coding RNA identification utilizing sequence intrinsic composition, structural information and physicochemical property. *Brief Bioinform.* **2019**, *20*, 2009–2027. [CrossRef]

58. Li, A.; Zhang, J.; Zhou, Z. PLEK: A tool for predicting long non-coding RNAs and messenger RNAs based on an improved k-mer scheme. *BMC Bioinform.* **2014**, *15*, 311. [CrossRef]

59. Sun, L.; Luo, H.; Bu, D.; Zhao, G.; Yu, K.; Zhang, C.; Liu, Y.; Chen, R.; Zhao, Y. Utilizing sequence intrinsic composition to classify protein-coding and long non-coding transcripts. *Nucleic Acids Res.* **2013**, *41*, e166. [CrossRef]

60. Ellinghaus, D.; Kurtz, S.; Willhoeft, U. LTRharvest, an efficient and flexible software for de novo detection of LTR retrotransposons. *BMC Bioinform.* **2008**, *9*, 18. [CrossRef]

61. Steinbiss, S.; Willhoeft, U.; Gremme, G.; Kurtz, S. Fine-grained annotation and classification of de novo predicted LTR retrotransposons. *Nucleic Acids Res.* **2009**, *37*, 7002–7013. [CrossRef]

62. Llorens, C.; Futami, R.; Covelli, L.; Dominguez-Escriba, L.; Viu, J.M.; Tamarit, D.; Aguilar-Rodriguez, J.; Vicente-Ripolles, M.; Fuster, G.; Bernet, G.P.; et al. The Gypsy Database (GyDB) of mobile genetic elements: Release 2.0. *Nucleic Acids Res.* **2011**, *39*, D70–D74. [CrossRef]

63. Neumann, P.; Novák, P.; Hoštáková, N.; Macas, J. Systematic survey of plant LTR-retrotransposons elucidates phylogenetic relationships of their polyprotein domains and provides a reference for element classification. *Mobile DNA* **2019**, *10*, 1. [CrossRef]

64. Katoh, K.; Rozewicki, J.; Yamada, K.D. MAFFT online service: Multiple sequence alignment, interactive sequence choice and visualization. *Brief. Bioinform.* **2017**, *20*, 1160–1166. [CrossRef]

65. Letunic, I.; Bork, P. Interactive Tree Of Life (iTOL) v4: Recent updates and new developments. *Nucleic Acids Res.* **2019**, *47*, W256–W259. [CrossRef]

66. Waterhouse, A.M.; Procter, J.B.; Martin, D.M.; Clamp, M.; Barton, G.J. Jalview Version 2—A multiple sequence alignment editor and analysis workbench. *Bioinformatics* **2009**, *25*, 1189–1191. [CrossRef]

67. Ge, S.X.; Jung, D.; Yao, R. ShinyGO: A graphical gene-set enrichment tool for animals and plants. *Bioinformatics* **2019**, *36*, 2628–2629. [CrossRef]

68. Kim, D.; Langmead, B.; Salzberg, S.L. HISAT: A fast spliced aligner with low memory requirements. *Nat. Methods* **2015**, *12*, 357–360. [CrossRef]

69. Patro, R.; Duggal, G.; Love, M.I.; Irizarry, R.A.; Kingsford, C. Salmon provides fast and bias-aware quantification of transcript expression. *Nat. Methods* **2017**, *14*, 417–419. [CrossRef]

70. Xiao, B.; Lu, H.; Li, C.; Bhanbhro, N.; Cui, X.; Yang, C. Carbohydrate and plant hormone regulate the alkali stress response of hexaploid wheat (*Triticum aestivum* L.). *Environ. Exp. Bot.* **2020**, *175*, 104053. [CrossRef]

71. Yang, Z.; Peng, Z.; Wei, S.; Liao, M.; Yu, Y.; Jang, Z. Pistillody mutant reveals key insights into stamen and pistil development in wheat (*Triticum aestivum* L.). *BMC Genom.* **2015**, *16*, 211. [CrossRef]

72. Lanciano, S.; Carpentier, M.C.; Llauro, C.; Jobet, E.; Robakowska-Hyzorek, D.; Lasserre, E.; Ghesquiere, A.; Panaud, O.; Mirouze, M. Sequencing the extrachromosomal circular mobilome reveals retrotransposon activity in plants. *PLoS Genet* **2017**, *13*, e1006630. [CrossRef]

73. Wickham, H. ggplot2. *Wiley Interdiscip. Rev. Comput. Stat.* **2011**, *3*, 180–185. [CrossRef]

74. Gu, Z.; Eils, R.; Schlesner, M. Complex heatmaps reveal patterns and correlations in multidimensional genomic data. *Bioinformatics* **2016**, *32*, 2847–2849. [CrossRef]

75. Furtado, A.; Bundock, P.C.; Banks, P.M.; Fox, G.; Yin, X.; Henry, R.J. A novel highly differentially expressed gene in wheat endosperm associated with bread quality. *Sci. Rep.* **2015**, *5*, 10446. [CrossRef]

76. Kirov, I.; Pirsikov, A.; Milyukova, N.; Dudnikov, M.; Kolenkov, M.; Gruzdev, I.; Siksin, S.; Khrustaleva, L.; Karlov, G.; Soloviev, A.J.A. Analysis of Wheat Bread-Making Gene (wbm) Evolution and Occurrence in Triticale Collection Reveal Origin via Interspecific Introgression into Chromosome 7AL. *Agronomy* **2019**, *9*, 854. [CrossRef]

77. Unver, T.; Tombuloglu, H. Barley long non-coding RNAs (lncRNA) responsive to excess boron. *Genomics* **2020**, *112*, 1947–1955. [CrossRef]

78. Wang, T.; Song, H.; Wei, Y.; Li, P.; Hu, N.; Liu, J.; Zhang, B.; Peng, R. High throughput deep sequencing elucidates the important role of lncRNAs in Foxtail millet response to herbicides. *Genomics* **2020**, *112*, 4463–4473. [CrossRef]

79. Li, X.; Shahid, M.Q.; Wen, M.; Chen, S.; Yu, H.; Jiao, Y.; Lu, Z.; Li, Y.; Liu, X. Global identification and analysis revealed differentially expressed lncRNAs associated with meiosis and low fertility in autotetraploid rice. *BMC Plant Biol.* **2020**, *20*, 82. [CrossRef]

80. Flórez-Zapata, N.M.V.; Reyes-Valdés, M.H.; Martínez, O. Long non-coding RNAs are major contributors to transcriptome changes in sunflower meiocytes with different recombination rates. *BMC Genom.* **2016**, *17*, 490. [CrossRef]

81. Gupta, P.K.; Kulwal, P.L.; Jaiswal, V. Chapter Two—Association mapping in plants in the post-GWAS genomics era. In *Advances in Genetics*; Kumar, D., Ed.; Academic Press: Cambridge, MA, USA, 2019; Volume 104, pp. 75–154.

82. Oberlin, S.; Sarazin, A.; Chevalier, C.; Voinnet, O.; Marí-Ordóñez, A. A genome-wide transcriptome and translatome analysis of Arabidopsis transposons identifies a unique and conserved genome expression strategy for Ty1/Copia retroelements. *Genome Res.* **2017**, *27*, 1549–1562. [CrossRef]

83. Chaparro, C.; Gayraud, T.; de Souza, R.F.; Domingues, D.S.; Akaffou, S.; Laforga Vanzela, A.L.; Kochko, A.; Rigoreau, M.; Crouzillat, D.; Hamon, S.; et al. Terminal-repeat retrotransposons with GAG domain in plant genomes: A new testimony on the complex world of transposable elements. *Genome Biol. Evol.* **2015**, *7*, 493–504. [CrossRef]

84. Campillos, M.; Doerks, T.; Shah, P.K.; Bork, P.J.T.i.G. Computational characterization of multiple Gag-like human proteins. *Trends Genetics* **2006**, *22*, 585–589. [CrossRef]

85. Nefedova, L.N.; Kuzmin, I.V.; Makhnovskii, P.A.; Kim, A.I. Domesticated retroviral GAG gene in Drosophila: New functions for an old gene. *Virology* **2014**, *450–451*, 196–204. [CrossRef]

86. Ashley, J.; Cordy, B.; Lucia, D.; Fradkin, L.G.; Budnik, V.; Thomson, T. Retrovirus-like Gag Protein Arc1 Binds RNA and Traffics across Synaptic Boutons. *Cell* **2018**, *172*, 262–274.e11. [CrossRef]

87. Makhnovskii, P.; Balakireva, Y.; Nefedova, L.; Lavrenov, A.; Kuzmin, I.; Kim, A.J.G. Domesticated gag Gene of Drosophila LTR Retrotransposons Is Involved in Response to Oxidative Stress. *Genes* **2020**, *11*, 396. [CrossRef]
88. Saha, A.; Mitchell, J.A.; Nishida, Y.; Hildreth, J.E.; Ariberre, J.A.; Gilbert, W.V.; Garfinkel, D.J. A trans-dominant form of Gag restricts Ty1 retrotransposition and mediates copy number control. *J. Virol.* **2015**, *89*, 3922–3938. [CrossRef]
89. Tucker, J.M.; Larango, M.E.; Wachsmuth, L.P.; Kannan, N.; Garfinkel, D.J. The Ty1 Retrotransposon Restriction Factor p22 Targets Gag. *PLoS Genet.* **2015**, *11*, e1005571. [CrossRef]
90. Kirov, I.; Omarov, M.; Merkulov, P.; Dudnikov, M.; Gvaramiya, S.; Kolganova, E.; Komakhin, R.; Karlov, G.; Soloviev, A. Genomic and Transcriptomic Survey Provides New Insight into the Organization and Transposition Activity of Highly Expressed LTR Retrotransposons of Sunflower (*Helianthus annuus* L.). *Int. J. Mol. Sci.* **2020**, *21*, 9331. [CrossRef]

Publisher's Note: MDPI stays neutral with regard to jurisdictional claims in published maps and institutional affiliations.

Article

Factors Influencing Genomic Prediction Accuracies of Tropical Maize Resistance to Fall Armyworm and Weevils

Arfang Badji [1,*], Lewis Machida [2], Daniel Bomet Kwemoi [3,*], Frank Kumi [4], Dennis Okii [1], Natasha Mwila [1], Symphorien Agbahoungba [5], Angele Ibanda [1], Astere Bararyenya [1], Selma Ndapewa Nghituwamhata [1], Thomas Odong [1], Peter Wasswa [1], Michael Otim [3], Mildred Ochwo-Ssemakula [1], Herbert Talwana [1], Godfrey Asea [3], Samuel Kyamanywa [1] and Patrick Rubaihayo [1]

[1] Department of Agricultural Production, Makerere University, Kampala P.O. Box 7062, Uganda; dennisokii@gmail.com (D.O.); mwilanatasha@yahoo.co.uk (N.M.); angeltanzito@gmail.com (A.I.); barastere@gmail.com (A.B.); snghituwamhata@gmail.com (S.N.N.); thomas.l.odong@gmail.com (T.O.); wasswa@caes.mak.ac.ug (P.W.); mknossemakula@gmail.com (M.O.-S.); haltalwana@gmail.com (H.T.); Skyamanywa@gmail.com (S.K.); prubaihayo@gmail.com (P.R.)
[2] Alliance Bioversity-CIAT, Africa-Office, Kampala P.O. Box 24384, Uganda; l.machida@cgiar.org
[3] National Crops Resource Research Institute, Kampala P.O. Box 7084, Uganda; otim_michael@yahoo.com (M.O.); grasea9@gmail.com (G.A.)
[4] Department of Crop Science, University of Cape Coast, Cape Coast PMB, Ghana; frankkumifk@gmail.com
[5] Laboratory of Applied Ecology, University of Abomey-Calavi, Cotonou 01BP 526, Benin; agbahoungbasymphorien@gmail.com
*   Correspondence: arbad2009@caes.mak.ac.ug (A.B.); kdbomet@gmail.com (D.B.K.)

Citation: Badji, A.; Machida, L.; Kwemoi, D.B.; Kumi, F.; Okii, D.; Mwila, N.; Agbahoungba, S.; Ibanda, A.; Bararyenya, A.; Nghituwamhata, S.N.; Odong, T.; et al. Factors Influencing Genomic Prediction Accuracies of Tropical Maize Resistance to Fall Armyworm and Weevils. *Plants* **2021**, *10*, 29. https://dx.doi.org/10.3390/plants10010029

Received: 13 July 2020
Accepted: 14 September 2020
Published: 24 December 2020

Publisher's Note: MDPI stays neutral with regard to jurisdictional claims in published maps and institutional affiliations.

Abstract: Genomic selection (GS) can accelerate variety improvement when training set (TS) size and its relationship with the breeding set (BS) are optimized for prediction accuracies (PAs) of genomic prediction (GP) models. Sixteen GP algorithms were run on phenotypic best linear unbiased predictors (BLUPs) and estimators (BLUEs) of resistance to both fall armyworm (FAW) and maize weevil (MW) in a tropical maize panel. For MW resistance, 37% of the panel was the TS, and the BS was the remainder, whilst for FAW, random-based training sets (RBTS) and pedigree-based training sets (PBTSs) were designed. PAs achieved with BLUPs varied from 0.66 to 0.82 for MW-resistance traits, and for FAW resistance, 0.694 to 0.714 for RBTS of 37%, and 0.843 to 0.844 for RBTS of 85%, and these were at least two-fold those from BLUEs. For PBTS, FAW resistance PAs were generally higher than those for RBTS, except for one dataset. GP models generally showed similar PAs across individual traits whilst the TS designation was determinant, since a positive correlation (R = 0.92***) between TS size and PAs was observed for RBTS, and for the PBTS, it was negative (R = 0.44**). This study pioneered the use of GS for maize resistance to insect pests in sub-Saharan Africa.

Keywords: prediction accuracy; mixed linear and Bayesian models; machine learning algorithms; training set size and composition; parametric and nonparametric models

1. Introduction

Insect damage on maize plants and stored grains potentially impedes food security in Africa [1–3]. The fall armyworm (FAW) and stem borers in the field, and the maize weevils (MWs) in storage facilities, are some of the most devastating insect pests on the continent. These insect pests cause yield losses ranging from 10–90% leading to loss of grain marketability, and consumer health concerns due to the possible contamination of the grain with mycotoxins, such as aflatoxins [3–6]. In Africa, tremendous efforts were made during the last two decades to build host plant resistance to insect pests in maize through traditional pedigree (phenotypic)-based selection (PS) with substantial desirable results. Several Africa-adapted maize lines were developed and successfully tested for resistance to MW damage on grains [7–12]. Some of the success stories are from the International Center for Maize and Wheat Improvement (CIMMYT) of Kenya through the Insect Resistant Maize for Africa project (IRMA) that produced several storage pest and stem borer resistant maize lines [7,8,13–15]. On the other hand, the FAW is a new pest on the continent, first reported in 2016 in West and Central African countries [16], from where it spread throughout the African continent [17]. Hence, although efforts to develop FAW resistant varieties are underway at several institutions, including CIMMYT, published reports of FAW resistant varieties are not yet available [18,19].

The complex nature of insect resistance traits makes PS slow and expensive, and thus, difficult to implement, especially for resource-constrained breeding programs [20]. Application of traditional marker-assisted selection (MAS) is hampered by the necessity to first discover resistance-associated genomic regions through genetic linkage and genome-wide association mapping methods, both with several shortcomings, especially for complex traits [21–23]. In addition, genetic linkage and genome-wide association mapping studies have seldom been explored in African germplasm [8,24], which further impedes the application of MAS in the development of insect resistance maize germplasm in Africa. In a previous study, we discovered several quantitative trait nucleotides and genes that are putatively associated with FAW and MW resistance, confirming the quantitative nature of these traits, hence the difficulty in improving these traits through MAS [25]. An alternative to both PS and MAS is genomic selection (GS), which uses whole-genome markers to perform genomic prediction (GP) of breeding values of unphenotyped genotypes, from which one can select superior candidate genotypes for crossing to produce hybrids or to advance to the next generation [26]. GS was reported to achieve up to threefold annual genetic gain in maize improvement when compared to MAS, due to a more efficient accounting of trait-associated quantitative trait loci (QTL), faster selection cycles, and lower phenotyping costs [27–33].

Several statistical and machine learning GP models with various strengths and weaknesses have been developed to adapt to different contexts that are partly influenced by the genetic architecture of traits (number and effect size of QTL, proportions of additive and non-additive genetic effects) and reproductive classes of plants (allogamous vs. autogamous vs. clonally propagated) [34–36]. Therefore, to effectively implement GS in crop improvement programs, it is necessary to employ a holistic approach to determine the best GP strategy for particular breeding targets for given crop species [31,37]. Statistical models for GS vary in their prior assumptions and treatment of marker effects [31]. Parametric models focus on parameter estimates rather than prediction, while nonparametric algorithms give priority to prediction and have fewer assumptions [38]. Some parametric methods assume the SNP effects follow a normal distribution with equal variance for all loci, which seems unrealistic in practice.

Representative parametric methods are ridge regression best linear unbiased predictors (RR-BLUP) [39] and genomic BLUP (GBLUP) [40]. GBLUP was the first GP method to be developed, and it replaced the traditional pedigree-based relationship matrix with a genomic information-based matrix to improve prediction accuracies (PAs) [41]. Parametric methods BayesA [26] and weighted Bayesian shrinkage regression (wBSR) [42], on the other hand, consider a prior distribution of effect with a higher probability of moderate to large effects. Regarding parametric models such as BayesB [43] and BayesCπ [44], assumptions are made that consider the effects of some SNPs to be zero. The Bayesian least absolute shrinkage and selection operator (Bayesian LASSO) assumes that the effects of all markers follow a double exponential distribution [45], whilst the Bayesian sparse linear mixed model (BSLMM), a parametric method developed by Zhou et al. [46], combines the hypotheses of both GBLUP and Bayesian methods and achieves higher PAs than BayesCπ and BayesLASSO. Nonparametric or semi-parametric approaches such as random forest and reproducing kernel Hilbert space (RKHS) [47,48] are better suited for accounting for non-additive genetic effects (37,38), in contrast with parametric genomic prediction models [23,38,47,49]. Several studies compared the performances of GP models under different conditions. In a simulation study, Meuwissen et al. [26] found that while GBLUP achieved PAs of up to 73.2%, BayesA and BayesB comparatively provided additional increases of around 9% and 16%, respectively. However, when a population is composed of close relatives and the target traits are controlled by several small effect genes, the different methods perform similarly [50–52]. On the contrary, BayesB and BayesCπ are better when dealing with distant relatives and traits affected by a small number of large-effect loci [23]. Kernel methods such as RKHS are robust in predicting non-additive effects and in solving complex multi-environment multi-trait models [53,54]. Compared to the above-mentioned parametric methods, deep learning techniques such as support vector regression (SVR), multilayer perceptron, and convolutional neural networks models performed poorly in some studies [55,56]. However, there are also instances where RKHS outperformed one or several of the parametric methods, for instance, GBLUP, rrBLUP, and Bayesian algorithms, in terms of several traits in several crops including maize [51,57–59]. These results were most likely because nonparametric GS models capture more adequately the non-additive genetic components which are an essential characteristic of complex traits [23,37,38] and hence could be good candidate tools for the prediction of FAW and MW-resistance traits which are controlled by both additive and non-additive gene action [21,23,31,41,60]. Therefore, since GP for maize resistance to insect pests such as FAW and MW is not yet well explored, it is pivotal to compare performances of several available prediction algorithms to inform better future GS programs. Therefore, the Genomic Prediction 0.2.6 plugin of the KDCompute 1.5.2. beta (https://kdcompute.igss-africa.org/kdcompute/home), an online database developed by Diversity Array Technologies (DArT, https://www.diversityarrays.com) for the analysis of DArT marker data, presents great interest for this purpose. It hosts a suite of parametric, semiparametric, and nonparametric GP methods that can be run simultaneously on genotype-phenotype datasets.

Additional factors that influence PAs are the different sizes of the training sets (TSs) and breeding sets (BS) and their genetic relationships, the number of markers used to estimate genomic estimated breeding values (GEBV) of lines, the population structure, and the extent of linkage disequilibrium [21,23,31,41,60]. Since phenotyping is the current bottleneck in plant breeding and one of the disadvantages of GP is the requirement of large TSs for high PAs to be achieved, determination of effective TS composition and size is critical for effective implementation of GS in crop improvement programs [21,61–64]. Additionally, the best TS determination will depend on the genetic architecture and the extent of population structure of the trait targeted for GP [63], two parameters that are substantially variable among plant breeding traits. Another factor that is a determinant of the predictive ability is the kinship between the TS and the BS (63). Several methods are used for TS optimization and these generally fall into two categories—namely, untargeted and targeted approaches. For the untargeted approach, the TS is determined independently of its genomic information, whereas the targeted method considers the genomic relationship between the TS and the BS as a means of maximizing PAs [65]. However, deciding on the best TS selection method is not straightforward and depends on context [66].

Furthermore, in maize, GPs were previously conducted using either genotypic best linear unbiased estimators (BLUEs) [67–69] or best linear unbiased predictors (BLUPs) [31, 41,70] as means of phenotypic correction [70]. BLUEs are obtained by treating the genotypic effect of a mixed linear model as fixed effects and provide an estimated mean for each individual of a population equal to its true value. On the other hand, BLUPs generated by considering the genotypic factors as random and allowing for the shrinkage of the means towards the population mean [71]. Whether to use BLUPs or BLUEs in GP is debatable. Phenotypic BLUEs allow avoiding double penalization which BLUPs suffer from. With phenotypic BLUPs, this double penalization is, however, compensated through maximization of the correlation between predicted and true line values, while phenotypic BLUEs do not rely on this shrinkage [70]. However, the shrinkage in the BLUP procedure accounts better for outliers and environmental variabilities [72], permitting better estimates of individual genetic effects than BLUEs [71], and therefore, it usually yields more accurate predictions of phenotypic performance [70,72,73]. Furthermore, BLUPs are better in handling unbalanced data, wherein, for example, the number of individuals is not the same in different locations or in the different replications of an experiment [49,70]. On that basis, the current study was conducted to evaluate the efficacies of different parametric, semiparametric, and nonparametric methods from both statistical and machine learning GP models in generating prediction accuracies (PAs) for maize resistance to FAW and MW in a diverse panel using both genotypic BLUEs and BLUPs.

2. Material and Methods

2.1. Genetic Material and Field Experiments

The panel used in this study consisted of 358 maize lines with diverse genetic and geographic backgrounds, and they were sourced from the National Crop Resources Research Institute (NaCRRI/Namulonge, Uganda), the International Institute for Tropical Agriculture (IITA/Ibadan, Nigeria), and The International Maize and Wheat Improvement Center (CIMMYT/Nairobi, Kenya). The panel consisted of 71 inbred lines developed for various purposes at NaCRRI; 28 and five stem borer (SB)-resistant inbred lines from CIM-MYT [6,13,14] and IITA, respectively; 19 storage pest (SP)-resistant inbred lines [7,8]; and a doubled haploid (DH) panel of 235 lines developed at CIMMYT using six parents—three of which were stem borer-resistant, one was a storage pest-resistant inbred line (these were also included in the population), and two were CML elite lines (one, CML132 was included in the panel) (Supplementary Materials Table S1).

The panel was planted and evaluated in three environments, at Mubuku Irrigation Experimental Station in Kasese, western Uganda in 2017 (316 lines) during the second rainy season (2017B) and the National Crop Resources Research Institute (NaCRRI), Namulonge, central Uganda in 2018 (92 lines) and 2019 (252 lines) both during the first rainy seasons (2018A and 2019A, respectively). Detailed information on these locations is presented in Table 1.

Table 1. Geographical, climatic, and soil characteristics of the planting locations [74].

Locations	Geographical Coordinates	Altitude above Sea Level	Minimum Rainfall	Soil Characteristics
Kasese	0°16′10″ N 30°6′9″ E	1330 m asl	1000 mm	Sandy loam soils with a pH of 5.68
Namulonge	0°31′30″ N 32°36′54″ E	1160 m asl	1300 mm	Oxisols with a pH of 5.8

Each combination of location and season was considered an environment, resulting in a total of three environments. An augmented experimental design was adopted in all three environments using six checks in 2017B, two in 2018A, and four in 2019A replicated in all the blocks. The experiments in 2017B, 2018A, and 2019A consisted of twelve, five, and ten blocks, respectively, containing the replicated checks and unreplicated lines and the experiment in 2018A was replicated twice.

2.2. Genotyping, Quality Control, and Imputation for Genomic Prediction Analyses

Genotyping of the panel and SNP quality were described in our previous study [25]. In brief, maize leaves at the sixth-leaf stage of development were harvested from 341 of the 358 lines of the panel (5–10 plants per line) in 2017B and 2018A (for lines not captured in 2017B). The leaf samples were oven-dried overnight at 36 degrees Celsius and shipped to the Biosciences east and central Africa (BecA) Laboratory of the International Livestock Research Institute (ILRI, Nairobi, Kenya) for DNA extraction and genotyping. Diversity Array Technology (DArT) genotyping facilities (44) were used to successfully identify 34,509 SNPs from 341 of the 358 lines composing the panel; hence, only these lines were considered for the GP analyses. Duplicate SNPs were first removed using the R package DartR (45), leaving 28,919 unique SNPs (DRSNP) distributed across all the 10 chromosomes of the entire maize genome.

The DRSNP dataset was imputed before GP using KDCompute 1.5.2. beta (https://kdcompute.igss-africa.org/kdcompute/home), an online database developed by Diversity Array Technologies (DArT, https://www.diversityarrays.com) for the analysis of DArT marker data. KDCompute uses a suit of imputation methods to impute the SNP dataset and scores the imputation results by calculating simple matching coefficient (SMC). The method with the highest SMC is considered as optimal and used to impute the original genotypic dataset.

2.3. FAW and MW Resistance Phenotyping

After germination, plants were left unprotected to allow sufficient natural pressure of fall armyworm (FAW) population to build up. FAW damage scoring in all the three environments was carried out two months after planting when adequate natural FAW infestation levels had manifested, and scoring was based on a visual assessment using a scale of 1 (no or minor leaf damage) to 9 (all leaves highly damaged) [75], illustrated in Figure S1 [18].

Rearing of and bioassays for MW were performed as described in previous experiments carried out at NaCRRI [76,77]. Weevils were reared prior to the MW bioassay to obtain enough insects aged between 0 to 7 days for infestation. During rearing, standard conditions were provided for weevils to ensure proper acclimatization during the experiment. Rearing was carried out by preparing a weevil-maize grain culture of 300–400 unsexed insects and 1.5 kg of grains contained in 3000 cm^3 plastic jars incubated for 14 days in the laboratory at a temperature of 28 $\pm$ 2 °C and relative humidity of 70% $\pm$ 5%, to enhance oviposition. The lids of the jars were perforated and a gauze-wire mesh with a pore size smaller than one mm was fitted on each lid to allow proper ventilation while preventing the weevils from escaping.

After harvesting and shelling, grains of each line were bulked across environments. Then, samples of 30 g were weighed from each grain bulk, aiming to produce three replicates per line for the MW bioassay experiment. However, due to the lack of an adequate amount of grains for most of the lines of the panel, only 64, 123, and 132 lines could generate three, two, and one replicates, respectively, and were therefore considered for the MW bioassay experiment. Each of these samples was wrapped in polythene bags and kept at −20 °C for 14 days to eliminate any weevil infestation prior to the start of the experiment. After this disinfestation process, samples were left to thaw and transferred into 250 cm^3 glass jars and infested with 32 unsexed weevils. After 10-days of incubation to allow oviposition, all dead and living adult insects were removed. One month after infestation (MAI), each sample was removed from its jar, and the grains and the flour were isolated and their weights were recorded. The total number of holes inflicted by the weevils on the grains was counted along with the number of grains affected by such damages. Additionally, the numbers of dead and living weevils were recorded. After these measurements were collected, the grains were returned to their respective jars and all the measurements were repeated at two and three MAI. The collected data were used to infer, for each sample, the cumulative grain weight loss (GWL), the cumulative number of emerged adult weevil progenies (AP), and the final number of damage-affected kernels (AK).

2.4. Statistical Analyses of the Phenotypic Data

Both best linear unbiased estimators (BLUEs) and predictors (BLUPs) were generated using the general linear model with only phenotype option of the software Trait Association through Evolution and Linkage (TASSEL) [78] and the *ranef* function of the R package [79] lme4 [80], respectively. The mixed linear model for generating BLUEs (all factors considered as fixed) and BLUPs (all factors considered as random) for MW traits (GWL, AP, AK, NH, and FP) was as follows:

$$Y = \mu + Replication + Genotype + Error$$

The mixed model for generating BLUEs (all factors considered as fixed) and BLUPs (with all factors considered as random) for FAW damage scores across environments model was:

$$Y = \mu + Location + Block + Genotype + Location : Genotype + Error$$

where μ in the two equations is the intercept.

2.5. Strategies for TS and BS Determination

2.5.1. MW Resistance Traits

Due to inadequate amount of seeds, only 37% (126 out of 341 that had genotypic data) of lines from the panel had phenotypic data on grain weight loss (GWL), adult progeny emergence (AP), and number of affected kernels (AK). Therefore, to estimate GP accuracies for MW resistance, these 126 lines were used as the TS and the remaining 215 lines with only genotypic data constituted the breeding set (BS).

2.5.2. FAW Damage Resistance

The GP analyses for FAW resistance were carried out on the 341 lines of the panel that were genotyped and phenotyped for FAW damage resistance. To determine TS and BS sizes and compositions for the evaluation of maize resistance to FAW damage, two strategies, namely, random-based TS (RBTS) and pedigree-based TS (PBTS), were used.

2.5.3. Random-Based TS Determination

For the RBTS, 126 (37%) lines used for GPs of MW-resistance traits were used as the TS for FAW to predict the GEBVs of the remaining 215 lines first. To build the second TS for FAW, the 215 (63%) lines used earlier as BS were considered as a TS. Then to determine the third and fourth TSs for FAW, random selections of 75 and 85% of the lines in the entire panel were performed through the Excel formula "=INDEX($A:$A,RANDBETWEEN(1,COUNTA ($A:$A)),1)" and dragging until the adequate number of lines for each percentage determined above was obtained.

2.5.4. Pedigree-Based TS Determination

The four datasets determined based on the pedigrees of the lines in the panel (PBTS strategy) are presented in Table 2. For the first dataset (FAW.Ped1), the 235 (68.91%) CIMMYT doubled haploid (DH) lines were used as a TS and the remainder (106 lines) as a BS. Regarding the second dataset, the TS and BS were switched to consider the TS in FAW.Ped1 as BS, and BS in FAW.Ped1 as the TS. The third dataset, FAW.Ped3, had a TS composed of the 294 that were neither stem borer (SB) resistant nor storage pest (SP)-resistant lines from CIMMYT, whilst the 28 SB and 19 SP-resistant lines from CIMMYT constituted the BS. In the last dataset, FAW.Ped4, the 235 DH lines, the 28 SB and 19 SP-resistant lines from CIMMYT, and the five SB-resistant lines from IITA amounting to 287 (84.16%) genotypes were considered as the TS and the remaining 54 lines from NaCRRI lines were considered as the BS (Table 2).

Table 2. Compositions of the pedigree-based test sets (TSs) for fall armyworm (FAW) datasets.

FAW Datasets	FAW.Ped1	FAW.Ped2	FAW.Ped3	FAW.Ped4
TS composition	235 DH CIMMYT lines	106 Non-DH lines	294 Non-CIMMYT SB and SP resistant lines	287 DH and CIMMYT and IITA SB and SP lines
TS/Panel (%)	68.91	31.09	86.22	84.16

DH = doubled haploid; FAW, fall armyworm; FAW.Ped1 to 4, FAW datasets 1–4 with TS based on pedigree information of the lines in the panel; SB, stem borer; SP, storage pest; TS, training set; CIMMYT, International Center for Maize and Wheat Improvement; IITA, International Institute for Tropical Agriculture.

2.6. Genomic Prediction Algorithms

The GP analyses were performed using the BLUEs and BLUPs of the phenotypes and the 28,919 DRSNPs. Sixteen algorithms available in 10 GP methods were implemented using the Genomic Prediction 0.2.6 plugin of the KDCompute 1.5.2. beta. The 10 methods were directly translated from functions of five R packages designed for GP analyses:

2.6.1. Bayesian Models

Bayesian models have different prior distributions with a general model that can be as follows: $y = 1_n\mu + Z\mu + \varepsilon$, where y is the vector of observations, Z is the design matrix for random effects, and μ is the vector of random effects [31].

The BLR (Bayesian Linear Regression) algorithms from the BLR R Package [81] are used to fit the Bayesian ridge regression. The marker effects are assumed to have a Gaussian prior distribution with mean 0 and variance σ^2, where σ^2 is unknown and assumed to have scaled x^2 distribution. In the KDCompute genomic prediction 0.2.6 plugin, the Gibbs sampler is run with 4000 iterations and 1000 iterations for burn-in period as default parameters.

The Bayesian Generalized Linear Regression (BGLR) package fits various types of parametric and semi-parametric Bayesian regressions. The parametric Bayesian algorithms used from this package rely on different prior distributions that induce different types of shrinkages of the marker effects [82], including: Gaussian (Bayesian ridge regression, BRR [83]), scaled-t (BayesA [26]), double-exponential (Bayesian LASSO, BL [84]), and two component mixtures with a point of mass at zero and a distribution with a slab that can be either Gaussian (BayesC [44]) or scaled-t (BayesB [43]). In the KDCompute genomic prediction 0.2.6 plugin tool, defaults parameters for running the Gibbs sampler were used: 4000 iterations and 1000 iterations for burn-in period.

Reproducing kernel Hilbert space (RKHS) [47,48] is a semiparametric Bayesian method from the BGLR package implemented on the KDCompute genomic prediction 0.2.6 plugin. The RKHS methods employs a kernel function to convert the molecular markers as a between pairs of observations distances, thereby, generating a square matrix that fits in a linear model. This non-linear regression method is expected to capture dominance and epistasis effects more efficiently. This approach can be modelled as:

$$y = W\mu + K_h\alpha + \varepsilon,$$

where μ represents the fixed effects vector and ε is a vector of random residuals. The parameters α and ε are assumed to have independent prior distributions $\alpha \sim N(0, K_{h\sigma_\alpha^2})$ and $\varepsilon \sim N(0, I_{\sigma_\varepsilon^2})$, respectively, and the matrix K_h relies on a reproducing kernel function with a smoothing parameter h. The parameter h measures the genomic distances among genotypes that can be interpreted as a correlation matrix and it controls the rate of decay of the correlation among genotypes [51]. To perform this analysis, the same number of iterations and burn-in parameters as for the other Bayesian methods described above were set on the KDCompute genomic prediction 0.2.6 plugin.

2.6.2. Mixed Models

The Sommer (solving mixed model equations in R) package [85] was used to implement the *mmer* (mixed model equations in R) function on the KDCompute genomic prediction 0.2.6 plugin. The package solves mixed model equations proposed by Henderson [86]. It works incidence matrices and known variance covariance matrices for each random effect using four algorithms: efficient mixed model association (EMMA) [87], average information (AI) [88], expectation maximization (EM) [89], and the default Newton–Raphson (NR) [90].

The model by Sommer can be formulated as [85]: $y = X\beta + Z\mu + \varepsilon$ with variance $V(y) = V(Z\mu + \varepsilon) = ZGZ' + R$ Additionally, the mixed model equations for this model are:

$$\begin{bmatrix} X'R^{-1}X & X'R^{-1}Z \\ Z'R^{-1}X & Z'R^{-1}Z + G^{-1} \end{bmatrix}^{-1} \begin{bmatrix} X'R^{-1}y \\ Z'R^{-1}y \end{bmatrix} = \begin{bmatrix} \beta \\ \mu \end{bmatrix}$$

where $G = K\sigma_\omega^2$ is the variance covariance matrix of the random effect μ, from a multivariate normal distribution $\mu \sim MVN(0, K\sigma_\mu^2)$, K is the additive or genomic relationship matrix ($\mathbf{A}$ or $\mathbf{A_g}$) in the genomics context, X and Z are incidence matrices for fixed and random effects, respectively, and R is the matrix for residuals (here $I\sigma_e^2$). A mixed model with a single variance component other than the error (σ_e^2) can be used to estimate the genetic variance (σ_μ^2) along with genotype BLUPs to exploit the genetic relationships between individuals coded in $\mathbf{K(A)}$. The genomic relationship matrix was constructed according to VanRaden where $K = ZZ'/2\sum p_i(1 - p_i)$ [91].

The ridge regression best linear unbiased predictor (rrBLUP) packages can either estimate marker effects by ridge regression, or alternatively, BLUPs can be calculated based on an additive relationship matrix or a Gaussian kernel. Additionally, using the rrBLUP package, **the mixed model solution (MMS)** that calculates the maximum-likelihood (ML) or restricted-ML (REML) solutions for mixed models to perform GP [92] was fitted in the KDCompute genomic prediction 0.2.6 plugin.

The mixed models fitted by rrBLUP can be formulated as:

$$y = X\beta + Z\mu + \varepsilon,$$

where β is a vector of fixed effects and μ is a vector of random effects with variance $Var[\mu] = K_{\sigma_\mu^2}$. The residual variance is $Var[\varepsilon] = I_{\sigma_\varepsilon^2}$. This class of mixed models, in which there is a single variance component other than the residual error, has a close relationship with ridge regression (ridge parameter $\lambda = \sigma_\varepsilon^2/\sigma_\mu^2$) (https://kdcompute.igss-africa.org/kdcompute/home).

2.6.3. Machine Learning Algorithms

The R package RandomForest that implements Breiman's random forest algorithm for classification and regression [93] was used on the KDCompute genomic prediction 0.2.6 plugin to fit the function missForest. Random forest is a non-linear machine learning algorithm that uses a two-layer randomization process to build decorrelated bootstrapped trees. As a first randomization layer, it builds multiple trees using a bootstrap sample of the marker data in the training. Then, a second randomization process is carried at the novel nodes to grow final trees. The random forest method selects at each node of each tree, a random subset of variables, and only those variables are used as candidates to find the best split for the node [94]. To predict the breeding value of a line in the TS, predictions over trees for which the given observation was not used to build the tree are averaged [51]. On the KDCompute 1.5.2. beta platform, both options for the *mtry*, square root and regression (*sqrt(p)* and *p/3*, respectively, where p is number of variables in x), for the classification of the number of variables randomly sampled as candidates at each split were implemented in this study. Additionally, the trees to grow (*ntree*) was set to 10, while *node size* (minimum size of terminal nodes) and *max nodes* (maximum number of terminal nodes trees in the forest can have) were set to 5 and 10, respectively. The 16 methods used in this study and their statistical characteristics are presented in Table 3.

Table 3. Genomic prediction methods used for the analysis of the different traits and datasets.

	GP Algorithms	Abbreviations	Method Type
1	Sommer with Average Information (AI)	mmer_AI	Parametric/Mixed model
2	Sommer with Expectation Maximization (EM)	mmer_EM	Parametric/Mixed model
3	Sommer with Efficient Mixed Model Association (EMMA)	mmer_EMMA	Parametric/Mixed model
4	Sommer with default Newton-Raphson (NR)	mmer-NR	Parametric/Mixed model
5	Ridge-regression Best linear unbiased Predictor	rrBLUP	Parametric/Mixed model
6	Mixed Model solution with Maximum Likelihood (ML)	mms_ML	Parametric/Mixed model
7	Mixed Model solution with Restricted Maximum Likelihood (REML)	mms_REML	Parametric/Mixed model
8	BayesB	BayesB	Parametric/Bayesian
9	BayesA	BayesA	Parametric/Bayesian
10	BayesC	BayesC	Parametric/Bayesian
11	Bayesian least absolute shrinkage and selection operator (LASSO)	BL	Parametric/Bayesian
12	Bayesian Ridge Regression	BRR	Parametric/Bayesian
13	Bayesian Linear Regression	BLR	Parametric/Bayesian
14	Reproducible kernel Hilbert space	RKHS	Semi-parametric/Bayesian
15	Random Forest with Square root	missForest_Sqt	Nonparametric/Machine Learning
16	Random Forest with Regression	missForest_Reg	Nonparametric/Machine Learning

2.7. Cross-Validations and PA Estimation

To calculate the predictive accuracies of each of the 17 methods, a cross-validation approach was performed using the data for the TS with 10 folds and five repetitions amounting to 50 replications. The PAs were estimated as the correlation coefficient (R^2) averaged across the 50 cross-validation replications between the observed phenotypic values and the predicted genomic-estimated breeding values (GEBV) (https://kdcompute. igss-africa.org/kdcompute/plugins).

3. Results

3.1. Higher PAs Achieved for FAW and MW-Resistance Traits with BLUPs when Compared to BLUEs across GP Algorithms

Both genotypic BLUEs and BLUPs for resistance to FAW and MW traits such as AK, AP, and GWL were used in GPs. In general, BLUPs produced better predictions than BLUEs by at least two orders of magnitude in terms of PAs (Figure 1). The PAs realized with BLUEs (Figure S2) varied from −0.246 for FAW (mms_ML) to 0.299 for AP (BayesB), while PAs for BLUPs ranged from 0.668 for AP (mmer_NR) to 0.823 for AP (missForest_Reg). The differences in terms of accuracies between BLUEs and BLUPs were high, despite the highly significant ($p < 0.001$) correlations between BLUEs and BLUPs for each trait ranging from 0.93 for FAW to close to 1 for AP, AK, and GWL (presented in Figure 1); therefore, only results for BLUPs will be presented hereafter.

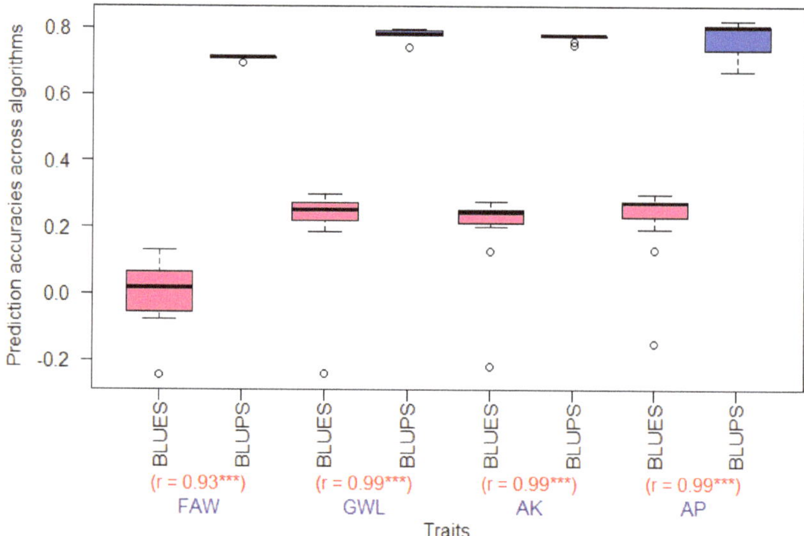

Figure 1. Boxplot of PAs (prediction accuracies) for best linear unbiased estimators (BLUEs) (in pink) and predictors (BLUPs) (in blue) of maize resistance to FAW and MW across prediction models and correlations (r) between BLUEs and BLUPs for each trait. FAW, fall armyworm; GWL, grain weight loss; AP, adult progeny emergence; AK, number of affected kernels. *** significant at $p < 0.001$.

3.2. PAs for MW Resistance Traits Using BLUPs

The PAs were generally high for the tested MW traits, mostly above 0.668 across the 12 GP models that were successfully run on the datasets (Figure 2); however, RKHS failed to work for AK. The highest PAs were achieved for AP with missForest_reg (0.823), followed by BRR (0.805), and RKHS (0.804), whilst mmer_NR algorithm had the lowest prediction accuracy of 0.667 (Figure 2). The PAs achieved for GWL ranged from 0.742 for missForest_Sqt to 0.795 for mmer_NR, while for AK, they varied from 0.749 for missForest_sqrt to 0.779 for BRR (Figure 2). In general, Bayesian models predicted better than both mixed model and machine learning methods, although the differences were small (Figure S3).

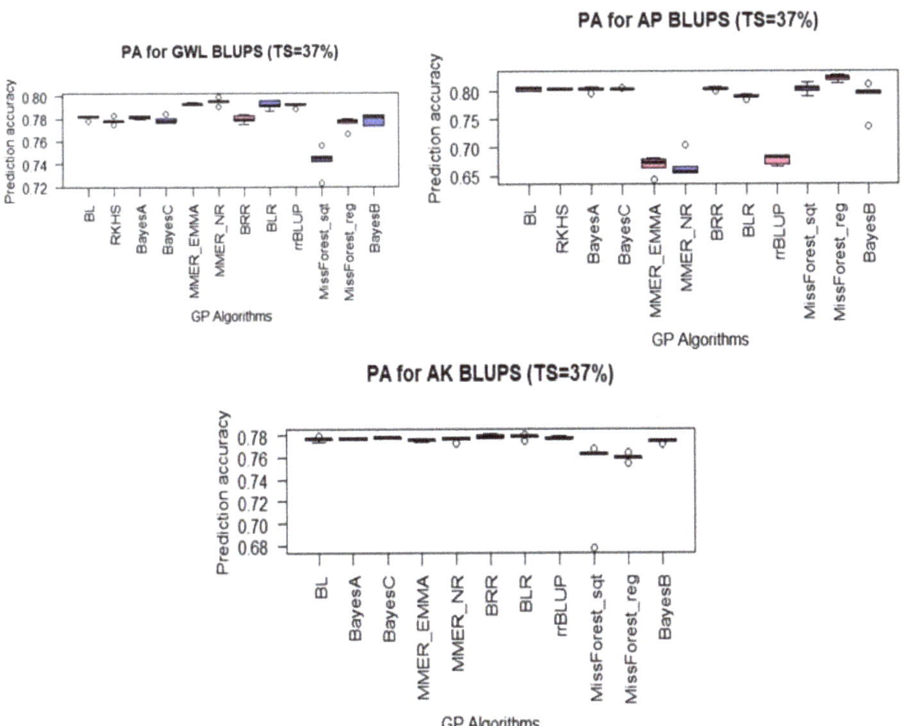

Figure 2. Boxplots of the genomic prediction accuracies of BLUPs for MW-resistance traits: GWL, grain weight loss; AP, adult progeny emergence; AK, number of affected kernels (See Table 3 for GP algorithms).

3.3. PA for FAW Resistance Using BLUPs

The different maize resistance to FAW datasets showed high predictive abilities with 10 of the 16 GP algorithms used in the study. For the RBTS approach, the PAs were lowest with the dataset that had a TS composed of 37% (lowest size) of the panel and highest with the largest TS (85% of the panel). Even with a TS of 37%, the PAs were still high, ranging from 0.694 to 0.714 for mms_ML and BLR methods, respectively (Figure 3). However, it should be noted that with equal TS sizes and same composition (37% of the panel), higher PAs were achieved for MW-resistance traits (GWL, AP, and AK) compared to FAW-resistance ones (Figure S3). The PA for the RBTS of 63% varied from 0.833 for BL method to 0.838 for the missForest_Sqt; thus, there was a small variation among different methods. Similarly, there was minimal variation among GP algorithms on the dataset with a 75% TS whose PAs varied from 0.838 for mms_REML to 0.843 for MissForest_Reg. The same trend was obtained on the dataset with a RBTS of 85% of the panel, with PAs ranging from 0.843 for the BRR model to 0.847 for the missForest_Reg method. Furthermore, there was a high and significant ($p < 2.2.10^{-16}$) positive correlation of 0.92 (Figure 4) between the PAs and TS sizes for FAW datasets for the RBTS denoting a steady improvement of the PAs as the TS size increased. However, the PAs for FAW resistance reached a plateau at TS size above 63% of the panel (Figure 5).

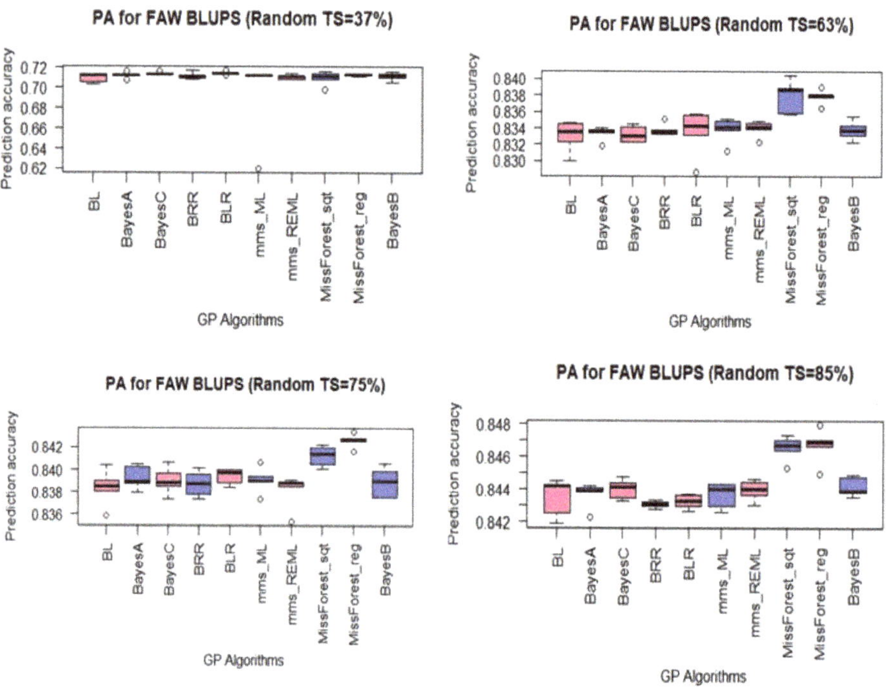

Figure 3. Boxplot of PAs for maize resistance to the fall armyworm (FAW) datasets with the RBTS approach with random selection of 37, 63, 75, and 87% of the entire panel (see Table 3 for GP algorithms).

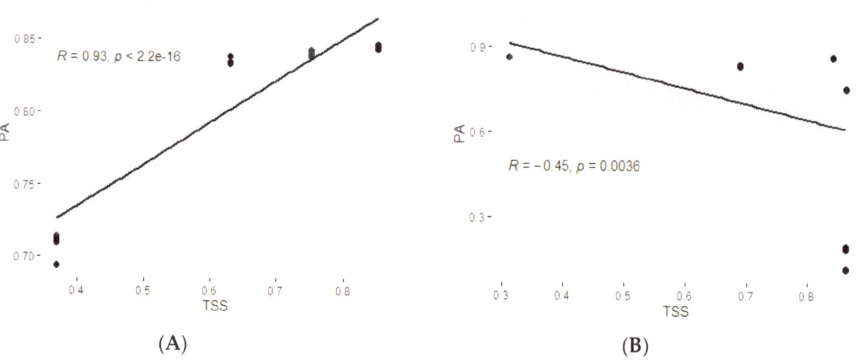

Figure 4. Pearson correlation between training set (TS) sizes and prediction accuracies (PAs) across the 10 genomic prediction algorithms conducted on RBTS (**A**) and PBTS (**B**) datasets for fall armyworm resistance (FAW) resistance.

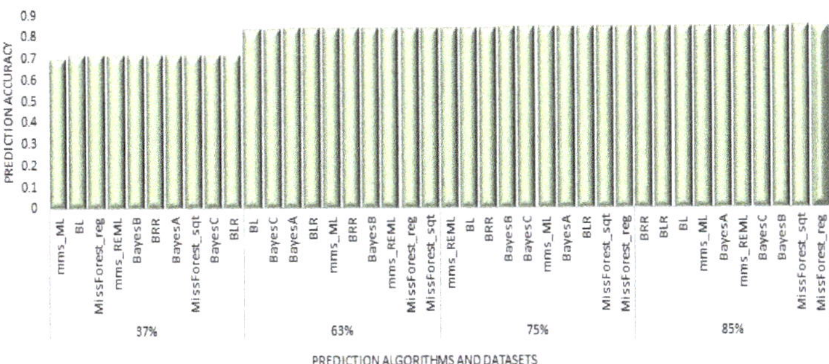

Figure 5. Prediction accuracies for FAW with RBTS across algorithms and training sets with different sizes in percent of the total panel.

Although the PAs did not vary much among GP algorithms, especially when the analyses involved larger TS sizes equal or bigger than 63% of the panel, the machine learning methods slightly outperformed other GP algorithms for all the traits, except for the TS of 37% where Bayesian methods such as BLR and BayesC showed a slight advantage over the machine learning methods (Figure S4). The PAs for FAW-resistance datasets with PBTS were generally high, mostly above 0.82 (Figure 6). For the first dataset (FAW.Ped1) with a TS of 68.91% of the panel (see Table 2), the PAs varied between 0.828 for BLR to 0.835 for missForest_Sqt. For FAW.Ped2 (TS = 31.09%), the PAs ranged from 0.862 for BayesC to 0.864 for mms_REML.

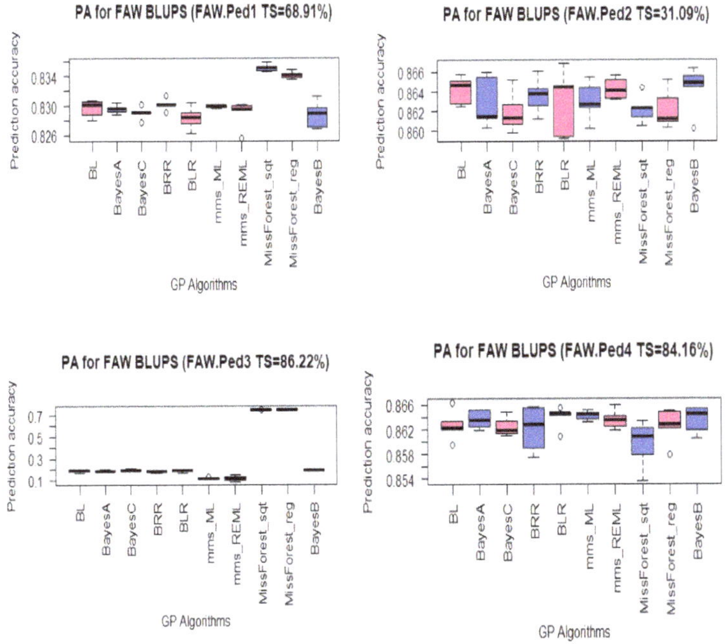

Figure 6. Boxplots of PAs for maize resistance to the fall armyworm (FAW) datasets using the PBTS approach (see Table 2 for the PBTS strategy and Table 3 for GP algorithms).

For FAW.Ped4, with a TS of 84.16%, PAs varied between 0.860 to 0.864 for missForest_Sqt and mms_ML, respectively. However, for FAW.Ped3 with the largest TS (86.22%), eight of the 10 algorithms achieved low PAs (below 0.20) and only missForest_Reg and missForest_Sqt attained PAs of 0.749 and 0.750, respectively. Thus, the Pearson correlation between the sizes of the PBTS datasets and the predictions accuracies for the 10 GP algorithms revealed a significant ($p > 0.0036$) negative relationship of $r = -0.45$ (Figure 4).

In the FAW datasets, the PAs were more influenced by the composition of the TS and its genetic relationship with the BS (see Table 2). Using the doubled haploid (DH) lines as TS (FAW.Ped1) and vice-versa (FAW.Ped2) or DH and stem borer (SB) and storage pest (SP)-resistant lines as TS (FAW.Ped4) permitted achieving relatively high PAs from all the 10 algorithms, which when considering the CIMMYT SB and SP-resistant lines as BS and the remainder as a TS (FAW.Ped3), only resulted in machine learning algorithms missForest_reg and missForest_Sqt achieving relatively high PAs. Furthermore, the composition of the TS and its relationship with the BS determined which GP methods achieved the highest Pas; machine learning algorithms worked best on FAW.Ped1 and FAW.Ped3, linear mixed model approaches outperformed Bayesian and machine learning algorithms on FAW.Ped2 and FAW.Ped4, and Bayesian methods ranked either second or third on all datasets (Figure S5). It should be noted that the PBTS strategy generally achieved better PAs than the RBTS irrespective of the size of the TS, except for the FAW.Ped3 dataset (Figures 3 and 6).

4. Discussion

Tropical maize germplasm is characterized by rapid linkage disequilibrium (LD) decay with high diversity [95]. These germplasm genetic characteristics make genomic selection (GS) a promising approach to integrate into African breeding programs [96]. However, genomic prediction (GP) models are very diverse and their differential performances depend on crops and trait architectures, besides other parameters such as the size of the training set (TS) and its genetic relationship with the breeding set (BS) [31,37]. Therefore, this study aimed at assessing the feasibility of genomic selection for maize resistance to FAW and MW through estimation of the genomic prediction accuracies achieved by parametric, semiparametric, and nonparametric (machine learning) genomic prediction (GP) algorithms using phenotypic BLUEs and BLUPs, and random and pedigree-based TS determination strategies.

4.1. Higher Pas Were Achieved for BLUPs Compared to BLUEs for Both FAW and MW-resistance Traits

With a RBTS of 37% of the panel, which was the smallest and expected to give the worst PAs, PAs were higher (at least two-fold) across both FAW and MW-resistance traits and for all GP models when trait BLUPs were used as phenotypes compared to BLUEs, although there were high Pearson correlations between these two categories of phenotypic data for each trait. In general, BLUPs were reported to have higher predictability than BLUEs owing to better accounting for outliers and environmental variabilities permitted by the shrinkage procedure in BLUPs, which results in more accurate estimates of individual genetic effects [70–73]. Furthermore, most of the predictive differences between BLUPs and BLUEs might have stemmed from BLUPs being more suitable than BLUEs in fitting data recorded from unbalanced experiments [49,70] as was the case for both FAW damage scores across environments and MW bioassay in this study. Therefore, for all subsequent analyses with higher RBTS sizes and the PBTS strategy for FAW, only BLUPs were focused at in this study and will be further discussed.

4.2. High PAs Were Achieved for FAW and MW-Resistance Traits Using Moderately Sized Training Sets

The obtained PAs were high for both MW and FAW-resistance traits even with TS of moderate sizes confirming the potential of genomic selection (GS) in Africa-adapted germplasms [28–30,33]. With a TS of 37% of the entire panel, high PAs (above 0.70) for MW-resistance traits, grain weight loss (GWL), adult progeny emergence (AP), the number of affected kernels (AK), and FAW resistance were achieved in agreement with the moderate to high heritability values for these traits as, reported earlier [21,31,41]. These results are significantly important considering that one of the disadvantages of GS is the requirement of large TS which negatively impacts the reduction of phenotyping cost [62,64].

The PAs increased up to above 0.85 in proportion to the increase in TS (RBTS approach) size for FAW resistance which was the only trait phenotyped for all the lines of the panel. It would be interesting to phenotype other lines of the panel that were not evaluated for MW-resistance traits to establish larger TS which may improve the PAs [31,65,97,98]. Very few reports of GP are available for maize resistance to biotic stresses. High PAs were achieved for maize resistance to chlorotic mottle virus (up to 0.95) and maize lethal necrosis (reaching 0.87) in tropical germplasm [67]. However, lower PAs of up to 0.59 were obtained in a study that assessed the predictability of maize resistance to the European corn borer [99] in temperate germplasm. Additionally, Gowda et al. (69) reported moderate PAs (close to 0.60) for maize resistance to a biotic stress, maize lethal necrosis in tropical maize populations.

4.3. GP Algorithms Performed Differently on FAW and MW Maize Resistance Traits

In this study, several GP models that included statistical and machine learning algorithms from parametric, semi-parametric, and nonparametric approaches were used to predict FAW and MW-resistance traits. These GP algorithms, as expected, performed differently on the different traits although the predictive variations were generally minimal, especially when large TS were involved, similarly to earlier model benchmarking reports [100,101]. Bayesian models (parametric: BLR and BRR, and semi-parametric: RKHS) performed better on MW traits, GWL, AP, and AK, while nonparametric machine learning algorithms (missForest, here), and to a lesser extent, the linear mixed model (especially in the PBTS approach), achieved the highest PAs on FAW datasets. The differential performances of the different GP algorithms on the insect resistance traits evaluated in this study could be due to differences in the genetic structures (extent of additive vs. non-additive gene action) of the respective traits [23,38,47,49]. Maize resistance to FAW, which was moderately heritable across environments [25], would be expected to be controlled by both additive and non-additive genetic factors, including epistasis [102–104], whereas, MW-resistance traits such as GWL, AP, and AK with heritability values above 90% [25] were most likely characterized by a prevalence of additive gene action [105,106] in the current panel.

This supposed genetic architecture difference between FAW and MW-resistance trait could be the reason for non-linear methods such as random forest performing better at predicting FAW resistance, since these are more capable of integrating epistasis in the statistical modelling [27,51]. However, the RKHS algorithm, also a non-linear GP approach known to efficiently handle epistatic genetic relation [51,59], did not successfully run on FAW dataset, although it was among the best models for predicting MW-resistance traits, except BLUPs for the number of affected kernels (AK), for which the RKHS algorithm did not run successfully. In this study, the reasons for some GP algorithms failing to run either on MW or FAW-resistance datasets are unclear, but this could be related to the BLUPS structure of the datasets that failed to run. It should be noted that all the algorithms ran successfully on phenotypic BLUEs datasets with the smallest TS (37% of the panel) being used to compare PAs between BLUPs and BLUEs in this study. However, the two to three-fold predictive ability gain with BLUPs compared to BLUEs would be an incentive to consider BLUPs in future GS activities for maize resistance to MW and FAW. Overall, future GS efforts for maize resistance to MW and FAW are recommended to focus more on Bayesian and machine learning algorithms such as random forest, BayesA, BayesB, BayesC, BRR, and BLR which outperformed mixed linear models for most datasets considered in the current study.

4.4. Influences of the Sizes and the Compositions of TS and BS on PAs

Two factors, the relative sizes of the TS and BS (RBTS approach) and their genetic relationship (PBTS approach), influenced the levels of PAs across FAW-resistance datasets, corroborating earlier reports [31,63,65,97,98,107,108]. A net increase in PAs for maize resistance to FAW was realized when the size of the TS was increased from 37% (0.694 to 0.714) to 63% (0.833 to 0.838), similar to earlier reports on wheat yield [109]. This increase was followed by a slight gain in predictability at 75% (0.837 to 0.843) and 85% (0.843 to 0.847), and thus, the PAs plateaued when TS sizes above 63% were considered in this study as reported earlier in other studies [21,64,109–111]. Thus, future GS programs for maize resistance to FAW could be designed around TS composed of a minimum of 60% of the entire breeding germplasm to achieve high genetic gains. These results were further supported by the highly significant ($p > 2.2.10^{-16}$) positive correlation (R = 0.92) between TS size and PAs. Similarly, positive correlations between the number of lines in the TS and the PAs, and plateau for the PAs were also reported by Edwards et al. [109].

The composition of the TS and its relationship with the BS are determinant factors for the genomic predictability of complex traits [63,112–114]. In the current study, using the PBTS approach, these two parameters were more important than the size of the TS since higher PAs were achieved in FAW.Ped2 (0.862 to 0.864) with a TS of 31.09% compared to all other FAW PBTS datasets, including FAW.Ped3 (0.114 to 0.750), with the largest TS of 66.22%. In fact, FAW.Ped3 achieved the lowest PAs among all the PBTS FAW datasets. These results were further illustrated by the significantly ($p < 0.0036$) negative correlation (R = −0.45) between the sizes of the PBTS and the achieved PAs.

However, it is not very clear why the predictions for the BS FAW.Ped3 (47 CIMMYT SB and SP-resistant lines) and the TS (DH, IITA SB, and NaCRRI lines) led to lower PAs for FAW.Ped3. A possible explanation could be that these two sets were distantly related since only two and one CIMMYT SB and SP-resistant lines, respectively, were used as parents to develop the DH lines. Spindel et al. [111] argued that high PAs can be achieved with small-sized TS when lines in the TS and the BS are closely related, since such TS would sample the full genetic diversity of the population. However, the more distantly related the TS and the BS are, the larger the required TS size to reach high PAs [111]. Using the CIMMYT SB and SP-resistant lines as a TS would most likely lead to lower PAs since such a TS would be additionally disadvantaged by its small size (47 lines). The DH lines in the current study are involved as a TS in most of the best performing GP datasets evaluated in the current study (both in the RBTS and PBTS approaches) and as unique lines in the BS of the best performing pedigree-based BS (FAW.Ped2). This DH population could be of interest in future breeding activities targeted at improving insect resistance in maize [23,115–117] and potentially useful for GS of complex traits with low to moderate heritability [118].

5. Conclusions

This study assessed prediction accuracies of genomic-estimated breeding values for fall armyworm (FAW) and maize weevil (MW)-resistance traits in a diverse Africa-adapted maize panel using several parametric, semi-parametric, and non-parametric genomic prediction models. Prediction accuracies for maize resistance to FAW and MW traits were relatively high, even with a moderate training set size. For FAW resistance, although the prediction accuracies were positively correlated with the size of the training set, the composition and the relationship of the training set with the breeding set were more influential in predicting line performance. Additionally, TS determination-related parameters were more important than the type of genomic prediction models in predicting FAW and MW-resistance traits. However, Bayesian models on MW-resistance traits and machine learning models on FAW damage resistance outperformed mixed linear models in almost all the datasets used in this study. Therefore, future genomic selection programs for maize resistance to insect pests such as FAW and MW in Africa should put more effort into designing effective training sets and use selected Bayesian and machine learning GP algorithms to improve genetic gains, shorten breeding cycles, and accelerate variety release.

Such programs could greatly benefit from using the genetically diverse maize panel used in this study as a base population, since it consists of lines adapted to several African agro-ecologies.

Supplementary Materials: The following are available online at https://www.mdpi.com/2223-774 7/10/1/29/s1, Figure S1: Rating of maize plants based on foliar damage by FAW, Figure S2: Boxplot of PA for best linear unbiased estimators (BLUEs) of maize resistance to the fall armyworm (FAW) and maize weevil (MW) with identical training set size (37%) and compositions, Figure S3: Comparisons of genomic prediction accuracies of the three best algorithms for best linear unbiased predictors (BLUPs) of maize weevil resistance traits: number of affected kernels (AK), adult progeny emergence (AP), and grain weight loss (GWL) vs., fall armyworm resistance dataset with identical TS, Figure S4: Genomic prediction accuracies of the three best algorithms for each fall armyworm resistance BLUPs datasets with RBTS of 37, 62, 75, and 85% of the entire dataset, Figure S5: Genomic prediction accuracies of the three best algorithms for each fall armyworm resistance BLUPs datasets with PBTS, Table S1: Descriptions of parents and crosses that constituted the doubled-haploid population.

Author Contributions: Conceptualization, A.B. (Arfang Badji), P.R., S.K., M.O., D.B.K., and L.M.; methodology, A.B. (Arfang Badji), D.B.K., and L.M.; investigation, A.B. (Arfang Badji) and D.B.K.; formal analysis, A.B. (Arfang Badji) and L.M.; resources, A.B. (Arfang Badji), G.A., M.O., D.B.K., and L.M.; visualization, A.B. (Arfang Badji); supervision, P.R., S.K., M.O., and L.M.; project administration, P.R. and M.O.; funding acquisition, A.B. (Arfang Badji), M.O., G.A., D.B.K., and P.R.; writing—original draft preparation, A.B. (Arfang Badji); writing—review and editing, all authors (A.B. (Arfang Badji), L.M., D.B.K., F.K., D.O., N.M., S.A., A.I., A.B. (Astere Bararyenya), S.N.N., T.O., P.W., M.O., M.O.-S., H.T., G.A., S.K., P.R.). All authors have read and agreed to the published version of the manuscript.

Funding: This research was funded by the capacity building competitive grant training the next generation of scientists provided by Carnegie Cooperation of New York through the Regional Universities Forum for Capacity Building in Agriculture (RUFORUM: RU/2016/Intra-ACP/RG/001). A. Badji received a Ph.D. scholarship from the Intra- ACP Academic mobility for Crop Scientists for Africa Agriculture (CSAA) project. Genotyping of the lines was carried out through a project of D.B.K. thanks to the Integrated Genotyping Service and Support (IGSS) coordinated by the International Livestock Research Institute (ILRI) and Bioscience east and central Africa (BecA), grant number: IGSS-DL0274. The National Crops Resources Research Institute (NaCRRI) of Namulonge, UGANDA through a grant of the USAID Feed-the-Future Uganda, Agriculture Research Activity/Maize paid the article processing charges. Further, NaCRRI financially and logistically supported field and laboratory activities of this research.

Acknowledgments: The authors thank all the technicians for experimental setup and data collection in the fields and laboratories of NaCRRI at Namulonge and Kasese, UGANDA. The authors acknowledge NaCRRI, the International Maize and Wheat Improvement Center (CIMMYT) of Nairobi, KENYA, and the International Institute of Tropical Agriculture (IITA) of Ibadan, NIGERIA for providing the original germplasm used for this research. The authors thank Clay SNELLER of the Ohio State University and all the personnel of ILRI and BecA who provided the genotyping support at BecA/ILRI.

Conflicts of Interest: The authors declare no conflict of interest.

References

1. Demissie, G.; Tefera, T.; Tadesse, A. Importance of husk covering on field infestation of maize by *Sitophilus zeamais* Motsch (Coleoptera: Curculionidea) at Bako, Western Ethiopia. *Afr. J. Biotechnol.* **2008**, *7*, 3777–3782.
2. Shiferaw, B.; Prasanna, B.M.; Hellin, J.; Bänziger, M. Crops that feed the world 6. Past successes and future challenges to the role played by maize in global food security. *Food Secur.* **2011**, *3*, 307–327. [CrossRef]
3. Awata, L.A.O.; Tongoona, P.; Danquah, E.; Ifie, B.E.; Suresh, L.M.; Jumbo, M.B.; Marchelo-D'ragga, P.W.; Sitonik, C. Understanding tropical maize (*Zea mays* L.): The major monocot in modernization and sustainability of agriculture in sub-Saharan Africa. *Int. J. Adv. Agric. Res.* **2019**, *7*, 32–77.
4. Nyukuri, R.W.; Wanjala, F.M.; Kirui, S.C.; Cheramgoi, E.; Chirchir, E.; Mwale, R. Damage of stem borer species to Zea mays L.,Sorghum bicolor L. and three refugia graminae. *Adv. Agric. Biol.* **2014**, *1*, 37–45.
5. Tefera, T.; Goftishu, M.; Ba, M.; Rangaswamy, M. *A Guide to Biological Control of Fall Armyworm in Africa Using Egg Parasitoids,* 1st ed.; ICIPE: Nairobi, Kenya, 2019.

6. Munyiri, S.W.; Mugo, S.N.; Otim, M.; Mwololo, J.K.; Okori, P. Mechanisms and sources of resistance in tropical maize inbred lines to *Chilo partellus* stem borers. *J. Agric. Sci.* **2013**, *5*, 51–60. [CrossRef]

7. Mwololo, J.K.; Mugo, S.; Okori, P.; Tefera, T.; Otim, M.; Munyiri, S.W. Sources of resistance to the maize weevil *Sitophilus zeamais* in tropical maize. *J. Agric. Sci.* **2012**, *4*, 206–215.

8. Mwololo, J.K. Resistance in Tropical Maize to the Maize Weevil and Larger Grain Borer. Ph.D. Thesis, Makerere University, Kampala, Uganda, 2013.

9. Kasozi, L.C.; Derera, J.; Tongoona, P.; Tukamuhabwa, P.; Muwonge, A.; Asea, G. Genotypic variation for maize weevil resistance in eastern and southern Africa maize inbred lines. *Uganda J. Agric. Sci.* **2016**, *17*, 83–97. [CrossRef]

10. Tende, R.; Derera, J.; Mugo, S.; Oikeh, S. Estimation of genetic diversity of germplasm used to develop insect-pest resistant maize. *Maydica* **2016**, *61*, 1–8.

11. Khakata, S.; Mbute, F.N.; Chemining'wa, G.N.; Mwimali, M.; Karanja, J.; Harvey, J.; Mwololo, J.K. Post-harvest evaluation of selected inbred lines to maize weevil *Sitophilus zeamais* resistance. *J. Plant Breed. Crop Sci.* **2018**, *10*, 105–114. [CrossRef]

12. Sodedji, F.K.A.; Kwemoi, D.B.; Kasozi, C.L.; Asea, G.; Kyamanywa, S. Genetic analysis for resistance to *Sitophilus zeamais* (Motschulsky) among provitamin-A maize germplasm. *Maydica* **2018**, *63*, 8. Available online: https://journals-crea.4science.it/index.php/maydica/article/view/1698 (accessed on 13 November 2019).

13. Munyiri, W.S.; Mugo, N.S.; Otim, M.; Tefera, T.; Beyene, Y.; Mwololo, K.J.; Okori, P. Responses of tropical maize landraces to damage by *Chilo partellus* stem borer. *Afr. J. Biotechnol.* **2013**, *12*, 1229–1235.

14. Munyiri, S.W.; Mugo, S.N.; Mwololo, J.K. Mechanisms and levels of resistance in hybrids, open pollinated varieties and landraces to *Chilo partellus* maize stem borers. *Int. Res. J. Agric. Sci. Soil Sci.* **2015**, *5*, 81–90.

15. Mwololo, J.K.; Munyiri, S.W.; Semagn, K.; Mugo, S. Genetic diversity analysis in tropical maize germplasm for stem borer and storage pest resistance using molecular markers and phenotypic traits. *Mol. Plant Breed.* **2015**, *6*, 1–22.

16. Goergen, G.; Kumar, P.L.; Sankung, S.B.; Togola, A.; Tamò, M. First report of outbreaks of the fall armyworm *Spodoptera frugiperda* (JE Smith) (Lepidoptera, Noctuidae), a new alien invasive pest in West and Central Africa. *PLoS ONE* **2016**, *11*, e0165632. [CrossRef]

17. Padhee, A.K.; Prassanna, B.M. The emerging threat of Fall Armyworm in India. *Indian Farming* **2019**, *69*, 51–54.

18. Prasanna, B.M.; Huesing, J.E.; Eddy, R.; Peschke, V.M.; Regina, E.; Virginia, M.P. *Fall Armyworm in Africa: A Guide for Integrated Pest Management*, 1st ed.; West Africa Regional Training of Trainers and Awareness Generation Workshop on Fall Armyworm Management, IITA, Cotonou, Bénin; Prasanna, B.M., Regina, E., Virginia, M.P., Eds.; CIMMYT: Queretaro, Mexico, 2018.

19. Gedil, M.; Menkir, A. An integrated molecular and conventional breeding scheme for enhancing genetic gain in maize in Africa. *Front. Plant Sci.* **2019**, *10*, 1–17. [CrossRef] [PubMed]

20. Murenga, M.; Derera, J.; Mugo, S.; Tongoona, P. A review of genetic analysis and response to selection for resistance to *Busseola fusca* and *Chilo partellus*, stem borers in tropical maize germplasm: A Kenyan perspective. *Maydica* **2016**, *61*, 1–11.

21. Crossa, J.; Pérez-Rodríguez, P.; Cuevas, J.; Montesinos-López, O.; Jarquín, D.; de los Campos, G.; Burgueno, J.; Gonzales-Camacho, J.M.; Perez-Elizade, S.; Beyene, Y.; et al. Genomic selection in plant breeding: Methods, models, and perspectives. *Trends Plant Sci.* **2017**, *22*, 1–15. [CrossRef]

22. Roorkiwal, M.; Jarquin, D.; Singh, M.K.; Gaur, P.M.; Bharadwaj, C.; Rathore, A.; Howard, R.; Srinivasan, S.; Jain, A.; Garg, V.; et al. Genomic-enabled prediction models using multi-environment trials to estimate the effect of genotype × environment interaction on prediction accuracy in chickpea. *Sci. Rep.* **2018**, *8*, 1–11. [CrossRef]

23. Robertsen, C.D.; Hjortshøj, R.L.; Janss, L.L. Genomic selection in cereal breeding. *Agronomy* **2019**, *9*, 95. [CrossRef]

24. Munyiri, S.W.; Mugo, S.N. Quantitative trait loci for resistance to spotted and African maize stem borers (*Chilo partellus* and *Busseola fusca*) in a tropical maize (*Zea mays* L.) population. *Afr. J. Biotechnol.* **2017**, *16*, 1579–1589. [CrossRef]

25. Badji, A.; Kwemoi, D.B.; Machida, L.; Okii, D.; Mwila, N.; Agbahoungba, S.; Kumi, F.; Ibanda, A.; Bararyenya, A.; Solemangey, M.; et al. Genetic basis of maize resistance to multiple-insect pests: Integrated genome-wide comparative mapping and candidate. *Genes* **2020**, *11*, 689. [CrossRef] [PubMed]

26. Meuwissen, T.H.E.; Hayes, B.J.; Goddard, M.E. Prediction of total genetic value using genome-wide dense marker maps. *Genetics* **2001**, *157*, 1819–1829. [PubMed]

27. Jannink, J.; Lorenz, A.J.; Iwata, H. Genomic selection in plant breeding: From theory to practice. *Brief. Funct. Genom.* **2010**, *9*, 166–177. [CrossRef]

28. Massman, J.M.; Jung, H.J.G.; Bernardo, R. Genomewide selection versus marker-assisted recurrent selection to improve grain yield and stover-quality traits for cellulosic ethanol in maize. *Crop Sci.* **2013**, *53*, 58–66. [CrossRef]

29. Beyene, Y.; Semagn, K.; Mugo, S.; Tarekegne, A.; Babu, R.; Meisel, B.; Sehabiague, P.; Makumbi, D.; Magorokosho, C.; Oukeh, S.; et al. Genetic gains in grain yield through genomic selection in eight bi-parental maize populations under drought stress. *Crop Sci.* **2015**, *55*, 154–163. [CrossRef]

30. Vivek, B.S.; Krishna, G.K.; Vengadessan, V.; Babu, R.; Zaidi, P.H.; Kha, L.Q.; Mandal, S.S.; Grudloyma, P.; Takalkar, S.; Krothapalli, K.; et al. Use of genomic estimated breeding values results in rapid genetic gains for drought tolerance in maize. *Plant Genome* **2017**, *10*, 1–8. [CrossRef]

31. Liu, X.; Wang, H.; Wang, H.; Guo, Z.; Xu, X.; Liu, J.; Wang, S.; Li, W.-X.; Zou, C.; Prasanna, B.M.; et al. Factors affecting genomic selection revealed by empirical evidence in maize. *Crop J.* **2018**, *6*, 341–352. [CrossRef]

32. Yuan, Y.; Scheben, A.; Batley, J.; Edwards, D. Using genomics to adapt crops to climate change. In *Sustainable Solutions for Food Security*; Springer International Publishing: Cham, Switzerland, 2019; pp. 91–109.

33. Zhang, X.; Pérez-Rodríguez, P.; Burgueño, J.; Olsen, M.; Buckler, E.; Atlin, G.; Prasanna, B.M.; Vargas, M.; San Vicente, F.; Crossa, J. Rapid cycling genomic selection in a multiparental tropical maize population. *G3 Genes Genomes Genet.* **2017**, *7*, 2315–2326. [CrossRef]

34. Hassen, M.B.; Bartholomé, J.; Valè, G.; Cao, T.V.; Ahmadi, N. Genomic prediction accounting for genotype by environment interaction offers an effective framework for breeding simultaneously for adaptation to an abiotic stress and performance under normal cropping conditions in rice. *G3 Genes Genomes Genet.* **2018**, *8*, 2319–2332. [CrossRef]

35. Muleta, K.T.; Pressoir, G.; Morris, G.P. Optimizing genomic selection for a sorghum breeding program in Haiti: A simulation study. *G3 Genes Genomes Genet.* **2019**, *9*, 391–401. [CrossRef]

36. Suontama, M.; Klápště, J.; Telfer, E.; Graham, N.; Stovold, T.; Low, C.; McKinley, R.; Dungey, H. Efficiency of genomic prediction across two *Eucalyptus nitens* seed orchards with different selection histories. *Heredity* **2019**, *122*, 370–379. [CrossRef] [PubMed]

37. Voss-Fels, K.P.; Cooper, M.; Hayes, B.J. Accelerating crop genetic gains with genomic selection. *Appl. Genet.* **2019**, *132*, 669–686. [CrossRef] [PubMed]

38. Kadam, D.C.; Lorenz, A.J. Evaluation of nonparametric models for genomic prediction of early-stage single crosses in maize. *Crop Sci.* **2019**, *59*, 1411–1423. [CrossRef]

39. Whittaker, J.C.; Thompson, R.; Denham, M.C. Marker-assisted selection using ridge regression. *Genet. Res.* **2000**, *75*, 249–252. [CrossRef] [PubMed]

40. Clark, S.A.; van der Werf, J. *Genomic Best Linear Unbiased Prediction (gBLUP) for the Estimation of Genomic Breeding Values*; Springer: Berlin/Heidelberg, Germany, 2013; pp. 321–330.

41. Zhang, H.; Yin, L.; Wang, M.; Yuan, X.; Liu, X. Factors affecting the accuracy of genomic selection for agricultural economic traits in maize, cattle, and pig populations. *Front. Genet.* **2019**, *10*, 1–10. [CrossRef] [PubMed]

42. Hayashi, T.; Iwata, H. EM algorithm for Bayesian estimation of genomic breeding values. *BMC Genet.* **2010**, *11*, 3. [CrossRef]

43. Meuwissen, T.H.; Solberg, T.R.; Shepherd, R.; Woolliams, J.A. A fast algorithm for BayesB type of prediction of genome-wide estimates of genetic value. *Genet. Sel. Evol.* **2009**, *41*, 1–10. [CrossRef]

44. Habier, D.; Fernando, R.L.; Kizilkaya, K.; Garrick, D.J. Extension of the bayesian alphabet for genomic selection. *BMC Bioinform.* **2011**, *12*, 186. [CrossRef]

45. De Los Campos, G.; Naya, H.; Gianola, D.; Crossa, J.; Legarra, A.; Manfredi, E.; Weigel, K.; Cotes, J.M. Predicting quantitative traits with regression models for dense molecular markers and pedigree. *Genetics* **2009**, *182*, 375–385. [CrossRef]

46. Zhou, X.; Carbonetto, P.; Stephens, M. Polygenic Modeling with Bayesian Sparse Linear Mixed Models. *PLoS Genet.* **2013**, *9*, e1003264. [CrossRef] [PubMed]

47. Gianola, D.; Fernando, R.L.; Stella, A. Genomic-assisted prediction of genetic value with semiparametric procedures. *Genetics* **2006**, *173*, 1761–1776. [CrossRef] [PubMed]

48. Gianola, D.; de los Campos, G.; González-Recio, O.; Long, N.; Okut, H.; Rosa, G.J.M.; Weigel, K.A.; Wu, X.-L. Statistical learning methods for genome-based analysis of quantitative traits. In Proceedings of the 9th World Congress on Genetics Applied to Livestock Production, Leipzig, Germany, 1–6 August 2010; Gesellschaft für Tierzuchtwissenschaften eV: Neustadt am Rübenberge, Germany, 2010; Volume 14, pp. 1–6.

49. Howard, R.; Carriquiry, A.L.; Beavis, W.D. Parametric and nonparametric statistical methods for genomic selection of traits with additive and epistatic genetic architectures. *G3 Genes Genomes Genet.* **2014**, *4*, 1027–1046. [CrossRef] [PubMed]

50. De los Campos, G.; Hickey, J.M.; Pong-Wong, R.; Daetwyler, H.D.; Calus, M.P.L. Whole-genome regression and prediction methods applied to plant and animal breeding. *Genetics* **2013**, *193*, 327–345. [CrossRef] [PubMed]

51. Heslot, N.; Yang, H.P.; Sorrells, M.E.; Jannink, J.L. Genomic selection in plant breeding: A comparison of models. *Crop Sci.* **2012**, *52*, 146–160. [CrossRef]

52. Maltecca, C.; Parker, K.L.; Cassady, J.P.; Maltecca, C.; Parker, K.L.; Cassady, J.P. Application of multiple shrinkage methods to genomic predictions. *J. Anim. Sci.* **2014**, *90*, 1777–1787. [CrossRef]

53. E Sousa, M.B.; Cuevas, J.; de Oliveira Couto, E.G.; Pérez-Rodríguez, P.; Jarquín, D.; Fritsche-Neto, R.; Burgueno, J.; Crossa, J. Genomic-enabled prediction in maize using Kernel models with genotype × environment interaction. *G3 Genes Genomes Genet.* **2017**, *7*, 1995–2014.

54. Cuevas, J.; Granato, I.; Fritsche-Neto, R.; Montesinos-Lopez, O.A.; Burgueño, J.; e Sousa, M.B.; Crossa, J. Genomic-enabled prediction Kernel models with random intercepts for multi-environment trials. *G3 Genes Genomes Genet.* **2018**, *8*, 1347–1365. [CrossRef]

55. Bellot, P.; de los Campos, G.; Pérez-Enciso, M. Can deep learning improve genomic prediction of complex human traits? *Genetics* **2018**, *210*, 809–819. [CrossRef]

56. Montesinos-López, O.A.; Martín-Vallejo, J.; Crossa, J.; Gianola, D.; Hernández-Suárez, C.M.; Montesinos-López, A.; Juliana, P.; Singh, R. A benchmarking between deep learning, support vector machine and Bayesian threshold best linear unbiased prediction for predicting ordinal traits in plant breeding. *G3 Genes Genomes Genet.* **2019**, *9*, 601–618. [CrossRef]

57. Crossa, J.; Beyene, Y.; Semagn, K.; Pérez, P.; Hickey, J.M.; Chen, C.; de los Campos, G.; Burgueno, J.; Windhausen, V.S.; Buckler, E.; et al. Genomic prediction in maize breeding populations with genotyping-by-sequencing. *G3 Genes Genomes Genet.* **2013**, *3*, 1903–1926. [CrossRef] [PubMed]

58. Pérez-Rodríguez, P.; Gianola, D.; González-Camacho, J.M.; Crossa, J.; Manès, Y.; Dreisigacker, S. Comparison between linear and non-parametric regression models for genome-enabled prediction in wheat. *G3 Genes Genomes Genet.* **2012**, *2*, 1595–1605. [CrossRef] [PubMed]

59. De Andrade, L.R.B.; Bandeira e Sousa, M.; Oliveira, E.J.; de Resende, M.D.V.; Azevedo, C.F. Cassava yield traits predicted by genomic selection methods. *PLoS ONE* **2019**, *14*, e0224920. [CrossRef] [PubMed]

60. Hickey, J.M.; Chiurugwi, T.; Mackay, I.; Powell, W. Genomic prediction unifies animal and plant breeding programs to form platforms for biological discovery. *Nat. Genet.* **2017**, *49*, 1297–1303. [CrossRef] [PubMed]

61. Akdemir, D.; Isidro-sánchez, J. Design of training populations for selective phenotyping in genomic prediction. *Sci. Rep.* **2019**, *9*, 1–15. [CrossRef]

62. Gosal, S.S.; Wani, S.H. (Eds.) *Accelerated Plant Breeding, Volume 1: Cereal Crops*; Springer Nature; Springer International Publishing: Cham, Switzerland, 2020; Volume 1.

63. Sarinelli, J.M.; Murphy, J.P.; Tyagi, P.; Holland, J.B.; Johnson, J.W.; Mergoum, M.; Mason, R.E.; Babar, A.; Harrison, S.; Sutton, R.; et al. Training population selection and use of fixed effects to optimize genomic predictions in a historical USA winter wheat panel. *Theor. Appl. Genet.* **2019**, *132*, 1247–1261. [CrossRef]

64. Cericola, F.; Jahoor, A.; Orabi, J.; Andersen, J.R.; Janss, L.L.; Jensen, J. Optimizing training population size and genotyping strategy for genomic prediction using association study results and pedigree information. A case of study in advanced wheat breeding lines. *PLoS ONE* **2017**, *12*, e0169606. [CrossRef]

65. Isidro, J.; Jannink, J.L.; Akdemir, D.; Poland, J.; Heslot, N.; Sorrells, M.E. Training set optimization under population structure in genomic selection. *Theor. Appl. Genet.* **2015**, *128*, 145–158. [CrossRef]

66. Andres, R.J.; Dunne, J.C.; Samayoa, L.F.; Holland, J.B. *Enhancing Crop Breeding Using Population Genomics Approaches*; Springer: Berlin/Heidelberg, Germany, 2020.

67. Sitonik, C.; Suresh, L.M.; Beyene, Y.; Olsen, M.S.; Makumbi, D.; Oliver, K.; Das, B.; Bright, J.M.; Mugo, S.; Crossa, J.; et al. Genetic architecture of maize chlorotic mottle virus and maize lethal necrosis through GWAS, linkage analysis and genomic prediction in tropical maize germplasm. *Theor. Appl Genet.* **2019**, *132*, 2381–2399. [CrossRef]

68. Nyaga, C.; Gowda, M.; Beyene, Y.; Muriithi, W.T.; Makumbi, D.; Olsen, M.S.; Suresh, L.M.; Bright, J.M.; Das, B.; Prasanna, B.M. Genome-wide analyses and prediction of resistance to mln in large tropical maize germplasm. *Genes* **2020**, *11*, 16. [CrossRef]

69. Gowda, M.; Das, B.; Makumbi, D.; Babu, R.; Semagn, K.; Mahuku, G.; Olsen, M.S.; Bright, J.M.; Beyene, Y.; Prasanna, B.M. Genome-wide association and genomic prediction of resistance to maize lethal necrosis disease in tropical maize germplasm. *Theor. Appl Genet.* **2015**, *128*, 1957–1968. [CrossRef] [PubMed]

70. Galli, G.; Lyra, D.H.; Alves, F.C.; Granato, Í.S.C.; e Sousa, M.B.; Fritsche-Neto, R. Impact of phenotypic correction method and missing phenotypic data on genomic prediction of maize hybrids. *Crop Sci.* **2018**, *58*, 1481–1491. [CrossRef]

71. Molenaar, H.; Boehm, R.; Piepho, H.P. Phenotypic selection in ornamental breeding: It's better to have the BLUPs than to have the BLUEs. *Front. Plant Sci.* **2018**, *871*, 1–14. [CrossRef] [PubMed]

72. Piepho, H.P.; Möhring, J.; Melchinger, A.E.; Büchse, A. BLUP for phenotypic selection in plant breeding and variety testing. *Euphytica* **2008**, *161*, 209–228. [CrossRef]

73. Piepho, H.P.; Möhring, J. Selection in cultivar trials—Is it ignorable? *Crop Sci.* **2006**, *46*, 192–201. [CrossRef]

74. Dramadri, I.O.; Nkalubo, S.T.; Kelly, J.D. Identification of QTL Associated with drought tolerance in Andean common bean. *Crop Sci.* **2019**, *59*, 1007–1020. [CrossRef]

75. Williams, W.P.; Buckley, P.M.; Davis, F.M. Combining ability for resistance in corn to fall armyworm and southwestern corn borer. *Crop Sci.* **1989**, *29*, 913–915. [CrossRef]

76. Sodedji, F.A.K.; Kwemoi, D.B.; Asea, G.; Kyamanywa, S. Response of provitamin—A maize germplasm to storage weevil *Sitophilus zeamais* (Motschulsky). *Int. J. Agron. Agric. Res.* **2016**, *9*, 1–13.

77. Kasozi, L.C.; Derera, J.; Tongoona, P.; Zziwa, S.; Foundation, M.G.; Box, P.O. Comparing the effectiveness of the "weevil warehouse" and "laboratory bioassay" as techniques for screening maize genotypes for weevil resistance. *J. Food Secur.* **2018**, *6*, 170–177.

78. Bradbury, P.J.; Zhang, Z.; Kroon, D.E.; Casstevens, T.M.; Ramdoss, Y.; Buckler, E.S. TASSEL: Software for association mapping of complex traits in diverse samples. *Bioinformatics* **2007**, *23*, 2633–2635. [CrossRef]

79. R Development Core Team. *R: A Language and Environment for Statistical Computing*; R Development Core Team, Ed.; R Foundation for Statistical Computing: Vienna, Austria, 2011; Volume 1, p. 409.

80. Bates, D.M.; Maechler, M.; Bolker, B.; Walker, S. Fitting linear mixed-effects models using lme4. *J. Stat. Softw.* **2015**, *67*, 1–48. [CrossRef]

81. De los Campos, G.; Pérez, P.; Vazquez, A.I.; Crossa, J. Genome-enabled prediction using the BLR (Bayesian linear regression) R-package. In *Methods in Molecular Biology*; Springer: Berlin/Heidelberg, Germany, 2013; pp. 299–320.

82. Pérez, P.; de los Campos, G. BGLR: A statistical package for whole genome regression and prediction. *Genetics* **2014**, *198*, 483–495. [CrossRef] [PubMed]

83. Pérez, P.; de los Campos, G.; Crossa, J.; Gianola, D. Genomic-enabled prediction based on molecular markers and pedigree using the bayesian linear regression package in R. *Plant Genome J.* **2010**, *3*, 106. [CrossRef]

84. Park, T.; Casella, G. The Bayesian Lasso. *J. Am. Stat. Assoc.* **2008**, *103*, 681–686. [CrossRef]

35. Covarrubias-Pazaran, G. Genome-assisted prediction of quantitative traits using the R package sommer. *PLoS ONE* **2016**, *11*, e0156744. [CrossRef]

36. Henderson, C.R.; Henderson, A.C.R. Best linear unbiased estimation and prediction under a selection model published by: International biometric society stable. *Biometrics* **1975**, *31*, 423–447. [CrossRef]

37. Kang, H.M.; Zaitlen, N.A.; Wade, C.M.; Kirby, A.; Heckerman, D.; Daly, M.J.; Eskin, E. Efficient control of population structure in model organism association mapping. *Genetics* **2008**, *178*, 1709–1723. [CrossRef]

38. Gilmour, A.R.; Thompson, R.; Cullis, B.R. Average information REML: An efficient algorithm for variance parameter estimation in linear mixed models. *Biometrics* **1995**, *51*, 1440–1450. [CrossRef]

39. Searle, S.R. Applying the EM algorithm to calculating ML and REML estimates of variance components. In Proceedings of the American Statistical Association Meetings, San Francisco, CA, USA, 8–12 August 1993; pp. 1–9.

40. Tunnicliffe, G.W. On the use of marginal likelihood in time series model estimation. *JRSS* **1989**, *51*, 15–27.

41. Vanraden, P.M. Efficient methods to compute genomic predictions. *J. Dairy Sci.* **2008**, *91*, 4414–4423. [CrossRef]

42. Endelman, J.B. Ridge regression and other kernels for genomic selection with R package rrBLUP. *Plant Genome J.* **2011**, *4*, 250. [CrossRef]

43. Stekhoven, D.J.; Bühlmann, P. Missforest—Non-parametric missing value imputation for mixed-type data. *Bioinformatics* **2012**, *28*, 112–118. [CrossRef] [PubMed]

44. Chen, X.; Ishwaran, H. Genomics random forests for genomic data analysis. *Genomics* **2012**, *99*, 323–329. [CrossRef] [PubMed]

45. Romay, M.C.; Millard, M.J.; Glaubitz, J.C.; Peiffer, J.A.; Swarts, K.L.; Casstevens, T.M.; Elshire, R.J.; Acharya, C.B.; Mitchell, S.E.; Flint-Garcia, S.A.; et al. Comprehensive genotyping of the USA national maize inbred seed bank. *Genome Biol.* **2013**, *14*, R55. [CrossRef]

46. Peiffer, J.A.; Flint-Garcia, S.A.; De Leon, N.; McMullen, M.D.; Kaeppler, S.M.; Buckler, E.S. The genetic architecture of maize stalk strength. *PLoS ONE* **2013**, *8*, e67066. [CrossRef]

47. Lorenz, A.J.; Smith, K.P.; Jannink, J.L. Potential and optimization of genomic selection for Fusarium head blight resistance in six-row barley. *Crop Sci.* **2012**, *52*, 1609–1621. [CrossRef]

48. Arruda, M.P.; Brown, P.J.; Lipka, A.E.; Krill, A.M.; Thurber, C.; Kolb, F.L. Genomic selection for predicting fusarium head blight resistance in a wheat breeding program. *Plant Genome* **2015**, *8*, 1–12. [CrossRef]

49. Foiada, F.; Westermeier, P.; Kessel, B.; Ouzunova, M.; Wimmer, V.; Mayerhofer, W.; Presterl, T.; Dilger, M.; Kreps, R.; Eder, J.; et al. Improving resistance to the European corn borer: A comprehensive study in elite maize using QTL mapping and genome-wide prediction. *Theor. Appl Genet.* **2015**, *128*, 875–891. [CrossRef]

100. Riedelsheimer, C.; Technow, F.; Melchinger, A.E. Comparison of whole-genome prediction models for traits with contrasting genetic architecture in a diversity panel of maize inbred lines. *BMC Genom.* **2012**, *13*, 452. [CrossRef]

101. Azodi, C.B.; Bolger, E.; McCarren, A.; Roantree, M.; de Los Campos, G.; Shiu, S.H. Benchmarking parametric and Machine Learning models for genomic prediction of complex traits. *G3 Genes Genomes Genet.* **2019**, *9*, 3691–3702. [CrossRef]

102. Drouaillet, B.E.; Mendez, C.A.R. Combinatorial aptitude and resistance to leaf damage of *Spodoptera frugiperda* (J.E. Smith) in maize germplasm native to Tamaulipas. *Rev. Mex. Cienc. Agríc.* **2018**, *9*, 81–93.

103. Alvarez, P.; Branco, J.; Filho, D.M. Diallel crossing among miaze populations for resistance to fall armyworm. *Sci. Agric.* **2002**, *59*, 731–741. [CrossRef]

104. Viana, P.A.; Guimarães, P.E.O. Maize resistance to the lesser cornstalk borer and fall armyworm in Brazil. In *Embrapa Milho e Sorgo-Artigo em Anais de Congresso (ALICE), Proceedings of the International Symposium on Insect Resistant Maize: Recent Advances and Utilization, Mexico City, Mexico, 27 November–3 December 1994*; Centro Internacional de Mejoramiento de Maiz y Trigo (CIMMYT): Mexico City, Mexico, 1997.

105. Musundire, L.; Dari, S.; Derera, J.; Co-Zimbabwe, S.; Wgt, P.O.B. Genetic analysis of grain yield performance and weevil [*Sitophilus zeamais* (Motschulsky)] resistance in southern African maize hybrids. *Maydica* **2015**, *60*, M35.

106. Dhliwayo, T.; Pixley, K.V.; Kazembe, V. Combining ability for resistance to maize weevil among 14 southern African maize inbred lines. *Crop Sci.* **2005**, *45*, 662–667. [CrossRef]

107. Asoro, F.G.; Newell, M.A.; Beavis, W.D.; Scott, M.P.; Jannink, J.-L. Accuracy and training population design for genomic selection on quantitative traits in elite North American oats. *Plant Genome J.* **2011**, *4*, 132. [CrossRef]

108. Lenz, P.R.N.; Beaulieu, J.; Mansfield, S.D.; Clément, S.; Desponts, M.; Bousquet, J. Factors affecting the accuracy of genomic selection for growth and wood quality traits in an advanced-breeding population of black spruce (*Picea mariana*). *BMC Genom.* **2017**, *18*, 1–17. [CrossRef]

109. Edwards, S.M.K.; Buntjer, J.B.; Jackson, R.; Bentley, A.R.; Lage, J.; Byrne, E.; Burt, C.; Jack, P.; Berry, S.; Flatman, E.; et al. The effects of training population design on genomic prediction accuracy in wheat. *Theor. Appl Genet.* **2019**, *132*, 1943–1952. [CrossRef]

110. Wang, W.; Cao, X.H.; Miclău, M.; Xu, J.; Xiong, W. The promise of agriculture genomics. *Int. J. Genom.* **2017**, *2017*. [CrossRef]

111. Spindel, J.; Iwata, H. Genomic selection in rice breeding. In *Rice Genomics, Genet Breed*; Springer: Singapore, 2018; pp. 473–496.

112. Crossa, J.; Pérez, P.; Hickey, J.; Burgueño, J.; Ornella, L.; Cerón-Rojas, J.; Zhang, X.; Dreisigacker, S.; Babu, R.; Li, Y.; et al. Genomic prediction in CIMMYT maize and wheat breeding programs. *Heredity* **2014**, *112*, 48–60. [CrossRef]

113. Ou, J.H.; Liao, C.T. Training set determination for genomic selection. *Theor. Appl. Genet.* **2019**, *132*, 2781–2792. [CrossRef]

114. Mangin, B.; Rincent, R.; Rabier, C.E.; Moreau, L.; Goudemand-Dugue, E. Training set optimization of genomic prediction by means of EthAcc. *PLoS ONE* **2019**, *14*, e0205629. [CrossRef] [PubMed]

115. Albrecht, T.; Wimmer, V.; Auinger, H.J.; Erbe, M.; Knaak, C.; Ouzunova, M.; Simianer, H.; Schon, C.-C. Genome-based prediction of testcross values in maize. *Theor. Appl. Genet.* **2011**, *123*, 339–350. [CrossRef] [PubMed]
116. Hickey, J.M.; Dreisigacker, S.; Crossa, J.; Hearne, S.; Babu, R.; Prasanna, B.M.; Grondona, M.; Zambelli, A.; Windhausen, V.S.; Mathews, K.; et al. Evaluation of genomic selection training population designs and genotyping strategies in plant breeding programs using simulation. *Crop Sci.* **2014**, *54*, 1476–1488. [CrossRef]
117. Krchov, L.-M.; Bernardo, R. Relative efficiency of genomewide selection for testcross performance of doubled haploid lines in a maize breeding program. *Crop Sci.* **2015**, *55*, 2091–2099. [CrossRef]
118. Mayor, P.J.; Bernardo, R. Genomewide selection and marker-assisted recurrent selection in doubled haploid versus F2 populations. *Crop Sci.* **2009**, *49*, 1719–1725. [CrossRef]

 *plants*

Article

Genetic Diversity of Selected Rice Genotypes under Water Stress Conditions

Mahmoud M. Gaballah [1], Azza M. Metwally [2], Milan Skalicky [3], Mohamed M. Hassan [4], Marian Brestic [3,5], Ayman EL Sabagh [6,*] and Aysam M. Fayed [2]

[1] Rice Research and Training Center (RRTC), Field Crops Research Institute, Agricultural Research Center, Kafr El-Sheikh 33717, Egypt; Mahmoudgab@yahoo.com

[2] Molecular Biology Department, Genetic Engineering and Biotechnology Institute, University of Sadat City, Sadat City 32897, Egypt; azzaher2000@yahoo.com (A.M.M.); aisamge2004@yahoo.com (A.M.F.)

[3] Department of Botany and Plant Physiology, Faculty of Agrobiology, Food, and Natural Resources, Czech University of Life Sciences Prague, Kamycka 129, 165 00 Prague, Czech Republic; skalicky@af.czu.cz (M.S.); marian.brestic@uniag.sk (M.B.)

[4] Department of Biology, College of Science, Taif University, P.O. Box 11099, Taif 21944, Saudi Arabia; m.khyate@tu.edu.sa

[5] Department of Plant Physiology, Slovak University of Agriculture, Nitra, Tr. A. Hlinku 2, 949 01 Nitra, Slovakia

[6] Department of Agronomy, Faculty of Agriculture, Kafrelsheikh University, Kafr El-Sheikh 33516, Egypt

* Correspondence: ayman.elsabagh@agr.kfs.edu.eg

Citation: Gaballah, M.M.; Metwally, A.M.; Skalicky, M.; Hassan, M.M.; Brestic, M.; EL Sabagh, A.; Fayed, A.M. Genetic Diversity of Selected Rice Genotypes under Water Stress Conditions. *Plants* 2021, 10, 27. https://dx.doi.org/10.3390/plants10010027

Received: 13 November 2020
Accepted: 21 December 2020
Published: 24 December 2020

Publisher's Note: MDPI stays neutral with regard to jurisdictional claims in published maps and institutional affiliations.

Abstract: Drought is the most challenging abiotic stress for rice production in the world. Thus, developing new rice genotype tolerance to water scarcity is one of the best strategies to achieve and maximize high yield potential with water savings. The study aims to characterize 16 rice genotypes for grain and agronomic parameters under normal and drought stress conditions, and genetic differentiation, by determining specific DNA markers related to drought tolerance using Simple Sequence Repeats (SSR) markers and grouping cultivars, establishing their genetic relationship for different traits. The experiment was conducted under irrigated (normal) and water stress conditions. Mean squares due to genotype × environment interactions were highly significant for major traits. For the number of panicles/plants, the genotypes Giza179, IET1444, Hybrid1, and Hybrid2 showed the maximum mean values. The required sterility percentage values were produced by genotypes IET1444, Giza178, Hybrid2, and Giza179, while, Sakha101, Giza179, Hybrid1, and Hybrid2 achieved the highest values of grain yield/plant. The genotypes Giza178, Giza179, Hybrid1, and Hybrid2, produced maximum values for water use efficiency. The effective number of alleles per locus ranged from 1.20 alleles to 3.0 alleles with an average of 1.28 alleles, and the He values for all SSR markers used varied from 0.94 to 1.00 with an average of 0.98. The polymorphic information content (PIC) values for the SSR were varied from 0.83 to 0.99, with an average of 0.95 along with a highly significant correlation between PIC values and the number of amplified alleles detected per locus. The highest similarity coefficient between Giza181 and Giza182 (Indica type) was observed and are susceptible to drought stress. High similarity percentage between the genotypes (japonica type; Sakha104 with Sakha102 and Sakha106 (0.45), Sakha101 with Sakha102 and Sakha106 (0.40), Sakha105 with Hybrid1 (0.40), Hybrid1 with Giza178 (0.40) and GZ1368-S-5-4 with Giza181 (0.40)) was also observed, which are also susceptible to drought stress. All genotypes are grouped into two major clusters in the dendrogram at 66% similarity based on Jaccard's similarity index. The first cluster (A) was divided into two minor groups A1 and A2, in which A1 had two groups A1-1 and A1-2, containing drought-tolerant genotypes like IET1444, GZ1386-S-5-4 and Hybrid1. On the other hand, the A1-2 cluster divided into A1-2-1 containing Hybrid2 genotype and A1-2-2 containing Giza179 and Giza178 at coefficient 0.91, showing moderate tolerance to drought stress. The genotypes GZ1368-S-5-4, IET1444, Giza 178, and Giza179, could be included as appropriate materials for developing a drought-tolerant variety breeding program. Genetic diversity to grow new rice cultivars that combine drought tolerance with high grain yields is essential to maintaining food security.

Keywords: rice; drought stress; genetic diversity; SSR markers; dendrogram

Plants **2021**, *10*, 27. https://dx.doi.org/10.3390/plants10010027

1. Introduction

Rice is the most diversified crop, which is grown under diverse ecological conditions. It is the staple food for more than 50% of the world's population and is the world's most important food in terms of a natural calorie source [1]. It occupies almost one-fifth of the total land area covered under cereals [2]. Due to the changing climate, the frequent occurrence of many extreme events contributes to different abiotic stresses, limiting the productivity of rice globally. Among them, drought is one of the most critical abiotic stresses that continually threatens the world's food security [3]. The severity of the drought depends on many factors, such as the occurrence and distribution of rainfall, evaporative demands, and moisture-retaining capacity of the soils [4]. Therefore, it is imperative to find out the genotypes that can grow under water-scarce conditions to expand rice-growing areas in water-limited lands. It can be helpful to meet the challenge of ever-increasing global food demand [5]. Different rice varieties of a distinct genetic structure promise a future improvement of rice cultivars against drought stress [6]. Hence, the assessment of genetic diversity becomes important in establishing relationships among different cultivars [7]. The first step towards determining the magnitude of these risks is to evaluate the genetic diversity in improved rice genotypes as the success of a crop improvement program depends on the importance of genetic variability and how the desirable characters are heritable [8]. This identification of genotypes and their inter-relationships is important. The development of new biotechnological techniques provides increased support to evaluate genetic variation in both phenotypic and genotypic levels. The results derived from analyses of genetic diversity at the DNA level could be used for designing effective breeding programs aiming to broaden the genetic basis of commercially grown varieties.

Molecular marker technology is a powerful tool for determining genetic variation in rice varieties. In contrast to morphological traits, molecular markers can reveal large differences among genotypes at the DNA level, providing a more direct, reliable, and efficient tool for germplasm characterization, conservation, and management and untouched by environmental influence [9]. SSR markers can detect a high level of allelic diversity, and they have been extensively used to identify genetic variation among rice subspecies [10]. Simple sequence repeats (SSR) markers are efficient in detecting genetic polymorphisms and discriminating among genotypes from germplasms of various sources; even they can notice the finer level of variation among closely related breeding materials within the same variety [11]. Several quantitative trait loci (QTLs) in rice with consistent effects on grain yield under water-limited conditions were reported [12]. Among them, DTY1.1, located in chromosome 1 of the rice genome, was identified from rice variety N22 and successfully transferred to susceptible genotypes line IR64 and MTU1010 [13]. Besides this, two other major-effect QTLs viz. DTY3.1 and DTY2.1 were also identified, which explains about 30% and 15% of the phenotypic variance, respectively [14]. Later, Shamshudin et al. [15] reported another two QTLs, DTY2.2 ad DTY12.1, for reproductive stage drought tolerance in rice. However, as mentioned earlier, all of these QTLs are derived from stable grain yield under drought conditions. Because of the low heritability of grain yield under drought stress, selection for secondary traits was more effective than grain yield traits. Due to the lack of a useful trait selection index related to drought tolerance, it is essential to find a molecular marker associated with drought tolerance in rice [16]. Many SSR markers have been reported to be linked to drought tolerance traits in rice, such as yield under drought [13,17,18]. Although several investigations have researched rice germplasm characterization and diversity analysis, variability studies of the common landraces and cultivars grown are limited. Therefore, the study was conducted based on three main goals; (i) morphological characterization of the genotypes for grain and agronomic parameters under normal and drought stress conditions; (ii) genetic differentiation of 16 rice cultivars by determination of specific DNA markers related to drought tolerance using SSR markers and; (iii) grouping of cultivars according to their genotypes and subsequent decision their genetic relationship for different traits.

2. Materials and Methods

2.1. Experiment Site

The field experiment was conducted in Rice Research and Training Center Farm, Sakha Research Station, Agricultural Research Center, Egypt, in consecutive two rice growing seasons during 2018 and 2019 to investigate morphological traits and genetic diversity of 16 rice genotypes. The soil's physical and chemical properties in the Sakha Research Station in 2018 and 2019 years are illustrated in Table 1.

Table 1. Physical and chemical properties of soil at Sakha Research Station in 2018 and 2019 years (the soil was collected before starting the field preparation in each season).

Soil Physical and Chemical Properties	Sakha, Kafr El-Sheikh	
	2018	**2019**
Clay %	55	55
Silt %	32.4	32.4
Sand %	12.6	12.6
Texture	Clayey	Clayey
Organic Matter	1.39	1.39
pH	8.1	8.2
Ec (Ds/m)	3.30	3.33
Total N (ppm)	512	518
Available P (ppm)	15.09	16.03
Co_3^-	-	-
Hco_3^-	5.55	5.56
Mg^{++}	4.3	5
Na^+	1.88	1.69
K^+	16	16
Fe^{++}	4.55	4.55
Mn^{++}	3.1	3.5

2.2. Treatments and Design

The sixteen rice genotypes origin, pedigree, salience, and feature are shown in Table 2. All selected rice materials were grown under full irrigated (normal) and water stress conditions (flush irrigation every twelve days and exposed after fifteen days from transplanting) in a randomized complete block design with three replications.

2.3. Experimental Procedures

Seeds of all cultivars were sown in a nursery on 5 May and transplanted into the main field after 30 days in both years (2018 and 2019). A single seedling of each genotype was transplanted in 5 rows having 20 by 20 cm space (between rows and within row distance). Data were recorded from 10 randomly selected plants from each genotype. For characterization of root structure, large iron cylinders of 20 cm diameters and 60 cm height were used. They were buried inside the soil with a hammer, dug out with a spade, and pulled out using hooks. The roots were separated from the soil by thorough washing in a special washing facility. After taking the quantitative data, the shoot was separated from the root using a sharp knife and dried in an oven at 70 °C for five days. Root length (cm) was measured by the length of the root from the base of the plant to the tip of the longest root, root volume was determined by measuring the volume of water displaced by the plant root system (mm^3), root thickness, the average diameter (mM) of the tip portion (about 1 cm from the tip) of three random secondary roots at the middle position of the root/plant, the number of roots/plant were estimated by the account roots at the maximum tillering stage and root: shoot ratio the ratio of the root dry weight (g) to the shoot dry weight (g) at maximum tillering stage was measured. Days to heading was recorded after flowering by the daily count of panicle exertion. The physiological maturity dates were recorded when 80% grains turn into golden yellow color. The leaf rolling scores were estimated by visual estimation, and the susceptible varieties and lines first started the rolling symptoms in the morning. Highly sensitive lines did not unroll at early morning hours and were recorded based on methods proposed by (De Data et al. 1988). The flag leaf area (cm^2) was measured using a leaf area meter (LI-3100 (LI-COR Inc., Lincoln, NE, USA), plant height (cm) was measured in (cm), from the soil surface to the tip of the tallest panicle of each plant, relative water content (%) was measured using the formula (Fw − Dw) × 100/(Tw − Dw) where, Fw is fresh leaf weight, Dw is leaf dry weight, and Tw is turgid leaf weight. A number of panicles/plant were recorded at harvest by counting the number of panicles/plant, 100-grain weight (g) was recorded as the weight of 100 random chosen filled grains/plant, sterility percentage (%) was calculated by the divided number of unfilled spikelets/panicle on a number of total spikelets/panicle, grain yield/plant (g) was recorded by collecting the filled grains from all the tillers in a single plant and their weight recorded, water use efficiency was calculated as economic yield/total water consumed during the crop growth period.

All cultural practices were applied as recommended. Nitrogenous fertilizer was used in three splits as top dressing; phosphorus and potash were applied in full dose at sowing. Insect and weed control was used as and when required.

2.4. Statistical Analysis

A combined analysis of variance for the two years was carried out for the yield and yield components. Phenotypic correlation between yield and yield-related traits was done following Steel et al. [19], and the data were analyzed using the Co-State software program.

2.5. Genomic DNA Extraction

Ten seeds of each advanced genotype were placed into a Petri dish with filter paper soaked in distilled water for germination under aseptic conditions. Then, germinated seeds were grown into labeled pots. Genomic DNA was extracted from the healthy portion of young leaves harvested from 21 days old seedlings. DNA isolation was carried out using a mini preparation modified CTAB (cetyltrimethylammonium bromide) method, which did not require liquid nitrogen, and only a minimal amount of tissue samples were needed [20]. Leaf tissues were cut into small pieces, homogenized, and digested with extraction buffer (1 M Tris, 0.5 M Na 2EDTA, 5 M NaCl, and distilled Hybrid 2O, pH 8.0) and 20% SDS. Following incubation of leaf extracts for 10 min at 65 °C in a water-bath, 100 µL of 5M NaCl was added and mixed well by gentle inversion. Then 100 µL 10 × CTAB was added and again incubated for 10 min at 65 °C in a water-bath. After that, 900 µL of a mixture of chloroform and isoamyl alcohol (24:1) was added and centrifuged for 8 min at 11,000 rpm in a microcentrifuge. Then, 500 µL of the upper aqueous layer was separated, and 600 µL of ice-cold isopropanol was added to it, mixed, and centrifuged for 12 min at 13,200 rpm. A small pellet was visible, and the supernatant was decanted. The pellet was then washed with 200 µL cold 70% ethanol and centrifuged at 13,200 rpm for 12 min. After removing ethanol followed by air drying, the DNA pellets were re-suspended into 100 µL of 1× TE buffer and dissolved the pellet by warming in a 65 °C water bath for up to 1 h (with frequent mixing or flicking the tube with finger). Then the pellet was stored at −20 °C in an ultra-freezer. The quality of DNA was estimated by agarose gel (0.8%) electrophoresis and visualized with UV light.

Table 2. Origin, pedigree, salience, and feature of sixteen rice genotypes.

Genotype	Pedigree	Salience and Feature
Giza 177	Giza 171/Yomjo No. 1//PiNo.4	Japonica type—sensitive to drought—short stature—early duration—resistance to blast
Giza 178	Giza175/Milyang 49	Indica/Japonica type, medium maturing, semi-dwarf, resistant to blast, medium grain, tolerant to drought, and high yield
Giza179	GZ6296/GZ1368	Indica/Japonica type—moderate to drought—short stature—early duration—resistant to blast
Giza181	(lR1626-203)	Indica type—sensitive to drought—short stature—late duration—resistance to blast
Giza182	Giza181/IR39422-161-1-3-1/Giza181	Indica type—sensitive to drought—short stature—early duration—resistant to blast
Sakha101	Giza 176/Milyang	Japonica type—sensitive to drought—short stature—long duration—sensitive to blast
Sakha 102	GZ4096-7-1/GZ4120-2-5-2 (Giza 177)	Japonica type—sensitive to drought—short stature, early duration—resistance to blast
Sakha 103	Giza177/Suweon349	Japonica type—sensitive to drought—short stature—early duration—resistance to blast
Sakha 104	GZ4096-8-1/GZ4100-9-1	Japonica type—sensitive to drought—long stature, moderate duration—sensitive to blast
Sakha 105	GZ5581-46-3/GZ4316-7-1-1	Japonica type—sensitive to drought—short stature—early duration—resistant to blast
Sakha 106	Giza176/Milyang79	Japonica type—sensitive to drought—long stature—early duration—resistant to blast
Hybrid 1	IR6962SA/Giza178	Indica type—moderate to drought—short stature—moderate
Hybrid 2	IR6962SA/Giza179	Indica type—moderate to drought—short stature—moderate duration—resistant to blast
E. Yasmine	Introduction	Indica type—moderate to drought—short stature late duration—resistant to blast
GZ 1368-S-5-4	IR 1615-31/BG 94-2349	Indica type—moderate to drought—short stature moderate duration—resistant to blast
IET1444	TN 1/CO 29	Indica type—tolerant to drought—short stature moderate duration—resistant to blast

2.6. SSR Markers and PCR Amplification

Ten SSR markers related to drought tolerance traits/QTLs were used. The sequences of primer pairs are found on the Web database (http://www.gramene.org). Primers' names, repeat motifs, chromosome number, and related trait/QTL are shown in Table 3. PCR amplification reactions were done in 10 µL reaction mixtures, containing 50 ng/µL of template DNA, 0.5 µL of each forward and reverse primer, 5 µL of PCR master mix (Ferments), and 3 µL dd H_2O. Thermal cycler was used with the following PCR profile: an initial denaturation step at 94 °C for 5 min, followed by 35 cycles of denaturation at 94 °C for 1 min, annealing at 55 °C for 30 s, and primer elongation at 72 °C for 1 min and then a final extension at 72 °C for 5 min. Amplified products were stored at −20 °C until further use.

Table 3. Genetic details of SSR markers were used in the study.

Primer	Chromo-Some Number	Repeat Motif	Allele Size Range	No of Amplified Alleles	Effective Number of Alleles	Common Alleles	Gene Diversity	Polymorphic Information Content	Related Traits or QTLs	References
RM23	1	(GA)15	76.109—172.029	13.00	2.41	0.18	1.00	0.99	RSR, PH, DTH	[21]
RM164	5	(GT)16TT (GT)4	222.307—255.184	8.00	2.18	0.27	0.99	0.96	FGN, TSW, DTH	[22]
RM223	8	G11(GA)16	149.686—206.429	9.00	2.25	0.25	0.99	0.98	PH	[21]
RM242	9	(CT)25	127.35—130.068	7.00	2.63	0.37	0.98	0.95	RT, RL, SW	[21]
RM246	1	(CT)26	205.964—272.973	4.00	2.00	0.5	0.97	0.91	RL, SN, SSP	[21]
RM263	2	(CT)20	334.846—353.507	8.00	3.00	0.37	0.99	0.97	PH, RDW, RFW	[23]
RM272	1	(CT)34	144.029—146.487	2.00	1.20	0.6	0.94	0.83	DTH, SP, HI,	[24]
RM276	6	(GA)9	127.065—316.725	10.00	2.00	0.2	0.99	0.98	PL, LL, LW, SN	[24]
RM289	5	(TC)15	101.834—144.426	9.00	2.81	0.31	0.99	0.97	RV, PL	[24]
RM518	4	(AG)8A3 (GA)33	125.898—277.807	15.00	2.37	0.16	1.00	0.99	DTM, PH	[25]

RSR (root to shoot ratio), PH (Plant height), DTH (Days to heading), FGN (Filled grain number), TSW (1000-seed weight), RT (Root Thickness), RL (Root Length), SW (Seed weight), SN (Seed number), SSP (seed sterility percentage), RDW (root dry weight, RFW (root fresh weight), SP (Seed Percentage), HI (Harvest index), PL (panicle length), LL (Leaf Length), LW (leaf width), SN (Spikelet number), RV (Root volume), DTM (days to maturity).

2.7. Electrophoretic Separation and Visualization of Amplified Products

Five µL of PCR amplified product was loaded into each well of 3% agarose gel supplemented with ethidium bromide. The TAE 1× was used as a running buffer, and a 50 bp DNA ladder (0.5 µg/µL, ferments) was used to estimate the molecular size of the amplified fragments. Electrophoresis was conducted at 60 Volts for 2 h. Gels were then visualized and photographed using a Biometra gel documentation unit (Bio-Doc, Biometra, Germany).

2.8. SSR Data Analysis

The amplified SSR DNA bands representing different alleles were scored as different genotypes. For each marker, allelic bands were compared against a 100 bp DNA ladder. Then, fragment data was converted into the binary encoded allelic data to apply the multivariate analyses. Genetic distance, the ratios of shared DNA bands, and genetic similarities were estimated from the allele binary formatted data set using Nei and Li's coefficient [26]. Genetic distance was calculated as follows:

$$GDn = 1 - [2N11/(2N11 + N10 + N01)]$$

where N (1,1) is the number of loci having bands present in both accession, N (1,-) is the number of loci having a band present in the first accession, N (-, 1) is the number of loci having a band present in the second accession.

The accessions were clustered based on the matrix of genetic similarities using the unweighted pair group method with arithmetic averages (UPGMA). Polymorphic information content (PIC) values were calculated for each microsatellite based on the allelic frequency detected in the accessions studied using this formula.

Where, Pij is the frequency of the j-th allele for the i-th marker, and summation extends over n alleles. Polymorphic loci were defined as those whose most frequent allele had a frequency of less than 0.95.

Genetic diversity of the entries/populations (based on a set of measured molecular data) was estimated using diversity parameters other than PIC [27]. These are calculated as follows: percentage of polymorphic loci (PPL):

$$P = (k/n) \times 100\%$$

where k is the number of polymorphic loci, n is the total number of loci investigated. The average number of alleles per locus (A):

$$A = \Sigma\, Ai/n$$

where Ai is the number of alleles at the i-th locus and n is the total number of loci investigated. The average number of alleles per polymorphic locus (Ap):

$$Ap = \Sigma Api/np$$

where, Api is the number of alleles at a certain polymorphic locus, np is the total number of polymorphic loci investigated.

Percentage of polymorphic alleles (PPA)

$$PPA = (\Sigma\, Api/\Sigma\, Ai) \times 100\%$$

The similarity matrix using [26] genetic distance for SSR characterization was also used for principal coordinate analysis (PCoA) with the Dcenter, Eigen, Output, and Mxplot subprograms in NTSYS-PC.

3. Results and Discussion

The analyses of variance in Table 4; Table 5 showed that the mean squares due to years were significant for major studied traits, except for days to heading, which would indicate overall vast differences among the genotypes studied annually. Abdallah et al. [28] observed the ordinary analysis of variance showed highly significant differences among environments, genotypes, and environments × genotype interaction for root and shoot traits in both treatments (normal and drought). The variation due to interaction between year and variety was not significant for all measures [29]. Mean squares due to environments were highly significant for all traits studied, indicating that all environments showed significant differences. Mean squares due to genotype × environment interactions were highly significant for all traits except, root thickness, flag leaf area, plant height, relative water content, number of panicles/plant, 100-grain weight, sterility percentage, and grain yield/plant, which indicated that the tested genotypes varied from the environment to environment and ranked differently from the normal condition. Raman et al. [30] recorded that the variance analysis for grain yield indicated a highly significant genotype x degree of stress severity interaction. Mean squares due to genotype × year interactions were significant for all traits studied, except root length, root volume, root thickness, number of roots/plants, root: shoot ratio, and leaf rolling.

Some genotypes surpassed the others once the mean squares of genotypes were highly significant than the interaction G × Y mean squares and identified the most superior genotypes. Genotype × environment × year mean squares were not significant for all the studied traits, except, days to heading and leaf rolling, indicating that each genotype's performance in one environment will be changed from one year to another. The significant differences among rice genotypes in this investigation revealed genetic variability in the studied material and provided an excellent yield improvement opportunity. Grain yield and other characteristics exhibited stability across the seasons as significant genotype × environment interaction, which indicated the differences among genotypes were apparent (Table 4; Table 5). This research shows that further improvement through the selection of all studied characteristics could be effective. Genotype characteristics that confer an advantage in some water stress environments may prove useless or may even be a liability in other environments. This is reflected in the large G × E interactions in drought trials and the difficulty of identifying drought-tolerant check cultivars Zhang et al. [31].

3.1. Performance Across Environments

The ordinary analysis of variance indicated highly significant differences among genotypes for all traits studied in Table 4; Table 5 in the combined data. The studied genotypes' mean performances at the combined data over environments are presented in Tables 6–8. For root length, the genotypes showing high values were IET1444, Sakha101, Sakha106, and Hybrid 2 (26.39, 25.88, 25.73, and 25.01cm, respectively), while the lowest values were obtained from Sakha102, Giza178, Sakha104, and Sakha103 (21.94, 21.42, 19.83 and 18.71cm, respectively). The genotypes Giza178, Hybrid 1, IET1444, and Sakha101 gave the superior values for root volume, 64.06, 57.66, 54.84, and 52.28 mm^3, respectively; otherwise, the genotypes Giza177, Giza182, Sakha105, and Sakha103 gave the lowest one 103, 35.00, 29.88, 27.27, and 16.50 mm^3, respectively. Concerning the root thickness, the genotypes E. Yasmine, Hybrid 1, IET1444, and Hybrid 2 had increased values of 1.08, 1.08, 1.07, and 0.97 mM. On the other hand, Giza178, Sakha102, Sakha104, and Giza182 resulted in the decreased values of 0.58, 0.56, 0.46, and 0.41 mM, respectively (Table 6).

Table 4. Analysis of variance for growth characteristics under normal and drought conditions.

Source of Variance	df	Root Length (cm)	Root Volume (mm^3)	Root Thickness (mm)	Number of Roots/Plant	Root:Shoot Ratio	Days to Heading (Day)	Leaf Rolling
Blocks	2	53.5051 **	2.584	0.0051	23.507	0.0157	1.02	1.51 **
Year (Y)	1	63.19 **	178.33 **	0.055 *	4555.10 **	0.0845 **	11.02 **	0.09
Env. (E)	1	842.23 **	15528.61 **	4.931 **	677517.6**	16.96 **	927.52 **	487.82 **
Gen. (G)	15	56.27 **	1801.26 **	0.465 **	22490.53 **	0.702 **	859.53 **	6.78 **
Y × E	1	1.256	15.019 *	0.0016	407.19 **	0.030	0.001	0.34
Y × G	15	0.033	1.2263	0.0008	14.055	0.007	4.75 **	0.34
E × G	15	15.06 **	178.02 **	0.0131	13297.93 **	0.384 **	4.30 **	4.68 **
Y × E × G	15	0.23	0.508	0.0002	7.30	0.004	1.91 **	0.46 *
Error	126	0.577	2.9094	0.00138	19.2036	0.0001	0.70 **	0.26
Total	191							

*, ** significant and high significant at probability 0.05 and 0.01, respectively. Env. (Environment), Gen. (Genotypes).

Table 5. Analysis of variance for grain yield and related traits under normal and drought in 2018 and 2019 rice growing seasons.

Source of Variance	df	Flag Leaf Area (cm^2)	Plant Height (cm)	Relative Water Content (%)	Number of Panicles/Plants	100-Grain Weight (g)	Sterility Percentage %	Grain Yield/Plant (g)	Water Use Efficiency
Blocks	2	1.04	4.48	1.138	9.48 **	0.19 **	2.83	0.07	0.00116
Year (Y)	1	1.51	6.27	314.75 **	5.27 *	0.31 **	5.60 *	3.69	0.0013
Env. (E)	1	1113.6 **	34978.5 **	18680.1 **	1566.37 **	2.34 **	6394.08 **	14,121 **	0.004 **
Variety (V)	15	84.41 **	369.10 **	2834.5 **	75.30 **	1.25 **	57.32 **	185.97 **	0.091 **
Y × E	1	3.52 **	0.01 **	11.746 *	2.80	0.00 **	1.84	0.04	0.00004
Y × V	15	5.24 **	4.42*	2.533	2.65 **	0.001	2.97 **	4.48 **	0.003 **
E × V	15	6.49 **	185.62 **	248.41 **	13.45 **	0.10 **	90.42 **	61.88 **	0.024 **
Y × E × V	15	1.57	1.99	0.6180	0.97	0.001	1.33	1.71	0.001 *
Error	126	0.94	2.50	2.0424	0.77	0.00	1.12	1.13	0.0006
Total	191								

*, ** significant and high significant at probability 0.05 and 0.01, respectively.

Table 6. Performance of Growth characteristics under normal and drought conditions and their combined data.

Variety	Root Length (cm)			Root Volume (mm^3)			Root Thickness (mm)			Number of Roots/Plant			Root:Shoot Ratio		
	d	n	C	d	n	C	d	n	C	d	n	C	d	n	C
Giza177	20.50	24.60	22.55	29.01	41.00	35.00	0.51	0.92	0.72	95.84	290.38	193.11	0.41	1.08	0.74
Giza178	19.27	23.58	21.42	51.25	76.88	64.06	0.44	0.72	0.58	197.83	281.57	239.70	0.62	1.56	1.09
Giza179	22.35	25.63	23.99	26.65	47.50	37.08	0.62	0.92	0.77	160.41	265.48	212.94	0.78	1.48	1.13
Giza181	20.30	26.65	23.47	26.96	46.13	36.54	0.41	0.82	0.62	117.88	153.75	135.81	0.60	1.85	1.22
Giza182	20.71	24.09	22.40	23.88	35.88	29.88	0.21	0.62	0.41	138.38	180.40	159.39	0.26	1.33	0.79
Sakha101	23.06	28.70	25.88	37.93	66.63	52.28	0.51	0.92	0.72	185.53	359.78	272.65	0.54	0.59	0.57
Sakha102	19.27	24.60	21.94	27.37	43.56	35.47	0.41	0.72	0.56	133.76	299.61	216.69	0.49	0.99	0.74
Sakha103	16.91	20.50	18.71	9.94	23.06	16.50	0.51	0.80	0.66	133.25	205.00	169.13	0.31	0.53	0.42
Sakha104	19.68	19.99	19.83	34.13	51.25	42.69	0.31	0.62	0.46	174.25	359.47	266.86	0.47	0.94	0.71
Sakha105	18.35	27.16	22.76	21.22	33.31	27.27	0.51	0.92	0.72	78.41	302.38	190.39	0.45	0.66	0.55
Sakha106	23.78	27.68	25.73	41.00	51.25	46.13	0.51	0.92	0.72	184.50	311.91	248.20	0.38	0.93	0.66
E. Yasmine	23.58	26.14	24.86	40.69	61.50	51.10	0.92	1.23	1.08	115.00	164.37	157.18	0.45	0.51	0.48
Hybrid 1	20.19	26.65	23.42	51.25	64.06	57.66	0.92	1.23	1.08	127.50	231.96	179.73	0.38	1.40	0.89
Hybrid 2	21.83	28.19	25.01	23.88	61.50	42.69	0.82	1.13	0.97	126.59	241.59	184.09	0.48	0.98	0.73
GZ1368-S-5-4	21.53	22.55	22.04	37.62	48.69	43.15	0.51	0.92	0.72	183.99	276.03	230.01	0.50	0.99	0.75
IET1444	26.14	26.65	26.39	46.13	63.55	54.84	0.91	1.23	1.07	175.28	365.00	270.14	0.72	1.51	1.11
LSD0.05	1.18	1.23	0.61	2.61	3.37	1.38	0.16	0.21	0.1	7.7	7.71	3.54	0.025	0.25	0.09
LSD0.01	1.71	1.78	0.86	3.78	4.88	1.95	0.23	0.30	0.14	11.15	11.16	5.01	0.04	0.36	0.13

d, n, and c are drought, normal, and combined data, respectively.

Regarding the number of roots/plant, Sakha101, IET1444, Sakha104, Sakha106 produced the greatest number 272.65, 270.14, 266.86, and 248.20, respectively; meanwhile, the lowest number was found to be with Sakha103, Giza182, E. Yasmine, Giza181, recorded 169.13, 159.39, 157.18, and 135.81, respectively. The genotypes Giza181, Giza179, IET1444, and Giza178 recorded the major values 1.22, 1.13, 1.11, and 1.09 for Sakha103, 0.57, 0.55, 0.48, and 0.42, respectively. The genotype Giza178 produced the highest values for root volume and number of roots/plant. Hybrid 1 has superior values for the root volume. Root thickness and IET1444 have major root volume and root: shoot ratio under drought stress compared to other genotypes, indicating these genotypes can avoid water stress and increase the ability to absorb water from the soil. Moreover, efforts to increase yield under drought conditions also focused on improving secondary traits such as root architecture (root length, root volume, root thickness, number of roots, and root: shoot ratio). Rice genotypes that can maintain water status through adapted root systems come under the drought avoidance mechanism category. These genotypes can minimize the yield losses caused by drought [32]. Rice genotypes that avoid drought usually have deep, coarse roots with a high ability for branching and soil penetration and a higher root to shoot ratio [33]. Gaballah [34] reported the rice genotypes Moroberekan, Giza 178, and Sakha104 had the highest values for root characters under water shortage. Abdallah et al. [16] found the genotypes GZ5121-5-2 and GZ1368-S-5-4 had thicker roots, higher root diameter, and higher root length density than those grown under normal conditions.

Table 7. Growth characteristics of genotype under normal, drought conditions, and their combined data.

Variety	Days to Heading (Days)			Leaf Rolling			Flag Leaf Area (cm^2)			Plant Height (cm)			Relative Water Content (%)		
	d	n	c	d	n	c	d	n	c	d	n	c	d	n	C
Giza 177	92.33	96.00	94.17	5.25	1.17	3.21	17.10	21.25	19.18	74.85	99.42	87.13	41.97	80.50	61.24
Giza 178	104.17	108.00	106.08	3.33	1.33	2.33	15.00	19.42	17.21	79.37	106.92	93.14	78.47	85.10	81.79
Giza 179	92.67	96.33	94.50	3.33	1.00	2.17	15.00	23.23	19.12	74.20	97.23	85.72	82.41	85.10	83.76
Giza 181	111.17	117.00	114.08	5.83	1.50	3.67	19.67	23.33	21.50	82.83	98.75	90.79	59.20	74.53	66.87
Giza 182	92.67	96.33	94.50	5.83	1.50	3.67	15.00	21.00	18.00	78.53	99.15	88.84	77.68	83.20	80.44
Sakha 101	101.50	107.00	104.25	5.50	1.50	3.50	18.33	22.80	20.57	70.93	99.92	85.43	82.00	87.10	84.55
Sakha 102	92.67	96.17	94.42	5.33	1.17	3.25	11.97	16.57	14.27	82.13	118.77	100.45	76.88	88.40	82.64
Sakha 103	92.67	96.00	94.33	5.33	1.33	3.33	14.30	18.68	16.49	77.75	109.07	93.41	51.25	66.30	58.78
Sakha 104	98.67	104.50	101.58	5.09	1.67	3.38	16.53	18.33	17.43	85.40	108.32	96.86	82.10	84.50	83.30
Sakha 105	92.67	95.17	93.92	5.42	1.33	3.38	14.00	17.83	15.92	75.53	97.67	86.60	29.21	78.00	53.61
Sakha 106	92.83	96.17	94.50	5.42	1.17	3.29	13.35	17.83	15.59	81.98	110.28	96.13	71.80	86.53	79.16
E. Yasmine	115.33	121.17	118.25	5.33	1.33	3.33	17.75	23.48	20.62	78.07	110.15	94.11	71.69	80.50	76.09
Hybrid 1	102.67	106.67	104.67	3.17	1.00	2.08	15.93	19.80	17.87	76.67	101.23	88.95	69.86	88.40	79.13
Hybrid 2	100.83	105.67	103.25	3.17	1.17	2.17	15.67	20.33	18.00	74.42	100.20	87.31	62.36	80.50	71.43
GZ1368-S-5-4	110.50	114.83	112.67	2.17	1.00	1.58	16.87	23.58	20.23	79.33	97.92	88.63	65.56	82.80	74.18
IET1444	109.17	115.83	112.50	1.83	1.17	1.50	22.05	28.10	25.08	79.90	128.83	104.37	81.18	86.25	83.72
LSD0.05	1.40	1.21	0.68	0.87	0.75	0.41	1.31	1.95	0.78	3.46	2.14	1.28	1.22	1.54	1.16
LSD0.01	2.03	1.75	0.96	1.26	1.09	0.58	1.90	2.82	1.10	5.01	3.10	1.81	1.77	2.23	1.64

d, n, and c are drought, normal, and combined data, respectively.

Table 8. Grain yield performance and related traits under normal, drought conditions and their combined data.

Variety	Number of Panicles/Plants			100-Grain Weight (g)			Sterility Percentage%			Grain Yield/Plant (g)			Water Use Efficiency		
	d	n	c	d	n	c	d	n	c	d	n	c	d	n	c
Giza 177	12.03	18.67	15.35	2.74	2.96	2.85	25.33	6.83	16.08	22.03	42.40	32.22	0.68	0.75	0.71
Giza 178	17.50	21.33	19.42	2.14	2.36	2.25	14.78	4.83	9.81	26.50	43.70	35.10	0.80	0.77	0.78
Giza 179	16.07	24.25	20.16	2.71	2.74	2.73	14.83	5.73	10.28	28.78	46.23	37.51	0.86	0.83	0.84
Giza 181	14.30	17.17	15.73	1.98	2.81	2.40	22.83	9.67	16.25	20.33	35.83	28.08	0.62	0.62	0.62
Giza 182	16.33	19.82	18.08	3.03	3.24	3.14	18.58	7.48	13.03	22.75	37.17	29.96	0.71	0.66	0.68
Sakha 101	13.07	20.23	16.65	2.73	3.05	2.89	19.92	6.78	13.35	25.17	45.95	35.56	0.77	0.80	0.78
Sakha 102	12.95	16.67	14.81	2.94	3.06	3.00	20.17	7.33	13.75	22.33	42.50	32.42	0.68	0.74	0.71
Sakha 103	14.40	19.00	16.70	2.52	2.58	2.55	20.17	7.10	13.63	20.62	41.43	31.03	0.62	0.73	0.67
Sakha 104	12.50	20.08	16.29	2.53	2.72	2.63	25.50	6.00	15.75	25.17	44.13	34.65	0.77	0.77	0.77
Sakha 105	13.33	18.33	15.83	2.69	2.84	2.77	20.97	6.03	13.50	20.58	42.33	31.46	0.59	0.74	0.66
Sakha 106	10.73	19.17	14.95	2.93	3.12	3.03	22.00	7.05	14.53	25.00	42.80	33.90	0.77	0.76	0.76
E. Yasmine	17.33	22.00	19.67	3.03	3.34	3.19	14.72	10.92	12.82	26.77	35.28	31.03	0.80	0.60	0.70
Hybrid 1	16.73	24.67	20.70	2.48	2.64	2.56	22.97	6.00	14.48	30.17	49.83	40.00	0.89	0.87	0.88
Hybrid 2	19.33	24.83	22.08	2.75	2.88	2.82	14.40	5.70	10.05	30.50	52.50	41.50	0.92	0.92	0.92
GZ1368-S-5-4	12.50	15.33	13.92	1.94	2.25	2.10	12.87	14.73	13.80	27.10	38.52	32.81	0.80	0.67	0.74
IET1444	15.70	24.67	20.18	2.26	2.34	2.30	13.00	6.17	9.58	23.33	30.95	27.14	0.71	0.53	0.62
LSD0.05	1.00	1.43	0.71	0.37	0.33	0.001	2.15	1.54	0.85	1.48	1.18	0.86	0.04	0.02	0.02
LSD0.01	1.45	2.07	1.00	0.54	0.48	0.00	3.11	2.23	1.20	2.14	1.71	1.22	0.06	0.03	0.03

d, n, and c are drought, normal, and combined data, respectively.

The genotypes such as Sakha 102, Sakha 103, Giza177, and Sakha 105 took the least number of days to head and have earliness values of 94.42, 94.33, 94.17, and 93.92 days, respectively; in contrast, it took significantly longer time for E. Yasmine, Giza 181, GZ1368-S-5-4 and IET1444 having the earliness values of 118.25, 114.08, 112.67 and 112.50 days, respectively. These differences among rice genotypes might be attributed to their genetic background. The opposite strategy was observed in other cultivars, which had a significant delay in maturity with drought. Heading delay is a typical drought response observed in rice (Gaballah and Abd Allah) [35], which is expected to confer a benefit in those environments where stress is temporary, if development and flowering resume after the stress are relieved. Gaballah [34] mentioned that Moroberakan, Giza178, and Sakha104 gave the highest values under normal and drought conditions and for days to heading.

The desirable mean values for leaf rolling were found to be with Hybrid 1, Hybrid 2, GZ1368-S-5-4, and IET1444 were 2.08, 2.17, 1.58, and 1.50, respectively, while, the undesirable mean values observed with 'Giza177', Giza178, Giza179 and 'Giza181'were 3.21, 2.33, 2.17 and 3.67, respectively (Table 7). In this investigation, drought tolerance can be assessed by visual scoring based on leaf rolling. A smaller degree of leaf rolling indicates a greater degree of dehydration avoidance by the development of deep roots. Gaballah [34] mentioned that the drought after 12 days increased leaf rolling in rice genotypes.

Concerning the flag leaf area, which is an important functional factor for photosynthesis, assimilation, and transpiration along the experimental plant life recorded the highest values with significant differences by the genotypes, IET1444, Giza181, E. Yasmine and Sakha101 were 25.08, 21.50, 20.62, and 20.57 cm^2, respectively. Otherwise, the lowest value was recorded with Sakha102 (14.27 cm^2). Reduced soil moisture levels produced a lower leaf area, which might inhibit cell division under water-starved conditions. Zubaer et al. [36] mentioned that the highest leaf area was found at 100% field capacity (FC) of the soil in all the rice genotypes. The leaf area was reduced by reducing moisture levels, but the degree reduction was higher in Basmati (14.7 for 70% FC and 53.2% for 40% FC) cm. There were significant differences in the plant height among the studied rice genotypes, suggesting that the growth rates were different in these genotypes. With respect to plant height in Table 7, the most desirable mean values towards dwarfism were obtained in the genotypes Sakha101, Giza 179, Sakha105, and Giza 177, which were 85.43, 85.72, 86.60, and 87.13cm, respectively. Similarly, the tallest values were obtained in Sakha106, Sakha104, Sakha102, and IET1444, 96.13, 96.86, 100.45, and 104.37 cm, respectively. Rice cultivars have tolerance by assessing plant height reduction under drought stress conditions. Lafitte et al. [37] indicated that the low land stress reduced height by only 4 cm (3%), ranging from a 43 cm reduction to a 22 cm increase in height. Regarding relative water content, the genotypes Sakha104, IET1444, Giza179, and Sakha101 resulted in the maximum values 83.30, 83.72, 83.76, and 84.55%, respectively, while the minimum values were recorded in Sakha105, Sakha103, 'Giza177' and Giza181 were 53.61, 58.78, 61.24, and 66.87%, respectively (Table 7). Plant responses to tissue water potential determine their level of drought tolerance. The traits, such as leaf turgor (RWC) maintenance and leaf rolling, have been used as selection criteria in rice [38], due to rice cultivar's ability to save water in the leaf tissue to overcome water shortage. In the present study, we could also find a similar mechanism of drought tolerance is operating in genotypes, such as Giza 179, Sakha 101, Sakha 104, and IET1444, which were able to maintain significantly higher RWC under drought condition. We found that these genotypes can be considered as tolerant to moderately tolerant rice genotypes for drought stress.

For a number of panicles/plant, the genotypes Giza179, IET1444, Hybrid 1, and Hybrid 2 recorded the highest mean values of 20.16, 20.18, 20.70, and 22.08, respectively. Otherwise, the lowest values 13.92, 14.81, 14.95, and 15.35 with the genotypes GZ1368-S-5-4, Sakha102, Sakha106, and Giza177, respectively (Table 8). The heaviest 100-grain weight were 3.00, 3.03, 3.14, and 3.19 g, achieved with Sakha102, Sakha106, Giza182, and E. Yasmine. Therefore, the genotypes GZ1368-S-5-4, Giza178, IET1444, and Giza181 gave the lightest values, 2.10, 2.25, 2.30, and 2.40 g, respectively. The desirable values for sterility percentage were confirmed with genotypes IET1444, Giza178, Hybrid 2 and Giza179, which were 9.58, 9.81, 10.05 and 10.28%, respectively, otherwise the genotypes Sakha106, Sakha104, Giza177, and Giza181 gave undesirable values 14.53, 15.75, 16.08, and 16.25%, respectively. Concerning grain yield/plant, the genotypes Sakha101, Giza179, Hybrid 1, and Hybrid 2 resulted in the greatest values 35.56, 37.51, 40.00, and 41.50 g/plant. On the other hand, the lowest values obtained with the genotypes IET1444, Giza181, Giza182, and E. Yasmine were 27.14, 28.08, 29.96, and 31.03 g/plant, respectively. Mukamuhirwa et al. [39] reported that the cultivar Intsindagirabigega was most tolerant to drought, while Zong geng was the most sensitive. The genotypes Giza178, Giza179, Hybrid 1, and Hybrid 2 gave maximum values 0.78, 0.84, 0.88, and 0.92 respectively, for water use efficiency, while the genotypes Giza181, IET1444, Sakha105, and Sakha103 recorded the minimum values were 0.62, 0.62, 0.66, and 0.67, respectively. The rice genotypes Giza 179, Hybrid 1and Hybrid 2 had higher values for a number of panicles/plants, 100-grain weight, grain yield/plant and water use efficiency under normal, drought stress, and their combined data. Gaballah and AbdAllah [35] mentioned that the water stress reduced plant height, induced leaf rolling in the susceptible rice genotypes. The reduction of grain yield, number of panicles/plants, 100-grain weight, and high sterility percentage resulted from water stress at flowering and ripening stages. Water stress during vegetative, panicle initiation, flowering has reduced grain yield/plant by 28%, 34%, and 40%, respectively. Drought mitigation, through the development of drought-tolerant varieties with higher yields suitable for water-limiting environments, will be the critical factor in improving stable rice production.

3.2. Number of Alleles and Allelic Diversity

The sixteen rice genotypes were used in the present study were subjected to DNA polymorphism screening and assessment using SSR markers, which offer excellent potential for generating large numbers of markers evenly distributed throughout the genome and have efficiently been used to give reliable and reproducible genetic markers. Ten SSR primer pairs related to drought tolerance with known map positions distributed in the rice genome were used to screen a set of sixteen selected indica, japonica, and tropical-japonica rice genotypes with different levels with mechanisms of drought.

Among 10 SSR markers, spread on seven chromosomes (1, 2, 4, 5, 6, 8, and 9) generated polymorphic alleles. Table 9 showed that a total number of 85 alleles were detected at the nine markers' loci across the sixteen rice genotypes. The number of alleles per locus generated by each marker varied from 2 to 15 alleles, with an average of 8.5 alleles per locus. The effective number of alleles per locus ranged from 1.20 alleles to 3.0 alleles, with an average of 2.28 alleles. The highest number and the effective number of alleles per locus were observed for RM263 (3.0), RM289 (2.81), and RM242 (2.63). Similar results for a low number of alleles per locus were also obtained by [40] (3.33) and [41] (2.5). On the contrary, a high number of alleles per locus was obtained by [42] (8.57). There was a significant positive correlation between the number of alleles detected at a locus and the number of repeats within the targeted microsatellite DNA (r = 0.57 **). Thus, the larger the repeat number in the microsatellite DNA, the larger the number of alleles detected. Moreover, it was reported that the dinucleotide repeat motif (GA) displayed a high level of variation among the rice genotypes [24]. On the other hand, [40] reported no correlations between the number of alleles detected and the number of SSR repeats.

Table 9. The correlation coefficient for polymorphic SSR markers.

	No of Amplified Alleles	Effective Number of Alleles	Common Alleles	Gene Diversity
Effective number of alleles	1			
Common alleles	0.57 **	1		
Gene diversity	−0.48 **	−0.96 **	1	
Polymorphic information content	0.75 **	0.92 **	-0.92 **	1

** is highly significant at probability 0.001.

3.3. Gene Diversity

The gene diversity or heterozygosity (H_e) of a locus is defined as the probability that an individual is heterozygous for the locus in the population [43]. Higher values of this measure tend to be more informative because there is more allelic variation. As shown in Table 9, the H_e values for all SSR markers used in this study varied from 0.94 to 1.00, with an average of 0.98. The findings were in agreement with the observation of [44–46]. The highest H_e value (1.00) was recorded for RM23 and 518. Meanwhile, the lowest H_e values (0.98) were achieved by RM518.

3.4. PIC Value

PIC value refers to the value of a marker for detecting polymorphism within a population, depending on the number of detectable alleles and the distribution of their frequency; thus, it provides an estimate of the discriminating power of the marker [47]. As shown in Table 3, the PIC values for the SSR used in this study varied from 0.83 to 0.99, with an average of 0.95. This result is consistent with Sajib et al. [40] who reported high variation in PIC values for all tested SSR loci (from 0.14 to 0.71 with an average of 0.48). Higher averages of PIC values were reported by Zeng et al. [48] (0.57) and Ram et al. [49] (0.707). The highest PIC values were observed for RM23 (0.99), RM518 (0.99), RM223 (0.98) and RM276 (0.98). A highly significant correlation coefficient was found between PIC values and the number of amplified alleles detected per locus (r = −0.75 **), as shown in Table 9. A significant correlation between PIC value and the effective number of alleles (r = 0.92 **) and a highly significant correlation was found between PIC and gene diversity (r = −0.92 **). Similar results were obtained by Kumar et al. [50].

Figure 1 shows the PCR amplified fragments produced by the highest polymorphic markers in the current study, RM23, RM518, RM223, and RM276. These markers revealed the highest PIC values and gave the same values 1.00 and the highest number of alleles ranging from 8 to 9 alleles per locus, suggesting that these markers could be used for molecular characterization of a large number of rice genotypes rather than mapping populations for drought tolerance. The results were similar to those obtained by [10,51,52].

3.5. Identified MAS Marker

Among ten polymorphic SSR markers, RM518 was able to divide the studied genotypes into seven groups depending on their drought tolerance potential. The first group showed the first allele with a molecular weight of 125.89 bp included the drought susceptible genotypes japonica type1 and 2. The second allele with a molecular weight of 172.37 bp appeared in the second group included the drought moderate indica–japonica genotype Sakha105, Hybrid 1, Giza178, Giza181, and Giza182. The third the allele molecular weight 163.038 has six genotypes IET1444, Sakha101, Sakha102, Sakha106, Sakha104, and Giza177. The fourth allele molecular weight 375.85 bp with on genotype GZ1368-S-5-4 indica japonica type and tolerant to drought stress. The fifth allele molecular weight was 169.65 with one genotype Hybrid 2 moderate tolerant to drought and the six-allele molecular weight 158.56 bp with on cultivar Sakha103 susceptible to drought stress. This result agreed with [53] who reported that RM472 was linked to maximum root length and root dry weight characters. This marker could be useful in MAS for these characters in rice. The results were similar to those obtained by [46,54].

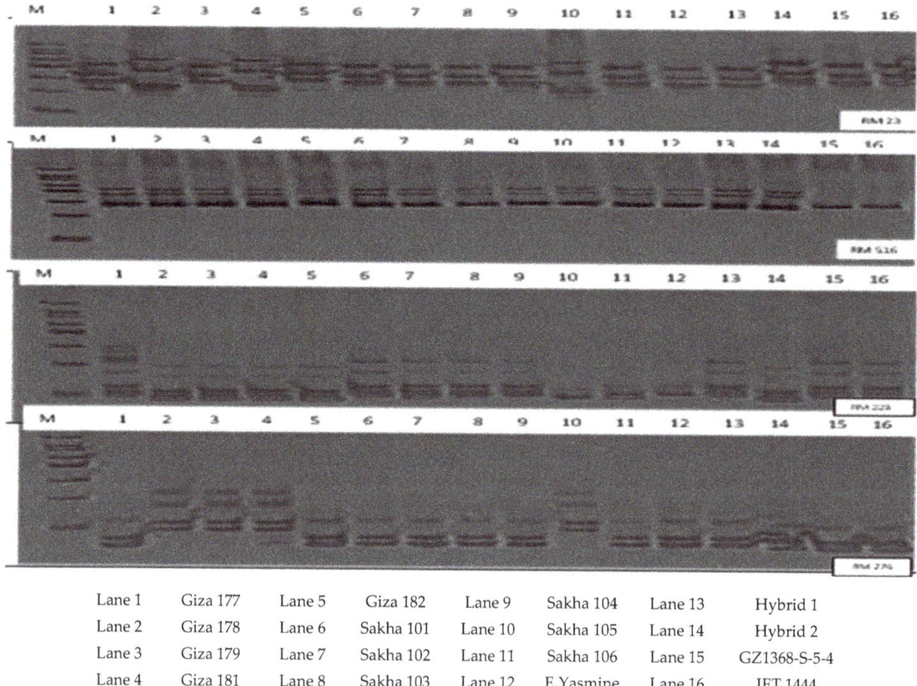

Lane 1	Giza 177	Lane 5	Giza 182	Lane 9	Sakha 104	Lane 13	Hybrid 1
Lane 2	Giza 178	Lane 6	Sakha 101	Lane 10	Sakha 105	Lane 14	Hybrid 2
Lane 3	Giza 179	Lane 7	Sakha 102	Lane 11	Sakha 106	Lane 15	GZ1368-S-5-4
Lane 4	Giza 181	Lane 8	Sakha 103	Lane 12	E.Yasmine	Lane 16	IET 1444

Figure 1. Agarose gel electrophoresis of PCR amplified fragments for the polymorphic SSR markers RM23, RM518, RM223, and RM276. M is a 100 bp DNA ladder.

3.6. Similarity

The maximum similarity coefficient (0.5) was recorded between Giza181 and Giza182, which are indica type and drought susceptible genotypes at the same time (Table 10). Moreover, a high similarity percentage was observed between the japonica type and susceptible to drought stress Sakha104 with Sakha102 and Sakha106 (0.45), Sakha101 with

Sakha102 and Sakha106 (0.40), Sakha105 with Hybrid 1 (0.40), Hybrid 1 with Giza178 (0.40) and GZ1368-S-5-4with Giza181 (0.40). On the other hand, no similarity percentage values were observed between the genotypes such as Sakha103, Giza179, IET1444, Sakha102, and Sakha104 with Giza178, GZ1368-S-5-4, Giza181, Giza182, E. Yasmine and Hybrid 2. These results were in harmony with [2], who reported a low similarity coefficient between japonica type and indica type genotypes, and [55] reported a relatively high level of similarity between closely related genotypes. Moreover, the findings are similar to those observed by [17,18].

3.7. Cluster Analysis

The genetic relationships among rice genotypes are presented in a dendrogram based on informative microsatellite alleles (Figure 2). All genotypes are grouped into two major clusters in the dendrogram at 66% similarity based on Jaccard's similarity index. Whereas Jaccard's similarity measure similarity for the two sets of data, it ranges from 0% to 100%. The higher the percentage, the more similar the two populations. Although it's easy to interpret, it is extremely sensitive to small sample sizes and may give erroneous results, especially with very small samples or data sets with missing observations. The first cluster (A) is divided into two minor groups A1 and A2. The A1 sub-cluster included two groups A1-1 and A1-2, of which A1-1 contains genotypes IET1444, GZ1386-S-5-4, and Hybrid1, which are drought stress-tolerant. On the other hand, the A1-2 cluster was further divided into A1-2-1 (having Hybrid2 genotype) and A1-2-2 (containing Giza179 and Giza178) at coefficient 0.91, represented moderate or drought stress.

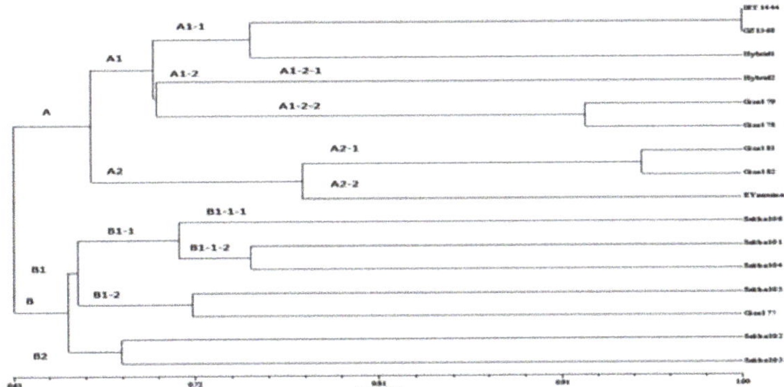

Figure 2. Dendrogram derived from unweighted pair group method with arithmetic averages (UPGMA) cluster analysis of sixteen rice genotypes based on Jaccard's similarity coefficient using 10 SSR markers.

The A2 cluster, also subdivided into A2-1, included Giza181 and Giza182 at coefficient 0.94, while the A2-2 cluster had Egyptian Yasmine genotype, which has coefficient 0.76 with cluster A2-1. The A2 cluster comprised of the sensitive genotypes and indica type. Similarly, the cluster B was divided into two minor groups B1 and B2. The cluster B1 included B1-1 and B1-2 at coefficient 0.65. The B1-1 divided into B1-1-1 and B1-1-2 at coefficient 0.72. In general, the B cluster japonica type, and all were sensitive to drought stress. El-Malky et al. [42] reported the ability of SSR markers to divide the varieties into two distinct groups, one included the indica varieties, and the other had the japonica varieties. Moreover, Zeng et al. [48] found that all genotypes grouped into two major branches in the dendrogram with less than 10% similarity based on Jaccard's similarity index, one unit represents the subspecies, japonica rice, and another unit represents the subspecies, indica, or the hybrids between japonica rice and indica rice.

Table 10. Similarity coefficient among studied genotypes based on SSR markers.

Genotypes	Sakha103	Giza179	IET1444	Sakha101	Sakha102	Sakha106	Sakha104	Giza177	Sakha105	Hybrid 1	Giza178	GZ1368-S-5-4	Giza181	Giza182	E. Yasmine
Giza179	0.35														
IET1444	0.18	0.13													
Sakha101	0.14	0.18	0.21												
Sakha102	0.19	0.18	0.26	0.40											
Sakha106	0.14	0.13	0.32	0.40	0.40										
Sakha104	0.18	0.23	0.25	0.32	0.45	0.45									
Giza177	0.05	0.04	0.08	0.18	0.18	0.30	0.35								
Sakha105	0.04	0.04	0.04	0.12	0.12	0.08	0.16	0.30							
Hybrid 1	0.04	0.04	0.04	0.08	0.08	0.04	0.07	0.13	0.40						
Giza178	0.00	0.00	0.00	0.04	0.04	0.04	0.00	0.04	0.17	0.40					
GZ1368-S-5-4	0.00	0.00	0.00	0.04	0.00	0.04	0.00	0.04	0.12	0.21	0.21				
Giza181	0.00	0.00	0.00	0.04	0.00	0.04	0.00	0.04	0.13	0.17	0.17	0.40			
Giza182	0.00	0.00	0.00	0.04	0.00	0.04	0.00	0.04	0.17	0.12	0.12	0.21	0.50		
E. Yasmine	0.00	0.00	0.00	0.00	0.00	0.00	0.00	0.04	0.13	0.18	0.18	0.17	0.25	0.30	
Hybrid 2	0.00	0.00	0.00	0.00	0.00	0.00	0.00	0.04	0.07	0.12	0.12	0.15	0.12	0.21	0.29

4. Conclusions

Genetic improvement for drought tolerance in rice can be achieved in this present study based on results obtained at phenotypic characterization. Extensive genetic diversity analyses were presented as valuable in selecting the truly promising drought-tolerant genotypes, which can be used to cross to development genotypes with increased water stress tolerance levels. The rice genotypes Giza179, IET1444, Hybrid1, and Hybrid2, achieved the greatest values for grain yield/plant and produced maximum values regarding water use efficiency. It could be summarized the genotypes GZ1368-S-5-4, IET1444, Giza 178, and Giza179 were suitable materials for developing drought breeding. Thus, this study's results indicate that incorporating genetic analyses with phenotypic data is very important to accelerate breeding programs by selecting suitable genotypes to improve target traits and could help exclude genotypes with bad performance.

Author Contributions: Conceptualization, M.M.G., and A.M.F.; methodology, M.M.G. and A.M.M.; software, A.M.M.; validation, A.E.S., M.M.G.; formal analysis, A.M.F.; investigation, M.M.G.; resources, M.M.G. and A.M.F.; data curation, M.M.G.; writing—original draft preparation, M.M.G., A.M.F., and A.M.M.; writing—review and editing, A.E.S., M.S., M.B., M.M.H.; visualization, A.M.F.; supervision, M.M.G.; project administration, A.M.F.; funding acquisition, M.S., M.B., M.M.H., A.E.S. All authors have read and agreed to the published version of the manuscript.

Funding: The current work was funded by Taif University Researchers Supporting Project number (TURSP—2020/59), Taif university, Taif, Saudi Arabia.

Acknowledgments: The authors sincerely acknowledge the contributions of Rice Research and Training Center (RRTC), Field Crops Research Institute, Agricultural Research Center, 33717, Sakha, Kafr Elsheikh, Egypt for providing necessary laboratory facility during the investigation. Besides, the authors extend their appreciation to Taif University for funding current work by Taif University Researchers Supporting Project number (TURSP—2020/59), Taif University, Taif, Saudi Arabia.

Conflicts of Interest: The authors declare no conflict of interest.

References

1. Long, S.P.; Ort, D.R. More than taking the heat: Crops and global change. *Curr. Opin. Plant Biol.* **2010**, *13*, 1–8. [CrossRef] [PubMed]
2. Chakravarthi, B.K.; Naravaneni, R. SSR marker-based DNA fingerprinting and diversity study in rice (*Oryza sativa* L.). *Afr. J. Biotechnol.* **2006**, *5*, 684–688.
3. Passioura, J.B. The drought environment: Physical, biological and agricultural perspectives. *J. Exp. Bot.* **2007**, *58*, 113–117. [CrossRef] [PubMed]
4. Wery, J.; Silim, S.N.; Knights, E.J.; Malhotra, R.S.; Cousin, R. Screening techniques and sources and tolerance to extremes of moisture and air temperature in cool season food legumes. *Euphytica* **1994**, *73*, 73–83. [CrossRef]
5. Anis, G.; Hassan, H.; Saneoka, H.; El Sabagh, A. Evaluation of new promising rice hybrid and its parental lines for floral, agronomic traits and genetic purity assessment. *Pak. J. Agric. Sci.* **2019**, *56*, 567–576.
6. Hossain, M.M.; Islam, M.M.; Hossain, H.; Ali, M.S.; Silva, T.D. Genetic diversity analysis of aromatic landraces of rice (*Oryza sativa* L.) by microsatellite markers. *Genes Genom. Genom.* **2012**, *6*, 42–47.
7. Sivaranjani, A.K.P.; Pandey, M.K.; Sudharshan, I.; Kumar, G.R.; Madhav, M.S. Assessment of genetic diversity among Basmati and non-Basmati aromatic rices of India using SSR markers. *Curr. Sci.* **2010**, *99*, 221–226.
8. Ravi, M.; Geethanjali, S.; Sameeyafarheen, F.; Maheswaran, M. Molecular marker based genetic diversity analysis in rice (*Oryza sativa* L.) using RAPD and SSR markers. *Euphytica* **2003**, *133*, 243–252. [CrossRef]
9. Singh, A.; Sengar, R.S. DNA fingerprinting based decoding of indica rice (*Oryza sativa* L) via molecular marker (SSR, ISSR, & RAPD) in aerobic condition. *Adv. Crop. Sci. Technol.* **2015**, *3*, 167.
10. Chukwu, S.C.; Rafii, M.Y.; Ramlee, S.I.; Ismail, S.I.; Oladosu, Y.; Kolapo, K.; Musa, I.; Halidu, J.; Muhammad, I.; Ahmed, M. Marker-Assisted Introgression of Multiple Resistance Genes Confers Broad Spectrum Resistance against Bacterial Leaf Blight and Blast Diseases in PUTRA-1 Rice Variety. *Agronomy* **2020**, *10*, 42. [CrossRef]
11. Lapitan, V.C.; Brar, D.S.; Abe, T.; Redofia, E.D. Assessment of genetic diversity of Philippine rice cultivars carrying good quality traits using SSR markers. *Breed Sci.* **2007**, *57*, 263–270. [CrossRef]
12. Suh, J.P.; Won, Y.J.; Ahn, E.K.; Lee, J.H.; Ha, W.G.; Kim, M.K.; Cho, Y.C.; Jeong, E.G.; Kim, B.K. Field performance and SSR analysis of drought QTL introgression lines of rice. *Plant Breed Biotechnol.* **2014**, *30*, 158–166. [CrossRef]

13. Vikram, P.; Swamy, B.P.M.; Dixit, S.; Cruz, S.; Ahmed, H.U.; Singh, A.K.; Kumar, A. qDTY1.1, a major QTL for rice grain yield under reproductive-stage drought stress with a consistent effect in multiple elite genetic backgrounds. *BMC Genet.* **2011**, *12*, 89. [CrossRef] [PubMed]

14. Venuprasad, R.; Dalid, C.O.; Del Valle, M.; Zhao, D.; Espiritu, M.; Sta Cruz, M.T.; Amante, M.; Kumar, A.; Atlin, G.N. Identification and characterization of large-effect quantitative trait loci for grain yield under lowland drought stress in rice using bulk-segregant analysis. *Theoret. Appl. Genet.* **2009**, *120*, 177–190. [CrossRef] [PubMed]

15. Shamshudin, N.A.; Swamy, B.M.; Ratnam, W.; Cruz, M.T.; Raman, A.; Kumar, A. Marker assisted pyramiding of drought yield QTLs into a popular Malaysian rice cultivar, MR219. *BMC genet.* **2016**, *17*, 3.

16. Abdallah, A.A.; Ammar, M.H.; Badawi, A.T. Screening rice genotypes for drought resistance in Egypt. *J. Plant Breed Crop. Sci.* **2010**, *2*, 205–215.

17. Mumtaz, M.Z.; Saqib, M.; Abbas, G.; Akhtar, J.; Ul-Qamar, Z. Drought stress impairs grain yield and quality of rice genotypes by impaired photosynthetic attributes and k nutrition. *Rice Sci.* **2020**, *27*, 5–9. [CrossRef]

18. Tabkhkar, N.; Rabiei, B.; Lahiji, H.S.; Chaleshtori, M.H. Identification of a new set of drought-related miRNA-SSR markers and association analysis under drought stress in rice (*Oryza sativa* L.). *Plant Gene* **2020**, *21*, 1–10. [CrossRef]

19. Steel, R.G.D.; Torrie, J.H.; Dickey, D.A. *Principles and Procedures for Statistics*; M.c. Graw Hill Book Co.: New York, NY, USA, 1997.

20. Zheng, K.; Huang, N.; Bennett, J.; Khush, G.S. *PCR-Based Marker Assisted Selection in Rice Breeding*; Irri Discussion Paper Series No 12; International Rice Research Institute: Manila, Philippines, 1995; pp. 9–11.

21. Chen, X.; Temnykh, S.; Xu, Y.; Cho, Y.G.; Mccouch, S.R. Development of a microsatellite framework map providing genome-wide coverage in rice (*Oryza sativa* L.). *Theoret. Appl. Genet.* **1997**, *95*, 553–567. [CrossRef]

22. Wu, K.S.; Tanksley, S.D. Abundance, polymorphism and genetic mapping of microsatellites in rice. *Mol. Gen. Genet.* **1993**, *241*, 225–235. [CrossRef]

23. Qu, S.; Desai, A.; Wing, R.; Sundaresan, V. A versatile transposon-based activation tag vector system for functional genomics in cereals and other monocot plants. *Plant Physiol.* **2008**, *146*, 189–199. [CrossRef] [PubMed]

24. Temnykh, S.; Park, W.D.; Ayres, N.M.; Cartinhour, S.; Hauck, N.; Lipovich, L.; Cho, Y.G.; Ishii, T.; Mccouch, S.R. Mapping and genome organization of microsatellite sequences in rice (*Oryza sativa* L.). *Theoret. Appld. Genet.* **2000**, *100*, 697–712. [CrossRef]

25. Temnykh, S.; Clerck, D.G.; Lukashova, A.; Lipovich, L.; Cartinhour, S.; Couch, M.S. Computational and experimental analysis of microsatellites in rice (*Oryza sativa* L.): Frequency, length variation, transposon associations, and genetic marker potential. *Genome Res.* **2001**, *11*, 1441–1452. [CrossRef] [PubMed]

26. Nei, M.; Li, W.H. Mathematical model for studying genetic variation in terms of restriction endonucleases. *Proc. Natl. Acad. Sci. USA* **1979**, *76*, 5269–5273. [CrossRef] [PubMed]

27. Sun, J.; Zhao, J.; Schwartz, M.A.; Wang, J.Y.; Weidmer, T.; Sims, P.J. c-Abl tyrosine kinase binds and phosphorylates phospholipid scramblase 1. *J. Boil. Chem.* **2001**, *276*, 28984–28990. [CrossRef]

28. Abdallah, A.; Gaballah, M.; Aml El-Saidy, M.; Ammar, M. Drought Tolerance of Anther Culture Derived Rice Lines. *J. Plant Prod.* **2014**, *11*, 115–124. [CrossRef]

29. Meng, T.Y.; Wei, H.H.; Li, C.; Dai, Q.G.; Xu, K.; Huo, Z.Y.; Wei, H.Y.; Guo, B.W.; Zhang, H.C. Morphological and Physiological Traits of Large-Panicle Rice Varieties with High Filled-Grain Percentage. *J. Integr. Agric.* **2016**, *15*, 1751–1762. [CrossRef]

30. Raman, A.; Verulkar, S.B.; Mandal, N.P.; Variar, M.; Shukla, V.D.; Dwivedi, J.L.; Singh, B.N.; Singh, O.N.; Swain, P.; Mall, A.K.; et al. Drought yield index to select high yielding rice lines under different drought stress severities. *Rice* **2012**, *5*, 31. [CrossRef]

31. Zhang, J.; Babu, R.C.; Pantuwan, G.; Kamoshita, A.; Blum, A.; Wade, L.; Sarkarung, S.; O'Toole, J.C.; Nguyen, H.T. Environments. In *Molecular Dissection of Drought Tolerance in Rice: From Physio-Morphological Traits to Field Performance*; Ito, O., O'Toole, J.C., Hardy, B., Eds.; International Rice Research Institute: Los Banos, Philippines, 1999; pp. 331–343.

32. Singh, S.; Pradhan, S.; Singh, A.; Singh, O. Marker validation in recombinant inbred lines and random varieties of rice for drought tolerance. *Aust. J. Crop. Sci.* **2012**, *6*, 606–612.

33. Wang, H.; Inukai, Y.; Yamauchi, A. Root development and nutrient uptake. *Crit. Rev. Plant Sci.* **2006**, *25*, 279–301. [CrossRef]

34. Gaballah, M.M. Studies on Physiological and Morphological Traits Associated with Drought Resistance in Rice (*Oryza sativa* L.). Ph.D. Thesis, Kafer El-Sheikh University, Kafer, Egypt, 2009.

35. Gaballah, M.M.; Abdallah, A.A. Effect of water irrigation shortage on some quantitative characters at different rice development growth stages. *World Rural Obs.* **2015**, *7*, 10–21.

36. Zubaer, M.A.; Chowdhury, A.K.; Islam, M.Z.; Ahmed, T.; Hasan, M.A. Effects of water stress on growth and yield attributes of Aman rice genotypes. *Int. J. Sustain. Crop. Prod.* **2007**, *2*, 25–30.

37. Lafitte, H.R.; Li, V.R.; Gao, C.H.M.; Shi, Y.M.; Xu, Y.; Fu, J.L.; Yu, B.Y.; Ali, S.B.; Domingo, A.J.; Maghirang, J.; et al. Improvement of rice drought tolerance through backcross breeding: Evaluation of donors and selection in drought nurseries. *Field Crop. Res.* **2006**, *97*, 77–86. [CrossRef]

38. Kumar, A.; Basu, S.; Ramegowda, V.; Pereira, A. Mechanisms of Drought Tolerance in Rice. *Burleigh Dodds Sci. Publ. Ltd.* **2018**, *12*, 1–34.

39. Mukamuhirwa, A.; Hovmalm, H.P.; Bolinsson, H.; Ortiz, R.; Nyamangyoku, O.; Johansson, E. Concurrent Drought and Temperature Stress in Rice-a Possible Result of the Predicted Climate Change: Effects on Yield Attributes, Eating Characteristics, and Health Promoting Compounds. *Int. J. Environ. Res. Public Health* **2019**, *16*, 1043. [CrossRef]

40. Sajib, A.M.; Hossain, M.M.; Mosnaz, A.T.M.; Hossain, H.; Islam, M.M.; Ali, M.S.; Prodhan, S.H. SSR marker-based molecular characterization and genetic diversity analysis of aromatic landreces of rice (*Oryza sativa* L.). *J. BioSci. Biotech.* **2012**, *1*, 107–116.

41. Vanniarajan, C.; Vinod, K.K.; Pereira, A. Molecular evaluation of genetic diversity and association studies in rice (*Oryza sativa*, L.). *J. Genet.* **2012**, *91*, 1–11. [CrossRef]

42. El-Malky, M.M.; Fahmy, A.I.; Kotb, A.A. Detection of genetic diversity using microsatellites in rice (*Oryza sative* L.). *Afr. Crop. Sci. Conf. Proc.* **2007**, *8*, 597–603.

43. Liu, B.H. *Statistical Genomics: Linkage, Mapping and QTL Analysis*; CRC Press: Boca Raton, FL, USA, 1998; p. 611.

44. Fasahat, P.; Muhammad, K.; Abdullah, A.; Wickneswari, R. Identification of introgressed alien chromosome segments associated with grain quality in *Oryza rufiogon* x MR219 advanced breeding genotypes using SSR markers. *Genet. Mol. Res.* **2012**, *11*, 3534–3546. [CrossRef]

45. Islam, M.D.S.; Guswami, A.A.P.; Sarid-Ullah, M.; Hossain, M.M.; Prodhan, H.F.M.S. Assessment of genetic diversity among moderately drought tolerant landraces of rice using RAPD markers. *J. Biosci. Biotech.* **2013**, *2*, 207–213.

46. El-Wahsh, S.M.; El-Refaee, Y.Z.; Emeran, A.A.; Mashaal, S.F.; Arafa, R.A. Genetic diversity of rice blast fungus populations (*Pyricularia grisea*) using molecular markers. *J. Agric. Chem. Biotechnol.* **2016**, *7*, 57–65. [CrossRef]

47. Nagy, S.; Poczai, P.; Cernak, I.; Gorji, A.M.; Hegedus, G.; Taller, J. PIC calc: An online program to calculate polymorphic information content for molecular genetic studies. *Biochem. Genet.* **2012**, *50*, 670–672. [CrossRef] [PubMed]

48. Zeng, L.; Kwon, T.R.; Liu, X.; Wilson, C.; Grieve, C.M.; Gregorio, G.B. Genetic diversity analyzed by microsatellite markers among rice (*Oryza sativa*, L.) genotypes with different adaptations to saline soils. *Plant Sci.* **2004**, *166*, 1275–1285. [CrossRef]

49. Ram, S.G.; Thiruvengadam, V.; Vinod, K.K. Genetic diversity among cultivars, landraces and wild relatives of rice as revealed by microsatellite markers. *J. Appl. Genet.* **2007**, *48*, 337–345. [CrossRef] [PubMed]

50. Kumar, V.; Kumar, S.; Chakrabarty, S.K.; Mohapatra, T.; Dadlani, M. Molecular characterization of farmers' varieties of rice (*Oryza sativa*). *Indian J. Agric. Sci.* **2015**, *85*, 118–124.

51. Aljumaili, S.J.; Rafii, M.Y.; Latif, M.A.; Sakimin, S.Z.; Arolu, I.W.; Miah, G. Genetic diversity of aromatic rice germplasm revealed by SSR markers. *BioMed Res. Int.* **2018**, *2018*, 7658032. [CrossRef]

52. Melaku, G.; Zhang, S.; Haileselassie, T. Comparative evaluation of rice SSR markers on different Oryza species. *J. Rice Res. Dev.* **2018**, *1*, 38–48.

53. Kanbar, A.; Shashidhar, H.E. Participatory selection assisted by DNA markers for enhanced drought resistance and productivity in rice (*Oryza sativa*, L.). *Euphytica* **2011**, *178*, 137–150. [CrossRef]

54. Pachauri, V.; Taneja, N.; Vikram, P.; Singh, N.K.; Singh, S. Molecular and morphological characterization of Indian farmers' rice varieties (*Oryza sativa* L.). *Aust. J. Crop. Sci.* **2013**, *7*, 923–932.

55. Kanawapee, N.; Sanitchon, J.; Srihaban, P.; Theerakulpisut, P. Genetic diversity analysis of rice cultivars (*Oryza sativa* L.) differing in salinity tolerance based on RAPD and SSR markers. *Electr. J. Biotechnol.* **2011**, *14*, 1–17.

Article

QMrl-7B Enhances Root System, Biomass, Nitrogen Accumulation and Yield in Bread Wheat

Jiajia Liu [1,2,†], Qi Zhang [1,2,†], Deyuan Meng [1,2], Xiaoli Ren [1,2], Hanwen Li [3], Zhenqi Su [3], Na Zhang [1], Liya Zhi [1], Jun Ji [1], Junming Li [1], Fa Cui [4,*] and Liqiang Song [1,5,*]

[1] Center for Agricultural Resources Research, Institute of Genetics and Developmental Biology, The Innovative Academy of Seed Design, Chinese Academy of Sciences, Shijiazhuang 050022, China; ljjzl2014@163.com (J.L.); zhqiemail@163.com (Q.Z.); mengdeyuan123@163.com (D.M.); renxiaoli2011@126.com (X.R.); zhangna@sjziam.ac.cn (N.Z.); xiaoya_19861028@163.com (L.Z.); jijun@sjziam.ac.cn (J.J.); ljm@sjziam.ac.cn (J.L.)

[2] University of Chinese Academy of Sciences, Beijing 100049, China

[3] College of Agronomy and Biotechnology, China Agricultural University, Beijing 100094, China; lihanwen@cau.edu.cn (H.L.); suzhenqi80@163.com (Z.S.)

[4] Key Laboratory of Molecular Module-Based Breeding of High Yield and Abiotic Resistant Plants in Universities of Shandong, School of Agriculture, Ludong University, Yantai 264025, China

[5] State Key Laboratory of North China Crop Improvement and Regulation, College of Agronomy, Hebei Agricultural University, Baoding 071000, China

[*] Correspondence: sdaucf@126.com (F.C.); lqsong@sjziam.ac.cn (L.S.); Tel.: +86-311-8588-7272 (L.S.); Fax: +86-311-8581-5093 (L.S.)

[†] The authors have contributed equally to this work.

Citation: Liu, J.; Zhang, Q.; Meng, D.; Ren, X.; Li, H.; Su, Z.; Zhang, N.; Zhi, L.; Ji, J.; Li, J.; et al. *QMrl-7B* Enhances Root System, Biomass, Nitrogen Accumulation and Yield in Bread Wheat. *Plants* **2021**, *10*, 764. https://doi.org/10.3390/plants10040764

Academic Editor: Igor G. Loskutov

Received: 16 March 2021
Accepted: 10 April 2021
Published: 13 April 2021

Publisher's Note: MDPI stays neutral with regard to jurisdictional claims in published maps and institutional affiliations.

Abstract: Genetic improvement of root systems is an efficient approach to improve yield potential and nitrogen use efficiency (NUE) of crops. *QMrl-7B* was a major stable quantitative trait locus (QTL) controlling the maximum root length in wheat (*Triticum aestivum* L). Two types of near isogenic lines (A-NILs with superior and B-NILs with inferior alleles) were used to specify the effects of *QMrl-7B* on root, grain output and nitrogen-related traits under both low nitrogen (LN) and high nitrogen (HN) environments. Trials in two consecutive growing seasons showed that the root traits, including root length (RL), root area (RA) and root dry weight (RDW), of the A-NILs were higher than those of the B-NILs at seedling stage (SS) before winter, jointing stage (JS), 10 days post anthesis (PA10) and maturity (MS), respectively. Under the LN environment, in particular, all the root traits showed significant differences between the two types of NILs ($p < 0.05$). In contrast, there were no critical differences in aerial biomass and aerial N accumulation (ANA) between the two types of NILs at SS and JS stages. At PA10 stage, the aerial biomass and ANA of the A-NILs were significantly higher than those of the B-NILs under both LN and HN environments ($p < 0.05$). At MS stage, the A-NILs also exhibited significantly higher thousand-grain weight (TGW), plot grain yield, harvest index (HI), grain N accumulation (GNA), nitrogen harvest index (NHI) and nitrogen partial factor productivity (NPFP) than the B-NILs under the corresponding environments ($p < 0.05$). In summary, the *QMrl-7B* A-NILs manifested larger root systems compared to the B-NILs which is favorable to N uptake and accumulation, and eventually enhanced grain production. This research provides valuable information for genetic improvement of root traits and breeding elite wheat varieties with high yield potential and NPFP.

Keywords: *Triticum aestivum* L.; *QMrl-7B*; root traits; grain yield; nitrogen use efficiency

1. Introduction

Bread wheat (*Triticum aestivum* L.) is one of the major crops worldwide; its production greatly affects food security and the global economy [1]. In general, high grain productivity largely depends on water and fertilizer input. However, over-application of fertilizers has led to not only natural resources exhaustion, but also soil, air and water quality degradation [2,3]. To resolve these environmental issues and ensure food security, breeding crops with efficient use of water and nutrients is urgently required for sustainable agriculture [4]. Roots are the primary organs that determine the acquisition efficiency of soil resources and have a direct impact on grain yield [5]; more and deeper roots may improve the water and mineral uptake from deeper soil layers and reduce nitrate leaching losses to the environment [6]. Although root traits are difficult to characterize and their breeding values are seldom assessed under field conditions, manipulating root system architecture to enhance nutrient uptake has been proposed to enable a very much needed new green revolution and further increase in yield potential [7].

Root traits can be dissected into root number, root length (RL), root weight, root surface area (RA), root volume, root thickness, and density of primary roots, lateral roots and adventitious roots as well as root/shoot dry weight ratio, etc. [8,9]. Since the 1990s, a large number of quantitative trait loci (QTLs) controlling root system architecture have been reported in rice and some of them have been successfully cloned [10]. In maize, several major QTLs involving root morphology have been detected, but no causal genes have been reported yet [11]. In recent years, a good many QTLs for root traits in wheat have been also documented [9,12–16]. However, most of these QTLs were identified at seedling stage in hydroponic culture. It was not clear whether these root-related QTLs were associated with yield-related traits in most cases. Considering that root is mainly grown in soil and root traits are plastic in adapting to environmental factors such as limitation of water [17] and nutrients [18], field experiments under diverse environments are necessary to elucidate the genetic effects of QTLs identified in hydroponic culture. With precise evaluation and verification at the population level, QTLs associated with root traits may be used in molecular wheat breeding practice.

Nitrogen (N), as the key element of proteins and other biomacromolecules, is quantitatively the most important mineral nutrient for plant growth and development. Application of enough synthetic N fertilizers at the appropriate time can overwhelmingly improve crop yield [19]. However, only 30%~40% of the applied N fertilizer is taken up from soil by crops. Therefore, improving nitrogen use efficiency (NUE) in crops can help minimize the detrimental impact of N fertilizers on the environment and be favorable for sustainable agriculture [20,21]. As a result, a number of NUE-improved cultivars of main cereal crops have been released. Numerous research studies on rice [22,23], maize [24,25] and barley [26] indicated that root traits are closely related with N uptake and genetically controlled by major QTLs [27,28]. In wheat, several studies also discovered the co-localization of QTLs for root traits, nitrogen uptake and grain productivity [8,9]. These results presented the common genetic basis of root traits and N utilization, suggesting the tremendous potential of root traits in improving grain yield and NUE. Nevertheless, more sufficient understanding of the role of the key loci conferring high NUE will facilitate its future application in molecular breeding.

Near-isogenic lines (NILs) are powerful tools to characterize the gene/QTL function for certain plant traits [29]. We [9] detected a major stable QTL, named *QMrl-7B*, controlling the maximum root length of wheat at seedling stage in hydroponic culture and developed a pair of *QMrl-7B* NILs with superior and inferior alleles, respectively. The objective of this study was to specify *QMrl-7B*'s genetic effects on root, above-ground biomass, grain yield and nitrogen accumulation, using the pair of *QMrl-7B* NILs as materials at the population level under different nitrogen environments, which would provide a valuable resource for molecular improvement of root traits.

2. Results

2.1. Root Morphology of QMrl-7B NILs

Field trials showed that the root traits of KN9204 and the *QMrl-7B* NILs displayed the tendency of rapid increase in the initial seedling stage and then gradual decrease with the advancement of the growth period of wheat, and the highest values of root length, root area and root dry weight of the three genotypes were recorded at the stage of 10 days post anthesis (Figure 1, Table 1). Identical changing trends in root traits were observed in both 2017~2018 and 2018~2019 growing seasons.

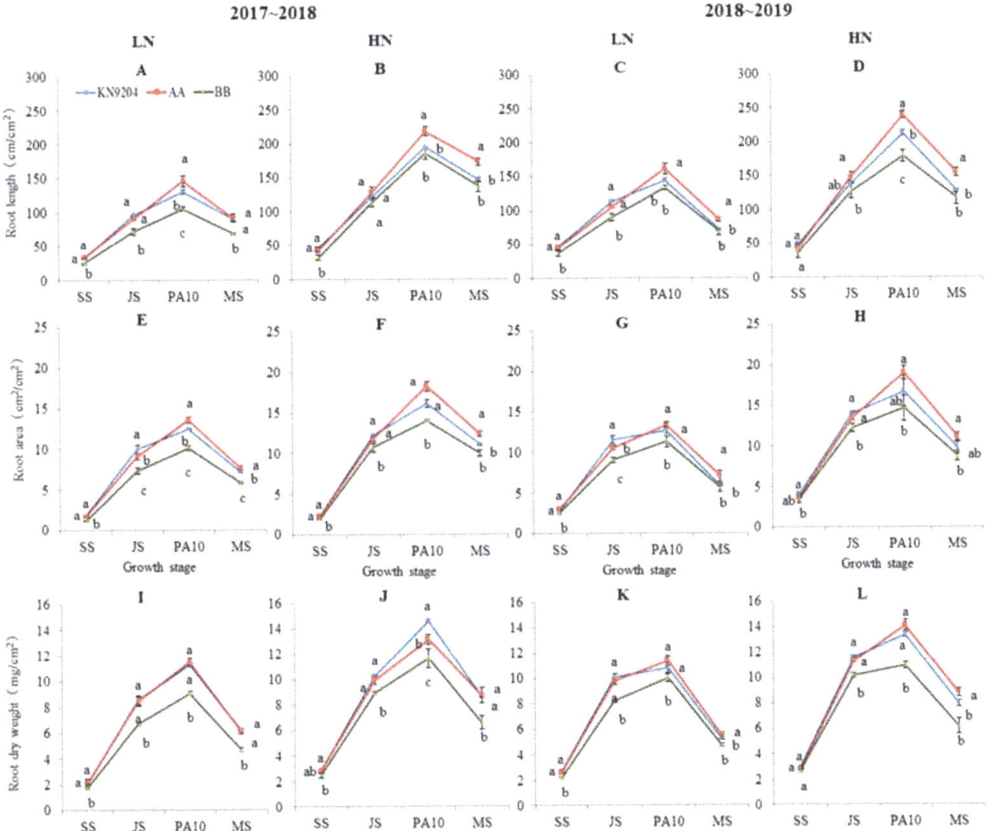

Figure 1. Root length (RL) (**A–D**), root surface area (RA) (**E–H**) and root dry weight (RDW) (**I–L**) of KN9204 and the *QMrl-7B* near isogenic lines (NILs) at different stages. Note: 2017~2018 and 2018~2019 indicate growing seasons; LN and HN indicate low nitrogen and high nitrogen environments, respectively; AA indicates *QMrl-7B* NILs with the superior alleles; BB indicates *QMrl-7B* NILs with the inferior alleles. SS, JS, PA10 and MS indicate seedling stage, jointing stage, 10 days post anthesis and maturity, respectively. Different lowercases indicate significant differences ($p < 0.05$) among the materials.

2.1.1. Root Length (RL)

In the 2017~2018 growing season, the mean RLs of A-NILs vs. B-NILs at SS, JS, PA10 and MS were 33.6 vs. 25.1, 90.2 vs. 72.1, 146.3 vs. 105.1 and 92.2 vs. 69.0 cm/cm^2 under the LN environment (Table 1, Figure 1A), and 42.8 vs. 32.5, 128.2 vs. 113.3, 218.4 vs. 185.6 and 173.4 vs. 137.2 cm/cm^2 under the HN environment (Table 1, Figure 1B), respectively; indicating that RLs of the A-NILs increased 33.9%, 25.1%, 39.2% and 33.6% under LN environment, and 31.7%, 13.2%, 17.7% and 26.4% under HN environment in comparison to those of the B-NILs at the comparable stages ($p < 0.05$). In the 2018~2019 growing season, the mean RLs of the A-NILs at the comparable stages were also significantly longer than those of the B-NILs under the corresponding nitrogen environments, except the RLs at JS stage under the HN environment (Table 1; Figure 1C,D).

Table 1. The RL, RA and RDW of KN9204 and the *QMrl-7B* NILs in different growing stages.

Root Trait	Growing Season	Material	LN				HN			
			SS	JS	PA10	MS	SS	JS	PA10	MS
RL (cm/cm^2)	2017	KN9204	32.6 ± 0.8 a	96.3 ± 0.9 a	131.0 ± 3.0 b	90.4 ± 3.0 a	46.6 ± 2.5 a	121.6 ± 0.8 a	194.5 ± 1.7 b	146.7 ± 3.2 b
	~	AA	33.6 ± 0.7 a	90.2 ± 1.2 a	146.3 ± 7.5 a	92.2 ± 5.7 a	42.8 ± 2.4 a	128.2 ± 8.4 a	218.4 ± 7.1 a	173.4 ± 5.1 a
	2018	BB	25.1 ± 1.2 b	72.1 ± 4.6 b	105.1 ± 3.2 c	69.0 ± 1.6 b	32.5 ± 4.4 b	113.3 ± 6.4 b	185.6 ± 8.2 b	137.2 ± 8.2 b
	2018	KN9204	46.3 ± 1.3 a	114.0 ± 1.1 a	144.7 ± 2.4 b	71.7 ± 1.8 b	49.8 ± 2.4 a	139.0 ± 0.8 ab	213.0 ± 3.6 b	127.9 ± 2.8 b
	~	AA	44.7 ± 3.2 a	105.4 ± 1.5 a	161.2 ± 8.4 a	86.4 ± 3.0 a	43.3 ± 5.2 a	147.9 ± 8.0 a	239.7 ± 5.7 a	154.8 ± 6.0 a
	2019	BB	36.8 ± 3.8 b	90.2 ± 5.9 b	134.0 ± 3.2 b	68.7 ± 5.7 b	37.4 ± 8.4 b	127.4 ± 10.6 b	179.1 ± 8.5 c	117.1 ± 10.2 b
RA (cm^2/cm^2)	2017	KN9204	1.9 ± 0.11 a	10.0 ± 0.5 a	12.4 ± 0.1 b	7.2 ± 0.19 b	2.4 ± 0.06 a	12.3 ± 0.5 a	16.1 ± 0.3 a	11.1 ± 0.11 ab
	~	AA	1.7 ± 0.04 a	9.1 ± 0.3 b	13.6 ± 0.4 a	7.6 ± 0.20 a	2.3 ± 0.18 a	11.8 ± 0.3 a	18.3 ± 0.4 a	12.3 ± 0.7 a
	2018	BB	1.2 ± 0.06 b	7.3 ± 0.4 c	10.1 ± 0.3 c	5.8 ± 0.10 c	2.0 ± 0.14 b	10.7 ± 0.3 b	14.0 ± 0.7 b	10.0 ± 0.5 b
	2018	KN9204	2.9 ± 0.03 a	11.6 ± 0.1 a	12.7 ± 0.5 a	5.9 ± 0.10 b	4.1 ± 0.28 a	14.0 ± 0.3 a	16.7 ± 1.9 ab	9.8 ± 0.78 ab
	~	AA	3.1 ± 0.20 a	10.5 ± 0.3 b	13.3 ± 0.6 a	7.1 ± 0.31 a	3.6 ± 0.07 ab	13.5 ± 0.3 a	19.0 ± 0.8 a	11.1 ± 0.5 a
	2019	BB	2.6 ± 0.22 b	9.1 ± 0.6 c	11.3 ± 0.2 b	5.6 ± 0.40 b	3.5 ± 0.47 b	12.2 ± 0.4 b	14.6 ± 1.5 b	8.8 ± 0.61 b
RDW (mg/cm^2)	2017	KN9204	2.2 ± 0.09 a	8.5 ± 0.37 a	11.3 ± 0.2 a	6.2 ± 0.11 a	2.9 ± 0.10 a	10.3 ± 0.1 a	14.6 ± 0.1 a	8.7 ± 0.10 a
	~	AA	2.2 ± 0.11 a	8.5 ± 0.37 a	11.5 ± 0.3 a	6.1 ± 0.21 a	2.8 ± 0.18 ab	10.0 ± 0.31 a	13.2 ± 0.4 b	8.8 ± 0.60 a
	2018	BB	1.8 ± 0.12 b	6.7 ± 0.13 b	9.1 ± 0.2 b	4.7 ± 0.18 b	2.5 ± 0.15 b	9.0 ± 0.15 b	11.7 ± 0.8 c	6.6 ± 0.55 b
	2018	KN9204	2.7 ± 0.10 a	10.1 ± 0.28 a	10.8 ± 0.4 a	5.2 ± 0.02 b	3.0 ± 0.10 a	11.7 ± 0.13 a	13.4 ± 0.2 a	8.0 ± 0.25 b
	~	AA	2.6 ± 0.05 a	9.8 ± 0.28 a	11.4 ± 0.1 a	5.6 ± 0.10 a	2.8 ± 0.05 a	11.4 ± 0.15 a	14.1 ± 0.4 a	8.8 ± 0.30 a
	2019	BB	2.2 ± 0.09 b	8.3 ± 0.15 b	10.1 ± 0.4 b	4.8 ± 0.08 b	2.7 ± 0.15 a	10.2 ± 0.18 b	11.0 ± 0.2 b	6.2 ± 0.58 c

Note: RL, RA and RDW indicate root length, root surface area and root dry weight, respectively; 2017~2018 and 2018~2019 indicate growing seasons; LN and HN indicate low nitrogen and high nitrogen environments, respectively; AA indicates *QMrl-7B* NILs with the superior alleles; BB indicates *QMrl-7B* NILs with the inferior alleles. Different lowercases indicate significant differences ($p < 0.05$) among materials at the same environment; SS, JS, PA10 and MS indicate seedling stage, jointing stage, 10 days post anthesis and maturity, respectively.

2.1.2. Root Surface Area (RA)

In the 2017~2018 growing season, likewise, the mean RAs of A-NILs vs. B-NILs at SS, JS, PA10 and MS were 1.7 vs. 1.2, 9.1 vs. 7.3, 13.6 vs. 10.1 and 7.6 vs. 5.8 cm^2/cm^2 under the LN environment (Table 1, Figure 1E), and 2.3 vs. 2.0, 11.8 vs. 10.7, 18.3 vs. 14.0 and 12.3 vs. 10.0 cm^2/cm^2 under the HN environment (Table 1, Figure 1F), respectively; indicating that the mean RAs of the A-NILs increased by 41.7%, 24.7%, 34.7% and 31.0% under the LN environment, and 15%, 10.3%, 30.7% and 23% under the HN environment higher than those of the B-NILs at the comparable stages ($p < 0.05$). In the 2018~2019 growing season, the unvarying trends in RA difference between the two types of NILs were observed at the comparable growth stages under the corresponding nitrogen environments, except the RAs at SS stage under the HN environment (Table 1; Figure 1G,H).

2.1.3. Root Dry Weight (RDW)

In the 2017~2018 growing season, similarly, the mean RDWs of A-NILs vs. B-NILs at SS, JS, PA10 and MS were 2.2 vs. 1.8, 8.5 vs. 6.7, 11.5 vs. 9.1 and 6.1 vs. 4.7 mg/cm^2, respectively, under the LN environment (Table 1, Figure 1I), indicating that the A-NILs were heavier than the B-NILs by 22.2%, 26.9%, 26.4% and 29.8% in RDW at the four growth stages ($p < 0.05$). Under the HN environment, the mean RDWs of A-NILs vs. B-NILs at SS, JS, PA10 and MS were 2.8 vs. 2.5, 10.0 vs. 9.0, 13.2 vs. 11.7 and 8.8 vs. 6.6 mg/cm^2, respectively (Table 1, Figure 1J), indicating that the A-NILs were 12.0%, 11.1%, 12.8% and 33.3% heavier than the B-NILs in RDW at the four growth stages ($p < 0.05$). In the 2018~2019 growing season, the mean RDWs of the A-NILs at the comparable stages were also significantly heavier than those of the B-NILs under the corresponding nitrogen environments, except the RDW at SS under the HN environment (Table 1; Figure 1K,L).

2.1.4. Root Vertical Distribution

To investigate the root distribution in soil, the RLD, RAD and RWD were measured every 10 cm of soil layer at MS, JS, PA10 and MS stages. The biggest values of RLDs, RADs and RWDs at each growth stage were recorded in the upper soil layer (0~10 cm and 10~20 cm), then gradual decrease of the root indices accompanied with the raised soil depth (Figure 2, Figures S1–S3). Noticeably, the root distribution in the 30~40 cm soil layer was much less than those in the neighboring soil layers (20~30 and 40~50 cm), which may result from the restriction effect of compact soil on root growth in this ploughed bottom layer. The A-NILs exhibited superior RLDs, RADs and RWDs over the B-NILs in each soil layer (except for 30~40 cm) at the most comparable stages ($p < 0.05$). Taking the 10~20 cm soil layer at PA10 stage as an example, the mean RLDs of the A-NILs were 3.6 cm/cm^3 in 2017~2018 and 4.0 cm/cm^3 in 2018~2019 growing seasons under the LN environment, respectively, which were 33.3% and 14.3% higher than those of the B-NILs (2.7 and 3.5 cm/cm^3), respectively (Figure 2A,C). Under the HN environment, the corresponding RLDs of the A-NILs were 5.2 and 4.9 cm/cm^3, respectively, which were 36.8% and 40.0% higher than those of the B-NILs (3.8 and 3.5 cm/cm^3) (Figure 2B,D). As expected, the RAD (Figure 2E–H) and RWD (Figure 2I–L) exhibited the consistent distribution pattern in different soil layers like RLD.

Besides, the root distribution in 0~30, 30~60, 60~100 and 100~150 cm groups of soil layers at PA10 stage was further analyzed (Table 2). The mean RL, RA and RDW of the A-NILs were significantly different from those of the B-NILs in most soil layers under the LN environment ($p < 0.05$), except for RL in the 0~30 cm soil layer and RA in the 60~100 cm soil layer in 2018~2019. Under the HN environment, significant differences in RL, RA and RDW between the two genotypes mainly took place in the 0~30 and 100~150 cm soil layers ($p < 0.05$). The ample roots of the A-NILs over the B-NILs in both upper soil and deeper soil would definitely improve water and mineral uptake, especially in water-deficient north China plain.

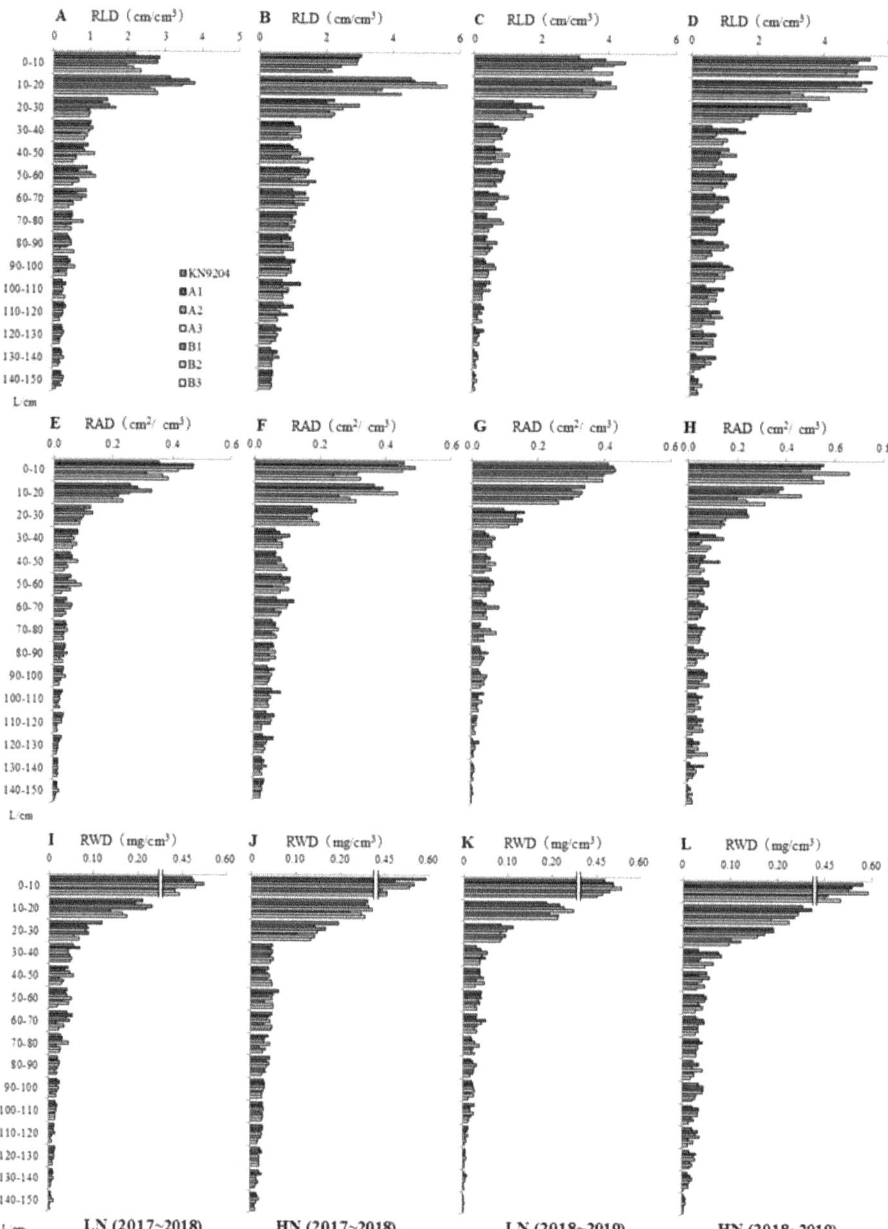

Figure 2. Root length density (RLD) (**A–D**), root area density (RAD) (**E–H**) and root weight density (RWD) (**I–L**) of KN9204 and the *QMrl-7B* near isogenic lines (NILs) in different soil layers at 10 days post anthesis. 2017~2018 and 2018~2019 indicate growing seasons; LN and HN indicate low nitrogen and high nitrogen environments, respectively; L indicates soil layer.

Table 2. The root traits of KN9204 and the *QMrl-7B* NILs in different soil layers at 10 days post anthesis.

Root Trait	Growing Season	Soil Layer (cm)	LN			HN			SV		
			KN9204	AA	BB	KN9204	AA	BB	E	G	E*G
RL (cm/cm^2)	2017 ~ 2018	0~30	67.4 ± 3.1 b	79.6 ± 1.5 a	58.3 ± 3.2 c	98.2 ± 0.3 a	106.2 ± 8.6 a	81.6 ± 4.7 b	**	**	ns
		30~60	29.0 ± 1.0 a	28.9 ± 4.0 a	20.4 ± 1.8 b	31.2 ± 0.5 a	37.8 ± 1.7 a	37.5 ± 9.6 a	**	ns	*
		60~100	22.4 ± 4.4 ab	23.7 ± 3.1 a	16.4 ± 3.6 b	37.4 ± 0.8 a	42.2 ± 5.3 a	39.4 ± 4.3 a	**	ns	ns
		100~150	12.2 ± 0.6 b	14.2 ± 0.4 a	10.0 ± 1.0 c	27.6 ± 2.2 b	32.2 ± 3.4 a	27.1 ± 2.0 b	**	*	ns
	2018 ~ 2019	0~30	81.5 ± 2.6 b	98.8 ± 4 a	86.4 ± 4.8 b	142.5 ± 2.8 a	132.6 ± 6.6 a	101.7 ± 4.8 b	**	**	**
		30~60	18.2 ± 0.5 b	25.8 ± 2.1 a	21.0 ± 3.0 ab	27.2 ± 0.5 a	34.6 ± 9.2 a	26.2 ± 3.7 a	**	*	ns
		60~100	15.7 ± 5.4 b	25.6 ± 2.8 a	19.9 ± 1.7 b	26.3 ± 0.1 b	41.0 ± 4.4 a	30.2 ± 3.3 b	**	**	ns
		100~150	9.3 ± 0.1 ab	10.9 ± 2.3 a	6.7 ± 1.6 b	16.9 ± 0.2 b	31.5 ± 3.7 a	21.0 ± 5.9 b	**	**	ns
RA (cm^2/cm^2)	2017 ~ 2018	0~30	7.39 ± 0.5 ab	8.58 ± 0.8 a	6.60 ± 0.4 b	9.97 ± 0.3 a	10.45 ± 0.3 a	7.56 ± 0.5 b	**	**	*
		30~60	2.04 ± 0.2 a	2.15 ± 0.3 a	1.56 ± 0.3 b	2.14 ± 0.5 a	2.70 ± 0.2 a	2.49 ± 0.5 a	**	ns	ns
		60~100	1.68 ± 0.0 a	1.76 ± 0.2 a	1.23 ± 0.2 b	2.22 ± 0.4 a	2.88 ± 0.3 a	2.36 ± 0.3 a	**	**	*
		100~150	1.31 ± 0.0 a	1.11 ± 0.1 b	0.72 ± 0.1 c	1.77 ± 0.0 ab	2.22 ± 0.5 a	1.60 ± 0.2 b	**	**	*
	2018 ~ 2019	0~30	8.45 ± 0.2 ab	8.89 ± 0.1 a	7.88 ± 0.6 b	11.78 ± 0.7 a	11.39 ± 0.5 a	9.12 ± 0.8 b	**	**	ns
		30~60	1.41 ± 0.1 b	1.80 ± 0.1 a	1.50 ± 0.2 b	1.82 ± 0.2 b	2.63 ± 0.9 a	1.86 ± 0.4 b	*	ns	ns
		60~100	1.10 ± 0.4 a	1.85 ± 0.6 a	1.43 ± 0.1 a	1.80 ± 0.4 b	2.83 ± 0.3 a	1.99 ± 0.3 b	**	**	ns
		100~150	0.73 ± 0.0 ab	0.79 ± 0.2 a	0.50 ± 0.1 b	1.30 ± 0.4 b	2.16 ± 0.7 a	1.67 ± 0.8 b	**	*	ns
RDW (mg/cm^2)	2017 ~ 2018	0~30	8.12 ± 0.1 a	8.05 ± 0.4 a	6.63 ± 0.2 b	10.42 ± 0.6 a	9.4 ± 0.5 a	8.19 ± 0.5 b	**	**	ns
		30~60	1.42 ± 0.1 ab	1.47 ± 0.1 a	1.15 ± 0.2 b	1.51 ± 0.2 a	1.29 ± 0.1 a	1.34 ± 0.2 a	ns	*	ns
		60~100	1.13 ± 0.1 a	1.32 ± 0.2 a	0.85 ± 0.1 b	1.61 ± 0.1 a	1.48 ± 0.1 ab	1.26 ± 0.2 b	**	**	ns
		100~150	0.64 ± 0.0 a	0.67 ± 0.1 a	0.45 ± 0.1 b	1.09 ± 0.0 a	1.04 ± 0.1 a	0.90 ± 0.1 b	**	**	ns
	2018 ~ 2019	0~30	7.51 ± 0.1 b	8.48 ± 0.2 a	7.70 ± 0.2 b	10.13 ± 0.5 a	9.77 ± 0.3 a	7.94 ± 0.3 b	**	**	ns
		30~60	1.09 ± 0.0 b	1.25 ± 0.1 a	1.10 ± 0.1 b	1.23 ± 0.1 a	1.55 ± 0.4 a	1.20 ± 0.3 a	ns	ns	ns
		60~100	0.91 ± 0.1 b	1.22 ± 0.2 a	0.94 ± 0.1 b	1.25 ± 0.2 b	1.59 ± 0.1 a	1.12 ± 0.1 b	**	**	ns
		100~150	0.51 ± 0.0 a	0.45 ± 0.1 a	0.32 ± 0.0 b	0.79 ± 0.0 b	1.22 ± 0.1 a	0.72 ± 0.2 b	**	**	**

Note: RL, RA and RDW indicate root length, root surface area and root dry weight, respectively; 2017~2018 and 2018~2019 indicate growing seasons; LN and HN indicate low nitrogen and high nitrogen environments, respectively; AA indicates *QMrl-7B* NILs with the superior alleles; BB indicates *QMrl-7B* NILs with the inferior alleles. Different lowercases indicate significant differences ($p < 0.05$) among materials at the same environment; SV indicate source of variation; E and G indicate environment and genotype, respectively; E*G indicate their interaction; "*" and "**" indicate significant differences at $p < 0.05$ and $p < 0.01$ levels, respectively; "ns" indicates no significant differences.

2.2. Aerial Biomass and Grain Yield of QMrl-7B NILs

2.2.1. Aerial Dry Weight (ADW)

Field trials showed that the ADWs of KN9204 and the *QMrl-7B* NILs increased gradually with the advancement of wheat development (Figure 3). In the 2017~2018 growing season, the mean ADWs of A-NILs vs. B-NILs at SS, JS, PA10 and MS were 48.1 vs. 41.0, 216.0 vs. 204.1, 573.8 vs. 525.2 and 882.2 vs. 832.7 g/m^2 under the LN environment (Figure 3A), and 100.8 vs. 89.4, 462.0 vs. 456.2, 1186.0 vs. 974.4 and 1475.2 vs. 1447.6 g/m^2 under the HN environment, respectively (Figure 3B). In the 2018~2019 growing season, the consistent trends in ADW difference between the two types of NILs were observed repeatedly at the comparable growth stages under the corresponding nitrogen environments (Figure 3C,D).

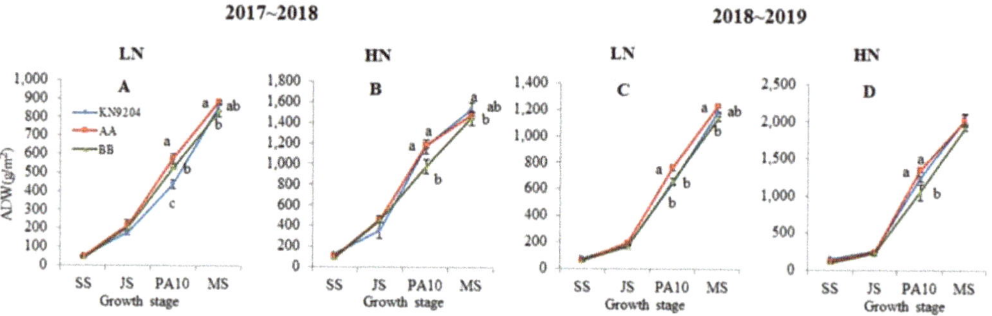

Figure 3. Aerial dry weight (ADW) of KN9204 and the *QMrl-7B* near isogenic lines (NILs) at different stages. 2017~2018 (**A,B**) and 2018~2019 (**C,D**) indicate growing seasons; LN (**A,C**) and HN (**B,D**) indicate low nitrogen and high nitrogen environments, respectively; AA indicates *QMrl-7B* NILs with the superior alleles; BB indicates *QMrl-7B* NILs with the inferior alleles; SS, JS, PA10 and MS indicate seedling stage, jointing stage, 10 days post anthesis and maturity, respectively; Different lowercases indicate significant differences ($p < 0.05$) among the genotypes at the same growth stage.

Unlike the findings in root traits, interestingly, no significant differences in ADW were found between the two types of NILs at SS and JS stages under both LN and HN environments. The biggest difference of ADW between the two types of NILs was recorded at the stage PA10 (Figure 3). The mean ADWs of A-NILs vs. B-NILs at this stage in 2018~2019 were 762.7 vs. 658.5 g/m^2 under the LN environment and 1349.1 vs. 1049.1 g/m^2 under the HN environment, respectively. This finding indicated that the A-NILs were heavier than the B-NILs in ADW by 9.3% and 15.8% under the LN environment, and 21.7% and 28.6% under the HN environment in the two trial years, respectively. Prior to harvest, no significant difference between the two types of NILs in ADW was observed under the HN environment in the two growing seasons. Under the LN environment; however, the mean ADWs of A-NILs vs. B-NILs at MS were 1231.0 vs. 1135.9 g/m^2 in 2018~2019, indicating that there were 6.76% and 8.37% phenotypic differences between the two types of NILs in the two years, respectively.

2.2.2. Grain Yield

The trends of annual variation in agronomic traits of KN9204 and the *QMrl-7B* NILs were basically the same between the two growing seasons. Under both LN and HN environments, there were no significant differences in plant height (PH), spike length (SL), total spikelets per spike (TSPS) and kernel number per spike (KNPS) between the two types of NILs, but the A-NILs manifested superior TGW and plot grain yield over the B-NILs (Table 3). Under the LN environment, the mean TGWs of the A-NILs were 38.8 g in 2017~2018 and 40.7 g in 2018~2019, respectively, which were 1.9 g (5.15%) and 3.3 g (8.82%) heavier than those of the B-NILs ($p < 0.05$). Under the HN environment, the mean TGWs of the A-NILs were 32.6 g in 2017~2018 and 37.9 g in 2018~2019, respectively, which were 5.50% and 6.76% higher than those of the B-NILs in the comparable growing seasons ($p < 0.05$). Consequently, the A-NILs yielded more than the B-NILs. Under the LN environment, GYs of the A-NILs were 4030.9 and 5735.4 kg/ha in 2017~2018 and 2018~2019, respectively; which were 454.8 kg/ha (12.72%) and 550.2 kg/ha (10.61%) heavier than those of the B-NILs ($p < 0.05$), respectively. Under the HN environment, GYs of the A-NILs were 6388.9 and 8426.8 kg/ha in 2017~2018 and 2018~2019, respectively; which were 6.40% (6004.1 kg/ha) and 9.99% (7661.4 kg/ha) higher than those of the B-NILs ($p < 0.05$), respectively. What is more, the mean HI of the A-NILs was also significantly higher than that of the B-NILs under the corresponding nitrogen environments.

2.3. Nitrogen Accumulation of QMrl-7B NILs

2.3.1. The Aerial N Content (ANC) and Accumulation (ANA)

Field trials revealed that the ANCs of KN9204 and the *QMrl-7B* NILs tended to decrease with the advancement of wheat development (Table S1, Figure 4A–D). The ANCs of A-NILs vs. B-NILs at SS, JS, PA10 and MS stages were 2.47% vs. 2.45%, 1.70% vs 1.71%, 1.33% vs. 1.22% and 1.25% vs. 1.13% in 2017~2018, and 3.53% vs 3.35%, 2.00% vs 2.01%, 1.64% vs. 1.57% and 1.80% vs. 1.62% in 2018~2019 under the LN environment, respectively. Under the HN environments, the ANCs of A-NILs vs. B-NILs at the comparable stages were 2.79% vs. 2.75%, 2.42% vs 2.40%, 1.79% vs. 1.76% and 1.65% vs. 1.50% in 2017~2018, and 3.69% vs 3.69%, 2.32% vs 2.30%, 2.05% vs. 1.95% and 1.98% vs. 1.91% in 2018~2019, respectively. The result showed that the A-NILs exhibited higher ANC than the B-NILs, but the differences were not significant in most cases. The significant differences were presented at SS and MS stages under the LN environment in 2018~2019 ($p < 0.05$).

The ANA tended to increase with the advancement of the growth period (Table S1, Figure 4E–H), but no significant differences were found between the two types of NILs at SS and JS. At PA10 and MS, on the other hand, the A-NILs exhibited significant higher ANA than the B NILs under both LN and HN environments ($p < 0.05$). At PA10 stage, the mean ANAs of the A-NILs vs. B-NILs were 7.62 vs. 6.35 and 12.51 vs. 10.37 g/m^2 under the LN environment, and 21.19 vs. 17.12 g/m^2 and 27.69 vs. 20.36 g/m^2 under the HN environment in the two growing seasons, respectively, indicating that the A-NILs were heavier than the B-NILs in ANA by 20.0%, 20.6%, 23.8% and 36.0% under the corresponding environments, respectively. At MS stage, the A-NILs also accumulated more N than the B-NILs, the mean ANAs of the A-NILs vs. B-NILs were 10.99 vs. 9.42 g/cm^2 and 22.12 vs. 18.43 g/m^2 under the LN environment, and 24.34 vs. 21.71 g/m^2 and 40.05 vs. 36.71 g/m^2 under the HN environment in the two growing seasons, respectively, indicating that ANAs of the A-NILs were higher than those of the B-NILs by 16.7% and 20.0% under the LN environment as well as 12.1% and 9.1% under the HN environment.

2.3.2. The Grain N Content (GNC) and Accumulation (GNA)

Compared to the B-NILs, the A-NILs had higher mean GNCs, but the differences were not significant (Table 4). The GNCs of A-NILs vs. B-NILs were 2.15% vs. 2.01%, 2.64% vs. 2.44% under the LN environment, and 2.51% vs. 2.23%, and 2.98% vs. 2.87% under the HN environment in the two trial years, respectively. In contrast, there were significant differences in GNAs between the two genotypes ($p < 0.05$). The GNAs of the A-NILs vs. B-NILs were 8.9 vs. 7.2 g/m^2 and 15.6 vs. 12.8 g/m^2 under the LN environment, and 16.4 vs. 13.5 g/m^2 and 25.3 vs. 22.1 g/m^2 under the HN environment in the two years, which were 23.6%, 21.9%, 21.5%, and 14.5% higher than those of the B-NILs under the corresponding environments, respectively (Table 4).

Table 3. Agronomic traits of KN9204 and the *QMrl-7B* NILs.

GS	E	Material	PH (cm)	SL (cm)	SN	TSPS	SSPS	KNPS	TGW (g)	GY (kg/ha)	HI
2017 ~ 2018	LN	KN9204	63.0 ± 3.3 b	6.6 ± 0.3 b	2.0 ± 0.0	18.1 ± 0.9 a	2.0 ± 0.8	36.9 ± 3.9 a	35.9 ± 0.7 b	3881.5 ± 19.4 b	0.46 ± 0.01 b
		AA	71.6 ± 1.4 a	7.9 ± 0.2 a	2.3 ± 0.2	15.9 ± 0.1 b	1.4 ± 0.1	33.1 ± 0.9 b	38.8 ± 0.2 a	4030.9 ± 58.9 a	0.47 ± 0.01 a
		BB	71.9 ± 1.3 a	8.0 ± 0.1 a	2.2 ± 0.1	16.2 ± 0.1 b	1.4 ± 0.2	33.2 ± 1.4 b	36.9 ± 0.3 b	3576.1 ± 76.2 c	0.43 ± 0.00 b
	HN	KN9204	67.2 ± 2.5 b	7.1 ± 0.4 b	6.4 ± 1.8	18.1 ± 1.1 a	2.1 ± 0.6	37.6 ± 5.5 a	30.1 ± 0.6 b	6851.9 ± 30.4 a	0.45 ± 0.02 a
		AA	78.0 ± 1.6 a	8.8 ± 0.2 a	6.5 ± 0.3	16.7 ± 0.3 b	1.7 ± 0.3	31.5 ± 1.4 b	32.6 ± 0.2 a	6388.9 ± 129.6 b	0.44 ± 0.01 a
		BB	76.8 ± 1.7 a	8.6 ± 0.1 a	6.3 ± 0.3	16.6 ± 0.2 b	1.8 ± 0.3	31.9 ± 0.9 b	30.9 ± 0.4 b	6004.1 ± 80.3 c	0.42 ± 0.01 b
2018 ~ 2019	LN	KN9204	70.3 ± 2.5 b	7.3 ± 0.5 b	2.8 ± 0.2	18.3 ± 0.9 a	3.4 ± 0.5	36.9 ± 1.9 a	37.8 ± 0.2 b	5257.1 ± 50.1 b	0.46 ± 0.00 b
		AA	85.0 ± 0.8 a	8.4 ± 0.2 a	3.0 ± 0.1	17.5 ± 0.3 b	2.4 ± 0.1	32.1 ± 0.9 b	40.7 ± 0.5 a	5735.4 ± 63.5 a	0.48 ± 0.01 a
		BB	83.5 ± 3.2 a	8.2 ± 0.1 a	2.9 ± 0.1	17.7 ± 0.1 b	3.0 ± 0.3	31.0 ± 1.1 b	37.4 ± 0.2 b	5185.2 ± 66.1 b	0.46 ± 0.01 b
	HN	KN9204	73.7 ± 2.5 b	8.2 ± 0.4 b	6.3 ± 2.2	19.9 ± 1.0 a	3.7 ± 0.5	36.6 ± 4.8 a	36.6 ± 0.6 ab	8054.9 ± 61.6 a	0.41 ± 0.02 ab
		AA	98.6 ± 1.3 a	8.9 ± 0.2 a	6.2 ± 0.2	19.0 ± 0.2 b	4.3 ± 0.2	29.3 ± 1.5 b	37.9 ± 0.5 a	8426.8 ± 195.2 a	0.42 ± 0.01 a
		BB	96.0 ± 0.9 a	8.5 ± 0.1 a	6.1 ± 0.1	18.4 ± 0.2 b	4.1 ± 0.5	27.7 ± 1.0 b	35.5 ± 0.5 b	7661.4 ± 244.1 b	0.40 ± 0.01 b

Note: GS indicates growing season; E indicates environment; AA indicates *QMrl-7B* NILs with the superior alleles; BB indicates *QMrl-7B* NILs with the inferior alleles; PH, SL, SN, TSPS, SSPS, KNPS, TGW, GY and HI indicate plant height, spike length, spike number, total spikelet per spike, sterile spikelet per spike, kernel number per spike, thousand-grain weight, grain yield and harvest index, respectively; different lowercases indicate significant differences ($p < 0.05$) among materials at the same environment by ANOVA.

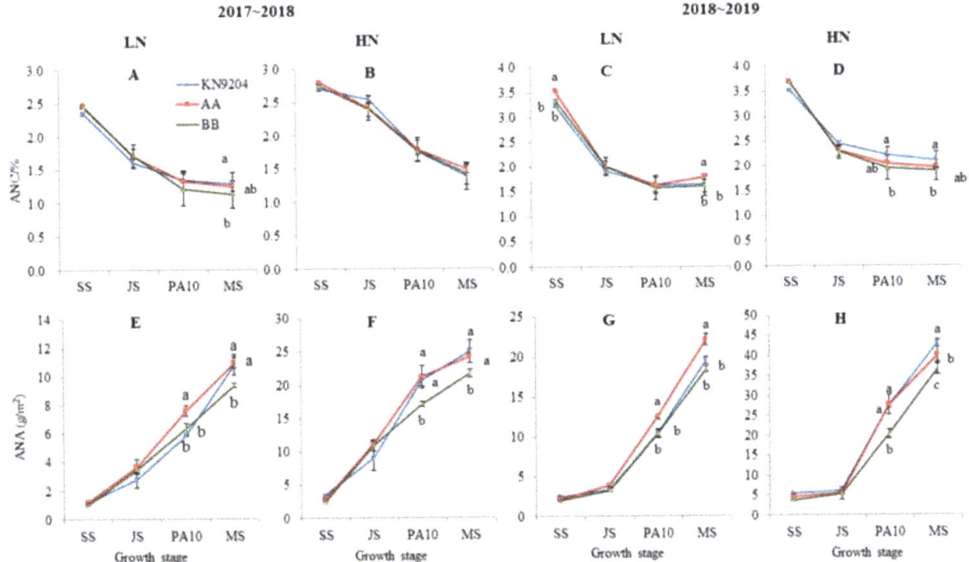

Figure 4. Aerial N content (ANC) (**A–D**) and accumulation (ANA) (**E–H**) of KN9204 and the *QMrl-7B* near isogenic lines (NILs) at different stages. 2017~2018 and 2018~2019 indicate growing seasons; LN and HN indicate low nitrogen and high nitrogen environments, respectively; AA indicates *QMrl-7B* NILs with the superior alleles; BB indicates *QMrl-7B* NILs with the inferior alleles. SS, JS, PA10 and MS indicate seedling stage, jointing stage, 10 days post anthesis and maturity, respectively; different lowercases indicate significant differences ($p < 0.05$) among the genotypes at the same growth stage.

Table 4. GNA, NHI and NPFP of KN9204 and the *QMrl-7B* NILs.

Growing Season	Environment	Material	GNC%	GNA (g/m²)	NHI	NPFP (kg kg⁻¹)
2017~2018	LN	KN9204	2.14 ± 0.17	8.5 ± 0.3 a	0.79 ± 0.02 b	–
		AA	2.15 ± 0.05	8.9 ± 0.3 a	0.81 ± 0.01 a	–
		BB	2.01 ± 0.08	7.2 ± 0.2 b	0.77 ± 0.01 b	–
	HN	KN9204	2.33 ± 0.05	16.1 ± 0.7 a	0.64 ± 0.04 ab	30.07 ± 0.8 a
		AA	2.51 ± 0.15	16.4 ± 0.5 a	0.68 ± 0.02 a	28.02 ± 0.6 a
		BB	2.23 ± 0.05	13.5 ± 0.5 b	0.62 ± 0.03 b	26.33 ± 0.4 b
2018~2019	LN	KN9204	2.43 ± 0.06	13.3 ± 0.7 b	0.68 ± 0.00 b	–
		AA	2.64 ± 0.10	15.6 ± 0.5 a	0.71 ± 0.01 a	–
		BB	2.44 ± 0.12	12.8 ± 0.2 b	0.69 ± 0.01 b	–
	HN	KN9204	2.95 ± 0.05	24.7 ± 0.9 a	0.57 ± 0.04 b	35.30 ± 0.5 a
		AA	2.98 ± 0.04	25.3 ± 0.2 a	0.63 ± 0.02 a	36.96 ± 0.9 a
		BB	2.87 ± 0.13	22.1 ± 0.6 b	0.60 ± 0.01 b	33.60 ± 0.9 b

2017~2018 and 2018~2019 indicate growing seasons; LN and HN indicate low nitrogen and high nitrogen environments, respectively; AA indicates *QMrl-7B* NILs with the superior alleles; BB indicates *QMrl-7B* NILs with the inferior alleles; GNC indicates grain N content; GNA indicates grain N accumulation; NHI indicates N harvest index; NPFP indicates partial factor productivity of applied N; different lowercases indicate significant differences ($p < 0.05$) among the genotypes at the same environment by ANOVA.

As expected, the A-NILs manifested significant higher mean NHIs in comparison to the B-NILs under both LN and HN environments ($p < 0.05$) (Table 4). The NHIs of A-NILs vs. B-NILs were 0.81 vs. 0.77 and 0.71 vs. 0.69 under the LN environment, and 0.68 vs. 0.62 and 0.63 vs. 0.60 under the HN environment in the two consecutive growing seasons, respectively, indicating that the NHIs of the A-NILs were higher than those of the B-NILs by 2.9 to 5.2% under the LN environment and 5.0 to 9.7 under the HN environment, respectively. Meanwhile, the NPFPs of the A-NILs vs. the B-NILs were 28.02 vs. 26.33 kg kg^{-1} in 2017~2018, and 36.96 vs. 33.60 kg kg^{-1} in 2018~2019, respectively, indicating that the NPFPs of the A-NILs were 6.4% to 10.0% higher than those of the B-NILs at the normal nitrogen management ($p < 0.05$) (Table 4).

3. Discussion

3.1. The Plasticity of Wheat Root Traits Is Affected by Both Genetic and Environmental Factors

A characteristic feature of plant development plasticity is that it does not follow a rigidly predefined plan but, instead, is continuously susceptible to modification by interactions with the environment [30,31]. Root architecture is a complicated trait not only controlled by endogenous genes/QTLs but also affected by soil environment. In Arabidopsis, for example, genes such as *MONOPTEROS (MP)* and *BODENLOSBDL* regulate root architecture through repressing primary root development [32,33]. In rice, Yao et al. [34] found that the short-root mutant, *srt5*, showed extreme inhibition of seminal root, crown root and lateral root elongation, as well as altered root hair formation at the seedling stage. The PIN1 family gene, *OsPIN1* and *ZmPIN1*, plays important roles in root growth in rice [35] and maize [36], respectively. In wheat, suppression of *LATERAL ROOT DENSITY (LRD)* expression in RNAi plants confers the ability to maintain root growth under water limitation and has a positive pleiotropic effect on grain size and number under optimal growth conditions [37]. Overexpression of *TaTRIP1* [38]) affects the growth of root in Arabidopsis. While knockdown of the transcription factor *TabZIP60* can increase the lateral root branching in wheat [39]. Uga et al. [22] reported that the *DRO1*, a rice quantitative trait locus controlling root growth angle, is involved in cell elongation in the root tip that causes asymmetric root growth and downward bending of the root in response to gravity. Maccaferri et al. [14] revealed 20 clusters of QTLs controlling root architecture such as root length, root number and root angle of wheat. *QMrl-7B*, a major stable QTL controlling maximum root length, was reported to regulate root development of wheat in hydroponic culture of different nitrogen conditions [9]. All the above findings indicated that root architecture is mainly controlled by both major genes as well as QTLs with moderate or minor effects.

Root plastic development is enormously influenced by environmental factors including soil water deficiency [40] and insufficient nutrient availability [18]. Developmental response to drought stress in crops is manifested through enhanced root growth and suppressed shoot growth resulting in increased root/shoot ratio [41]. According to the description of Zhang et al. [17], the root growth of bread wheat in the north China plain was even before winter, remained almost static in the winter, increased rapidly between jointing and grain filling stage, and then decreased at maturity. In the present study, a pair of *QMrl-7B* near isogenic lines experienced similar root growth patterns, the root traits including root length, root surface area and root dry weight expressed plasticity to varied soil nitrogen supplies. Interestingly, there were significant differences in root traits between the two types of *QMrl-7B* NILs from seedling till mature under both low and high nitrogen environments, indicating that the *QMrl-7B* played a vital role in the maintenance of root traits (Tables 1 and 2). *QMrl-7B* allele from KN9204 had significant positive effect on wheat root growth and development. For root vertical distribution, it was noticed that there was always significant difference between the two types of NILs, especially in deep soil, no matter what nitrogen environment there was (Tables 1 and 2). This result further supported the permanent effect of *QMrl-7B* on root development.

3.2. The Association of Root System with Nitrogen Accumulation

As an integral part of plants, roots are involved in the acquisition of water and nutrients, affecting efficiency of nitrogen uptake and utilization. Several studies in maize revealed that a larger root system contributed to effective N accumulation in N-efficient cultivars in comparison with N-inefficient cultivars [42,43]. Different wheat varieties responded to low N supply by expanding their root traits, such as root length, but manifested varied N accumulation [44]. Ehdaie et al. [45] suggested that positive and significant correlation coefficients existed between root biomass and plant N content, between root biomass and grain yield in wheat. KN9204, the donor of *QMrl-7B* superior alleles, is an efficient nitrogen use wheat cultivar [46] with long roots and large root system [47]. In the current study, the two types of NILs of *QMrl-7B* did not show a significant difference in aerial nitrogen accumulation before jointing stage, but the A-NILs, with huge root systems, exhibited enhanced N accumulation in both aerial vegetative organ at anthesis and grain over the B-NILs with small root systems, particularly under LN environment (Figure 4, Table 4). These results demonstrated that *QMrl-7B* has a prolonged positive effect on N accumulation during later vegetative growth and reproductive development of wheat.

Saengwilai et al. [48] found that maize genotypes with few crown roots in six RILs had 45% greater rooting depth in low-N soils, which further enhanced N acquisition, biomass and grain yield. Li et al. [28] detected 331 QTLs for root and NUE-related traits in maize and found about 70% of QTLs for NUE-related traits co-located in a cluster with those for root traits, suggesting genetic associations between root and NUE-related traits in most cases. Some reports in wheat revealed the linkage or co-localization of root trait QTLs with N uptake QTLs [8,12]. Using the KN9204-derived RIL population, Fan et al. [9] detected a list of QTLs for root architecture and NUE-related traits, and found most of them were mapped in nine clusters. In the present study, the pleiotropic effects of *QMrl-7B* were shown by the prolonged larger root system (Figures 1 and 2), higher N accumulation in the above-ground part and grain in the A-NILs (Table 3). In rice, Obara et al. [49] detected five QTLs for root system architecture and found that the most effective QTL increased the maximum root length and total root length 15.2–24.6%, in a near-isogenic line (NIL) over a wide range of nitrogen concentrations. Other studies showed that root and NUE-related traits might be regulated by the same gene. For example, overexpression of *TaNAC2-5A* enhanced root growth and nitrate influx rate in wheat, increased the root's ability to acquire nitrogen and nitrogen accumulation in aerial parts, and eventually allocated more nitrogen in grains [50].

3.3. The Ideal Root System Enhances Biomass, Grain Yield and NUE

Up to now, studies principally supported the theory that larger root system is positively correlated with the enhanced nutrient uptake, biomass accumulation and yield formation [51]. In the present study, the A-NILs with superior alleles at *QMrl-7B* exhibited extremely huge root systems over the B-NILs with inferior alleles from seedling till harvest. The seedling aerial biomass of the A-NILs, interestingly, were not significantly different from those of the B-NILs (Figure 1); this insignificant difference between the two types of NILs maintained till jointing stage. At PA10 stage, the aerial biomass of the A-NILs increased dramatically and surpassed that of the B-NILs remarkably ($p < 0.05$) (Figure 1). Till mature, the root dry weight of the two types of NILs paralleled the aerial biomass and grain yield linearly. These results illustrated that there is no correlation between root biomass and the aboveground biomass in early vegetative growth of the very wheat genotype, but the huge root system formed during seedling stage potentially associates with the final biomass and grain yield. The *QMrl-7B* donor parent KN9204, as a nitrogen efficient cultivar [46], bears a larger root system, but moderate tiller number and vegetative biomass in early seedling stage compared to the well-known 1RS-1BL cultivar 'Lovrin 10' [47]. Comprehensively, we proposed that the luxuriant root system, rather than abundant above-ground biomass before jointing, may be essential characteristics of modern wheat cultivars with high yield and NUE.

In wheat, deep root systems contribute to greater yield potential under drought conditions [52]. The drought-adapted genotype SeriM82 showed longer root systems in deep soil layers and higher potential grain yield [41], KN9204 with its robust root system showed high grain yield and high NUE [46]. Similarly, the A-NILs with large root systems also showed a higher aerial biomass prior to harvest than the B-NILs (Figure 3), demonstrating higher potential grain yield. Some researches pointed out that abundant roots in deep soil are essential for wheat growth and final yield, especially in deficient water and nutrient stresses [53,54]. The A-NILs manifested large root systems in the 100~150 cm soil layers under both LN and HN environments, and also showed significant higher grain yield than the B-NILs. These results demonstrated that *QMrl-7B* has a positive effect on enhanced aboveground biomass and grain yield.

Among the root system architecture traits, the maximum root length decides the root depth in soil and is considered as the most important root traits to impact crop yield [55]. Cane et al. [56] detected a QTL controlling root length on chromosome 7B co-located with grain weight in durum wheat. Fan et al. [9] found the cluster C7B had striking effect on TGW and the loci *QMrl-7B* with KN9204 allele could improve TGW by 4 g (10.64%). In the present study, the mean TGW of the A-NILs was significantly higher than that of the B-NILs by 5.15% to 8.82% under the LN environment and 5.50% to 6.76% under the HN environment, respectively (Table 3), when they were planted at the population level. But the significance was much less than the effect obtained at the individual level when the RILs (10.64%) and *QMrl-7B* NILs (9.19%) were planted in a large row [9]. It seems that planting density has vital influence on the precise evaluation of the genetic effect of *QMrl-7B*. What is more, the increased TGW devoted by *QMrl-7B* greatly contributed to plot grain yield of the A-NILs, over the B-NILs by 10.61% to 12.72% under LN environment and 6.40% to 9.99% under HN environment, respectively (Table 3). These results at the population level further showed that *QMrl-7B* is of great value in elevated grain weight and grain yield.

In conclusion, NILs with superior alleles of *QMrl-7B* not only manifested a luxuriant root system, but also had positive effects on aboveground biomass, grain yield and NPFP, indicating that *QMrl-7B* could facilitate genetic improvement of wheat root system. Therefore, this study provides a valuable case that improving root system via genetic manipulation can contributes directly to increased yield and NPFP.

4. Materials and Methods

4.1. Plant Materials and Experimental Design

A major stable QTL *QMrl-7B* (controlling the maximum root length) was identified by hydroponic culture using the recombinant inbred line population derived from the cross between KN9204 and J411 (KJ-RIL) [9]. This QTL was located in the interval 89.50–92.50 cM and the candidate physical region preliminarily ranged from 580.13 to 590.13 Mb (IWGSC1.0) [9]. A residual heterozygous line KJ-RIL239, which was heterozygous within the confidence interval of *QMrl-7B* detected by twelve PCR markers across this interval [9], was selected from F_6 progeny and self-pollinated for four generations till F_{10} progeny. Of which, two types of *QMrl-7B* NILs respectively, with superior alleles from KN9204 (A-NILs) and inferior alleles from J411 (B-NILs), were developed. In this study, the superior parent KN9204, three A-NILs (namely A1, A2 and A3) and three B-NILs (namely B1, B2 and B3) were used as materials.

The seven materials were evaluated under two different nitrogen environments in a split-plot design with three replicates at Luancheng (37°53′ N, 114°41′ E, 54 m altitude), Hebei province, China for 2017~2018 and 2018~2019 growing seasons, respectively (two years × two controlled-environments × three replicates). The low nitrogen (LN) environment was located on a long-term positioned experimental site where no nitrogen fertilizer but 600 kg ha^{-1} of superphosphate (around 16% P_2O_5) were applied throughout the growing period. In the high nitrogen (HN) environment, 300 kg ha^{-1} of diamine phosphate and 225 kg ha^{-1} of urea were applied before sowing, and 150 kg ha^{-1} of urea was applied at the elongation stage with irrigation every year. The field was irrigated twice at elongation and anthesis respectively to keep convenient soil hydraulic status for wheat growth. The soil fertility within the top tillage soil layer (0~20 cm) in each environment were measured after harvest (Table S2).

The plot was 6.3 m^2 ($7.0 \text{ m} \times 0.9 \text{ m}$) containing 6 rows 0.18 m apart, and 280 seeds were evenly planted in each row. All of the recommended agronomic practices were followed in each of the trials except for the nitrogen fertilization treatment as described above.

4.2. Root Sampling and Measurement

Roots were sampled at seedling stage before winter (SS), jointing stage (JS), 10 days post anthesis (PA10) and maturity (MS) under both LN and HN environments during the 2017~2018 and 2018~2019 growing seasons. After removing the above-ground part of the plants, the corer of 10 cm diameter was used to take the soil cores from the rows in each plot. The depth of sampling was 60, 100, 150 and 160 cm at the SS, JS, PA10 and MS at intervals of 10 cm, respectively. The soil cores were taken to the laboratory and the root samples were obtained as described by Zhang et al. [18]. The root samples were stored at $-20 \,^\circ\text{C}$ to prevent decay. On quantifying the root length (RL, cm) and root surface area (RA, cm^2), the root samples were tiled in a transparent dish to be scanned using ScanMaker i800 Plus Scanner (600 DPI) and analyzed by LA-S software (Hangzhou Wanshen Detection Technology Co., Ltd., Hangzhou, China, www.wseen.com). After being scanned, the roots were collected, oven-dried at $105 \,^\circ\text{C}$ for an hour and then kept at $80 \,^\circ\text{C}$ until constant weight to determine root dry weight (RDW, mg). The root length density (RLD, cm/cm^3), root area density (RAD, cm^2/cm^3) and root weight density (RWD, mg/cm^3) were calculated using RL, RA and RDW divided by the soil core volume.

4.3. Yield-Related Trait Evaluation

Ten representative plants in the center of the plot were randomly sampled at physiological maturity to evaluate the yield-related traits. The plant height (PH, cm), spike number per plant (SN), spike length (SL, cm), total spikelets per spike (TSPS), sterile spikelet number per spike (SSPS), kernel number per spike (KNPS) were determined. Thousand-grain weight (TGW, g) was evaluated after harvest using the Seed Counting and Analysis System of WSeen SC-G Instrument (Hangzhou Wanshen Detection Technology Co., Ltd., Hangzhou, China, www.wseen.com). The grain yield per plot (GY, kg/ha) was measured after harvest.

4.4. Measurement of Nitrogen-Related Traits

Ten representative plants in each plot were randomly sampled at the stages of SS, JS, PA10 and MS, respectively, and the aerial part was oven-dried at $105 \,^\circ\text{C}$ for an hour and then kept at $80 \,^\circ\text{C}$ until constant weight to determine dry matter accumulation (DW, g). The aerial part at the MS was further divided into shoot and grain parts. The dry matter accumulation was corrected to the aerial dry weight per unit area (aerial dry weight, ADW, g/m^2) and grain dry weight per unit area (grain dry weight, GDW, g/m^2), according to the number of plants per unit area. The dried samples were ground and sifted through a 0.5 mm sieves to determine the total aerial N content (ANC, %) and total grain N content (GNC, %) using a standard Kjeldahl procedure. Based on grain yield, dry weight and total N content, a suite of traits were calculated as follows:

Harvest index (HI) = GDW/ADW

Aerial N accumulation (ANA, g/m^2) = ANC $\times$ ADW

Grain N accumulation (GNA, g/m^2) = GNC $\times$ GDW

N harvest index (NHI) = GNA/ANA

Partial factor productivity of applied N (NPFP, kg kg^{-1}) = GY/N applied amount

Statistical analyses were conducted using the SPSS 20.0 (SPSS, Chicago, IL, United States) and the ANOVA was used to test the difference of the above traits among the genotypes at $p < 0.05$.

Supplementary Materials: The following are available online at https://www.mdpi.com/article/10.3390/plants10040764/s1, Table S1: Aerial N content (ANC) and accumulation (ANA) of KN9204 and the *QMrl-7B* NILs at different stages; Table S2: Summary of the major macronutrients in top tillage soil layer (0~20 cm) during the two growing seasons; Figure S1: Root length density (RLD), root surface area density (RAD) and root weight density (RWD) of KN9204 and the *QMrl-7B* NILs at seedling stage before winter (SS); Figure S2: Root length density (RLD), root surface area density (RAD) and root weight density (RWD) of KN9204 and the *QMrl-7B* NILs at jointing stage (JS); Figure S3: Root length density (RLD), root surface area density (RAD) and root weight density (RWD) of KN9204 and the *QMrl-7B* NILs at maturity (MS).

Author Contributions: L.S. and F.C. planned the research project. J.L. (Jiajia Liu), Q.Z., F.C. and L.S. made genotyping of the materials. J.L. (Jiajia Liu), Q.Z., L.S., J.J., N.Z., D.M., X.R., L.Z. and J.L. (Junming Li) conducted the field experiments. J.L. (Jiajia Liu), H.L., L.S. analyzed the data and wrote the manuscript. F.C., Z.S. and J.L. (Junming Li) revised the manuscript. All authors have read and agreed to the published version of the manuscript.

Funding: This research was jointly funded by the Natural Science Foundation of Hebei Provence (C2019503066), the National Key Research and Development Program of China (2016YFD0100706) and China Agriculture Research System (CARS-03).

Institutional Review Board Statement: Not applicable.

Informed Consent Statement: Not applicable.

Acknowledgments: We are grateful to our colleague Xiying Zhang for her critical review of the manuscript.

Conflicts of Interest: The authors declare that the research was conducted in the absence of any commercial or financial relationships that could be construed as potential conflicts of interest.

References

1. West, P.C.; Gerber, J.S.; Engstrom, P.M.; Mueller, N.D.; Brauman, K.A.; Carlson, K.M.; Cassidy, E.S.; Johnston, M.; Macdonald, G.K.; Ray, D.K.; et al. Leverage points for improving global food security and the environment. *Science* **2014**, *345*, 325–328. [CrossRef]
2. Galloway, J.N.; Townsend, A.R.; Erisman, J.W.; Bekunda, M.; Cai, Z.; Freney, J.R.; Martinelli, L.A.; Seitzinger, S.P.; Sutton, M.A. Transformation of the Nitrogen Cycle: Recent Trends, Questions, and Potential Solutions. *Science* **2008**, *320*, 889–892. [CrossRef]
3. Guo, J.H.; Liu, X.J.; Zhang, Y.; Shen, J.L.; Han, W.X.; Zhang, W.F.; Christie, P.; Goulding, K.W.T.; Vitousek, P.M.; Zhang, F.S. Significant Acidification in Major Chinese Croplands. *Science* **2010**, *327*, 1008–1010. [CrossRef]
4. Fillery, I.R.P. Plant-based manipulation of nitrification in soil: A new approach to managing N loss? *Plant Soil* **2007**, *294*, 1–4. [CrossRef]
5. Lynch, J.P. Roots of the second green revolution. *Aust. J. Bot.* **2007**, *55*, 493–512. [CrossRef]
6. Gastal, F.; Lemaire, G. N uptake and distribution in crops: An agronomical and ecophysiological perspective. *J. Exp. Bot.* **2002**, *53*, 789–799. [CrossRef]
7. Herder, G.D.; Van Isterdael, G.; Beeckman, T.; De Smet, I. The roots of a new green revolution. *Trends Plant Sci.* **2010**, *15*, 600–607. [CrossRef]
8. Atkinson, J.A.; Wingen, L.U.; Griffiths, M.; Pound, M.P.; Gaju, O.; Foulkes, M.J.; Le Gouis, J.; Griffiths, S.; Bennett, M.J.; King, J.; et al. Phenotyping pipeline reveals major seedling root growth QTL in hexaploid wheat. *J. Exp. Bot.* **2015**, *66*, 2283–2292. [CrossRef]
9. Fan, X.; Zhang, W.; Zhang, N.; Chen, M.; Zheng, S.; Zhao, C.; Han, J.; Liu, J.; Zhang, X.; Song, L.; et al. Identification of QTL regions for seedling root traits and their effect on nitrogen use efficiency in wheat (*Triticum aestivum* L.). *Theor. Appl. Genet.* **2018**, *131*, 2677–2698. [CrossRef]
10. Mai, C.D.; Phung, N.T.; To, H.T.; Gonin, M.; Hoang, G.T.; Nguyen, K.L.; Do, V.N.; Courtois, B.; Gantet, P. Genes controlling root development in rice. *Rice* **2014**, *7*, 30. [CrossRef]
11. Bray, A.L.; Topp, C.N. The Quantitative Genetic Control of Root Architecture in Maize. *Plant Cell Physiol.* **2018**, *59*, 1919–1930. [CrossRef]
12. An, D.; Su, J.; Liu, Q.; Zhu, Y.-G.; Tong, Y.; Li, J.; Jing, R.; Li, B.; Li, Z. Mapping QTLs for nitrogen uptake in relation to the early growth of wheat (*Triticum aestivum* L.). *Plant Soil* **2006**, *284*, 73–84. [CrossRef]
13. Ren, Y.; He, X.; Liu, D.; Li, J.; Zhao, X.; Li, B.; Tong, Y.; Zhang, A.; Li, Z. Major quantitative trait loci for seminal root morphology of wheat seedlings. *Mol. Breed.* **2012**, *30*, 139–148. [CrossRef]
14. Maccaferri, M.; El-Feki, W.; Nazemi, G.; Salvi, S.; Canè, M.A.; Colalongo, M.C.; Stefanelli, S.; Tuberosa, R. Prioritizing quantitative trait loci for root system architecture in tetraploid wheat. *J. Exp. Bot.* **2016**, *67*, 1161–1178. [CrossRef]

15. Ren, Y.; Qian, Y.; Xu, Y.; Zou, C.; Liu, D.; Zhao, X.; Zhang, A.; Tong, Y. Characterization of QTLs for Root Traits of Wheat Grown under Different Nitrogen and Phosphorus Supply Levels. *Front. Plant Sci.* **2017**, *8*, 2096. [CrossRef]

16. Li, T.; Ma, J.; Zou, Y.; Chen, G.; Ding, P.-Y.; Zhang, H.; Yang, C.-C.; Mu, Y.; Tang, H.; Liu, Y.; et al. Quantitative trait loci for seeding root traits and the relationships between root and agronomic traits in common wheat. *Genome* **2020**, *63*, 27–36. [CrossRef]

17. Zhang, X.; Pei, D.; Chen, S. Root growth and soil water utilization of winter wheat in the North China Plain. *Hydrol. Process.* **2004**, *18*, 2275–2287. [CrossRef]

18. Rasmussen, I.S.; Dresbøll, D.B.; Thorup-Kristensen, K. Winter wheat cultivars and nitrogen (N) fertilization—Effects on root growth, N uptake efficiency and N use efficiency. *Eur. J. Agron.* **2015**, *68*, 38–49. [CrossRef]

19. Raun, W.R.; Johnson, G.V. Improving Nitrogen Use Efficiency for Cereal Production. *Agron. J.* **1999**, *91*, 357–363. [CrossRef]

20. Hirel, B.; Tetu, T.; Lea, P.J.; Dubois, F. Improving Nitrogen Use Efficiency in Crops for Sustainable Agriculture. *Sustainability* **2011**, *3*, 1452–1485. [CrossRef]

21. Wu, K.; Wang, S.; Song, W.; Zhang, J.; Wang, Y.; Liu, Q.; Yu, J.; Ye, Y.; Li, S.; Chen, J.; et al. Enhanced sustainable green revolution yield via nitrogen-responsive chromatin modulation in rice. *Science* **2020**, *367*, 1–9. [CrossRef] [PubMed]

22. Uga, Y.; Sugimoto, K.; Ogawa, S.; Rane, J.; Ishitani, M.; Hara, N.; Kitomi, Y.; Inukai, Y.; Ono, K.; Kanno, N.; et al. Control of root system architecture by DEEPER ROOTING 1 increases rice yield under drought conditions. *Nat. Genet.* **2013**, *45*, 1097–1102. [CrossRef] [PubMed]

23. Kitomi, Y.; Hanzawa, E.; Kuya, N.; Inoue, H.; Hara, N.; Kawai, S.; Kanno, N.; Endo, M.; Sugimoto, K.; Yamazaki, T.; et al. Root angle modifications by theDRO1homolog improve rice yields in saline paddy fields. *Proc. Natl. Acad. Sci. USA* **2020**, *117*, 21242–21250. [CrossRef] [PubMed]

24. Lynch, J.P. Steep, cheap and deep: An ideotype to optimize water and N acquisition by maize root systems. *Ann. Bot.* **2013**, *112*, 347–357. [CrossRef]

25. Trachsel, S.; Kaeppler, S.; Brown, K.; Lynch, J. Maize root growth angles become steeper under low N conditions. *Field Crop. Res.* **2013**, *140*, 18–31. [CrossRef]

26. Jia, Z.; Liu, Y.; Gruber, B.D.; Neumann, K.; Kilian, B.; Graner, A.; Von Wirén, N. Genetic Dissection of Root System Architectural Traits in Spring Barley. *Front. Plant Sci.* **2019**, *10*, 400. [CrossRef]

27. Liu, J.; Li, J.; Chen, F.; Zhang, F.; Ren, T.; Zhuang, Z.; Mi, G. Mapping QTLs for root traits under different nitrate levels at the seedling stage in maize (*Zea mays* L.). *Plant Soil* **2008**, *305*, 253–265. [CrossRef]

28. Li, P.; Pengcheng, L.; Cai, H.; Liu, J.; Pan, Q.; Liu, Z.; Hongguang, C.; Mi, G.; Zhang, F.; Yuan, L. A genetic relationship between nitrogen use efficiency and seedling root traits in maize as revealed by QTL analysis. *J. Exp. Bot.* **2015**, *66*, 3175–3188. [CrossRef]

29. Mu, X.; Chen, F.; Wu, Q.; Chen, Q.; Wang, J.; Yuan, L.; Mi, G. Genetic improvement of root growth increases maize yield via enhanced post-silking nitrogen uptake. *Eur. J. Agron.* **2015**, *63*, 55–61. [CrossRef]

30. Forde, B.; Lorenzo, H. The nutritional control of root development. *Plant Soil* **2001**, *232*, 51–68. [CrossRef]

31. Hodge, A. The plastic plant: Root responses to heterogeneous supplies of nutrients. *New Phytol.* **2004**, *162*, 9–24. [CrossRef]

32. Hardtke, C.S.; Berleth, T. The Arabidopsis gene MONOPTEROS encodes a transcription factor mediating embryo axis formation and vascular development. *EMBO J.* **1998**, *17*, 1405–1411. [CrossRef]

33. Hamann, T.; Mayer, U.; Jürgens, G. The auxin-insensitive bodenlos mutation affects primary root formation and apical-basal patterning in the Arabidopsis embryo. *Development* **1999**, *126*, 1387–1395.

34. Yao, S.-G.; Taketa, S.; Ichii, M. A novel short-root gene that affects specifically early root development in rice (*Oryza sativa* L.). *Plant Sci.* **2002**, *163*, 207–215. [CrossRef]

35. Xu, M.; Zhu, L.; Shou, H.; Wu, P. A PIN1 Family Gene, OsPIN1, involved in Auxin-dependent Adventitious Root Emergence and Tillering in Rice. *Plant Cell Physiol.* **2005**, *46*, 1674–1681. [CrossRef]

36. Li, Z.; Zhang, X.; Zhao, Y.; Li, Y.; Zhang, G.; Peng, Z.; Zhang, J. Enhancing auxin accumulation in maize root tips improves root growth and dwarfs plant height. *Plant Biotechnol. J.* **2017**, *16*, 86–99. [CrossRef]

37. Placido, D.F.; Sandhu, J.; Sato, S.J.; Nersesian, N.; Quach, T.; Clemente, T.E.; Staswick, P.E.; Walia, H. The LATERAL ROOT DENSITY gene regulates root growth during water stress in wheat. *Plant Biotechnol. J.* **2020**, *18*, 1955–1968. [CrossRef]

38. He, X.; Fang, J.; Li, J.; Qu, B.; Ren, Y.; Ma, W.; Zhao, X.; Li, B.; Wang, D.; Li, Z.; et al. A genotypic difference in primary root length is associated with the inhibitory role of transforming growth factor-beta receptor-interacting protein-1 on root meristem size in wheat. *Plant J.* **2014**, *77*, 931–943. [CrossRef]

39. Yang, J.; Wang, M.; Li, W.; He, X.; Teng, W.; Ma, W.; Zhao, X.; Hu, M.; Li, H.; Zhang, Y.; et al. Reducing expression of a nitrate-responsive bZIP transcription factor increases grain yield and N use in wheat. *Plant Biotechnol. J.* **2019**, *17*, 1823–1833. [CrossRef]

40. Manschadi, A.M.; Christopher, J.; Devoil, P.; Hammer, G.L. The role of root architectural traits in adaptation of wheat to water-limited environments. *Funct. Plant Biol.* **2006**, *33*, 823–837. [CrossRef]

41. Sharp, R.E.; Poroyko, V.; Hejlek, L.G.; Spollen, W.G.; Springer, G.K.; Bohnert, H.J.; Nguyen, H.T. Root growth maintenance during water deficits: Physiology to functional genomics. *J. Exp. Bot.* **2004**, *55*, 2343–2351. [CrossRef]

42. Wang, Y.; Mi, G.; Chen, F.; Zhang, J.; Zhang, F. Response of Root Morphology to Nitrate Supply and its Contribution to Nitrogen Accumulation in Maize. *J. Plant Nutr.* **2005**, *27*, 2189–2202. [CrossRef]

43. Liu, J.; Chen, F.; Olokhnuud, C.; Glass, A.D.M.; Tong, Y.; Zhang, F.; Mi, G. Root size and nitrogen-uptake activity in two maize (*Zea mays*) inbred lines differing in nitrogen-use efficiency. *J. Plant Nutr. Soil Sci.* **2009**, *172*, 230–236. [CrossRef]

44. Melino, V.J.; Fiene, G.; Enju, A.; Cai, J.; Buchner, P.; Heuer, S. Genetic diversity for root plasticity and nitrogen uptake in wheat seedlings. *Funct. Plant Biol.* **2015**, *42*, 942–956. [CrossRef]

45. Ehdaie, B.; Merhaut, D.J.; Ahmadian, S.; Hoops, A.C.; Khuong, T.; Layne, A.P.; Waines, J.G. Root System Size Influences Water-Nutrient Uptake and Nitrate Leaching Potential in Wheat. *J. Agron. Crop. Sci.* **2010**, *196*, 455–466. [CrossRef]

46. Cui, Z.; Zhang, F.; Chen, X.; Li, F.; Tong, Y. Using In-Season Nitrogen Management and Wheat Cultivars to Improve Nitrogen Use Efficiency. *Soil Sci. Soc. Am. J.* **2011**, *75*, 976–983. [CrossRef]

47. Tong, Y.; Li, J.; Li, Z. Genotypic variations of nitrogen use efficiency in winter wheat (*Triticum aestivum* L.) II. Factors affecting nitrogen uptake efficiency. *Acta Bot. Boreali-Occident. Sin.* **1999**, *19*, 393–401.

48. Saengwilai, P.; Tian, X.; Lynch, J.P. Low Crown Root Number Enhances Nitrogen Acquisition from Low-Nitrogen Soils in Maize. *Plant Physiol.* **2014**, *166*, 581–589. [CrossRef] [PubMed]

49. Obara, M.; Ishimaru, T.; Abiko, T.; Fujita, D.; Kobayashi, N.; Yanagihara, S.; Fukuta, Y. Identification and characterization of quantitative trait loci for root elongation by using introgression lines with genetic background of Indica-type rice variety IR64. *Plant Biotechnol. Rep.* **2014**, *8*, 267–277. [CrossRef]

50. He, X.; Qu, B.; Li, W.; Zhao, X.; Teng, W.; Ma, W.; Ren, Y.; Li, B.; Li, Z.; Tong, Y. The nitrate inducible NAC transcription factor TaNAC2-5A controls nitrate response and increases wheat yield. *Plant Physiol.* **2015**, *169*, 1991–2005. [CrossRef] [PubMed]

51. Meister, R.; Rajani, M.; Ruzicka, D.; Schachtman, D.P. Challenges of modifying root traits in crops for agriculture. *Trends Plant Sci.* **2014**, *19*, 779–788. [CrossRef]

52. Reynolds, M.; Foulkes, M.J.; Slafer, G.A.; Berry, P.; Parry, M.A.J.; Snape, J.W.; Angus, W.J. Raising yield potential in wheat. *J. Exp. Bot.* **2009**, *60*, 1899–1918. [CrossRef]

53. Xu, C.; Tao, H.; Tian, B.; Gao, Y.; Ren, J.; Wang, P. Limited-irrigation improves water use efficiency and soil reservoir capacity through regulating root and canopy growth of winter wheat. *Field Crop. Res.* **2016**, *196*, 268–275. [CrossRef]

54. Liu, W.; Ma, G.; Wang, C.; Wang, J.; Lu, H.; Li, S.; Feng, W.; Xie, Y.; Ma, D.; Kang, G. Irrigation and Nitrogen Regimes Promote the Use of Soil Water and Nitrate Nitrogen from Deep Soil Layers by Regulating Root Growth in Wheat. *Front. Plant Sci.* **2018**, *9*, 32. [CrossRef]

55. Severini, A.D.; Wasson, A.P.; Evans, J.R.; Richards, R.A.; Watt, M. Root phenotypes at maturity in diverse wheat and triticale genotypes grown in three field experiments: Relationships to shoot selection, biomass, grain yield, flowering time, and environment. *Field Crop. Res.* **2020**, *255*, 107870. [CrossRef]

56. Canè, M.A.; Maccaferri, M.; Nazemi, G.; Salvi, S.; Francia, R.; Colalongo, C.; Tuberosa, R. Association mapping for root architectural traits in durum wheat seedlings as related to agronomic performance. *Mol. Breed.* **2014**, *34*, 1629–1645. [CrossRef]

Article

Genome-Wide Association Mapping of Salinity Tolerance at the Seedling Stage in a Panel of Vietnamese Landraces Reveals New Valuable QTLs for Salinity Stress Tolerance Breeding in Rice

Thao Duc Le [1], Floran Gathignol [2], Huong Thi Vu [1], Khanh Le Nguyen [3], Linh Hien Tran [1], Hien Thi Thu Vu [4], Tu Xuan Dinh [5], Françoise Lazennec [2], Xuan Hoi Pham [1], Anne-Aliénor Véry [6], Pascal Gantet [2,7,*] and Giang Thi Hoang [1,*]

1   National Key Laboratory for Plant Cell Biotechnology, Agricultural Genetics Institute, LMI RICE-2, Hanoi 00000, Vietnam; leducthao@agi.vaas.vn (T.D.L.); pcj.love96@gmail.com (H.T.V.); hienlinh.tran@gmail.com (L.H.T.); xuanhoi.pham@gmail.com (X.H.P.)
2   UMR DIADE, Université de Montpellier, IRD, 34095 Montpellier, France; floran.gathignol@gmail.com (F.G.); francoise.lazennec@umontpellier.fr (F.L.)
3   Faculty of Agricultural Technology, University of Engineering and Technology, Hanoi 00000, Vietnam; nl.khanh@vnu.edu.vn
4   Department of Genetics and Plant Breeding, Faculty of Agronomy, Vietnam National University of Agriculture, Hanoi 00000, Vietnam; vtthien@vnua.edu.vn
5   Incubation and Support Center for Technology and Science Enterprises, Hanoi 00000, Vietnam; dinhxt@gmail.com
6   UMR BPMP, Univ Montpellier, CNRS, INRAE, Institut Agro, 34060 Montpellier, France; anne-alienor.very@cnrs.fr
7   Department of Molecular Biology, Centre of the Region Haná for Biotechnological and Agricultural Research, Palacký University Olomouc, Šlechtitelů 27, 783 71 Olomouc, Czech Republic
*   Correspondence: pascal.gantet@umontpellier.fr (P.G.); nuocngamos@yahoo.com (G.T.H.); Tel.: +33-467-416-414 (P.G.); +84-397-600-496 (G.T.H.)

Citation: Le, T.D.; Gathignol, F.; Vu, H.T.; Nguyen, K.L.; Tran, L.H.; Vu, H.T.T.; Dinh, T.X.; Lazennec, F.; Pham, X.H.; Véry, A.-A.; et al. Genome-Wide Association Mapping of Salinity Tolerance at the Seedling Stage in a Panel of Vietnamese Landraces Reveals New Valuable QTLs for Salinity Stress Tolerance Breeding in Rice. *Plants* **2021**, *10*, 1088. https://doi.org/10.3390/plants10061088

Academic Editor: Igor G. Loskutov

Received: 14 April 2021
Accepted: 25 May 2021
Published: 28 May 2021

Publisher's Note: MDPI stays neutral with regard to jurisdictional claims in published maps and institutional affiliations.

Abstract: Rice tolerance to salinity stress involves diverse and complementary mechanisms, such as the regulation of genome expression, activation of specific ion-transport systems to manage excess sodium at the cell or plant level, and anatomical changes that avoid sodium penetration into the inner tissues of the plant. These complementary mechanisms can act synergistically to improve salinity tolerance in the plant, which is then interesting in breeding programs to pyramidize complementary QTLs (quantitative trait loci), to improve salinity stress tolerance of the plant at different developmental stages and in different environments. This approach presupposes the identification of salinity tolerance QTLs associated with different mechanisms involved in salinity tolerance, which requires the greatest possible genetic diversity to be explored. To contribute to this goal, we screened an original panel of 179 Vietnamese rice landraces genotyped with 21,623 SNP markers for salinity stress tolerance under 100 mM NaCl treatment, at the seedling stage, with the aim of identifying new QTLs involved in the salinity stress tolerance via a genome-wide association study (GWAS). Nine salinity tolerance-related traits, including the salt injury score, chlorophyll and water content, and K^+ and Na^+ contents were measured in leaves. GWAS analysis allowed the identification of 26 QTLs. Interestingly, ten of them were associated with several different traits, which indicates that these QTLs act pleiotropically to control the different levels of plant responses to salinity stress. Twenty-one identified QTLs colocalized with known QTLs. Several genes within these QTLs have functions related to salinity stress tolerance and are mainly involved in gene regulation, signal transduction or hormone signaling. Our study provides promising QTLs for breeding programs to enhance salinity tolerance and identifies candidate genes that should be further functionally studied to better understand salinity tolerance mechanisms in rice.

Keywords: rice; GWAS; salinity tolerance; Vietnamese landraces; QTL

1. Introduction

More than one third of cultivated lands are polluted by excess of salt (NaCl) [1]. Sodium is a toxic element for plants and this is particularly true for rice, which is often cultivated in river delta areas where irrigation water is increasingly frequently contaminated by sea water [2]. Rice is the most important food crop, feeding more than three billion people in the world [3]. In Vietnam, rice occupies 85% of the total agricultural area [4]. However, with 3620 km of coastline spreading from north to south, Vietnam has been ranked among the top five countries likely to be most affected by climate change [5]. Vietnam is one of the first rice exporters in the world and consequently plays an important role in food supply security, particularly in Asian countries [6,7]. The Mekong River Delta and Red River Delta are the main areas of rice production in Vietnam; the Mekong River Delta represents 50% of the total rice production area and supplies 90% of the rice exported by the country [7]. The Mekong River Delta is increasingly menaced by an elevation in salinity due to sea water intrusion that results from different climatic and anthropic factors [8]. According to data from the Ministry of Science and Technology of Vietnam, at the end of 2015 and the first months of 2016, saline intrusion in the Mekong River Delta reached the highest level measured during the past 100 years. In addition to global management of the causes leading to increases in salinity, it is important to breed new varieties of rice tolerant to salinity, which necessitates the identification of genetic determinants conferring salinity tolerance [9]. Several salinity tolerance QTLs (quantitative trait loci) have been identified in rice using association genetics approaches, and the mechanisms undelaying rice salinity tolerance start to be well known (for reviews see [10,11]). The mechanisms involved in rice salinity tolerance act at different levels and combine transcriptional and posttranscriptional or posttranslational regulation events that lead to sodium exclusion or compartmentation in specific cell infrastructures, osmolyte production or anatomical changes that avoid sodium penetration into the internal tissues of the plant [10]. These mechanisms act in different complementary ways that synergistically allow salinity tolerance [10]. For these reasons, it is interesting to combine genetic sources with different and complementary salinity tolerance to increase resistance to salinity, which can also buffer the susceptibility of QTL effects to environmental conditions [12,13]. To identify such complementary sources of salinity tolerance, it is necessary to look for them in the widest and most diverse panels of varieties possible.

To contribute to this goal, in this study, we screened a genotyped panel of 179 Vietnamese landrace varieties of indica and japonica rice collected in different agrosystems from North to South Vietnam for salinity tolerance [14]. Vietnamese landrace varieties are often underrepresented in the studied international panels such as the 3K panel developed by the International Rice Research Institute (IRRI), even though, they potentially constitute an original source of valuable alleles [15,16]. We already used this panel to identify valuable QTLs associated with root, leaf or panicle traits and water deficit tolerance by genome wide association study (GWAS) [17–20]. The plants were screened for salinity stress tolerance at an early developmental stage using a hydroponic culture system in the presence of 100 mM NaCl. The phenotypic standard evaluation score (SES) [21], chlorophyll and relative water content and the concentrations of K^+ and Na^+ ions were measured in leaves. GWAS revealed 26 QTLs including 10 QTLs associated with several traits. Most of these QTLs contain candidate genes that may explain their effect on salinity tolerance, and the function of the genes are further discussed.

2. Results

2.1. Phenotypic Variation and Heritability of Salinity Tolerance-Related Traits

The phenotyping experiment was conducted for three consecutive years, from 2017 to 2019. The observed salt tolerance diversity in different accessions was reproducible. The data from the last trial were chosen for performing GWAS, for which the screening protocol was improved and standardized for the Vietnamese rice landrace panel and for the parameter measurement as described in the Materials and Methods section. In this trial, on the tenth day after the start of salinization, 25 plots of 19 accessions were monitored to have simultaneously reached a score of 7, which included the susceptible check IR29. A total of 9 salinity tolerance-related traits were evaluated, three of which (leaf water content (WC), chlorophyll a to chlorophyll b ratio in leaves (Chla_b), and ratio of Na^+/K^+ in leaves (Na_K)) were computed from the directly measured traits. Statistical analysis was conducted for the full panel and the indica and japonica subpanels (Table 1). Within the full panel and the indica subpanel, significant replication and genotypic effects were observed for most of the traits, with the exception of Chla_b. Meanwhile, the genotypic effect for the chlorophyll traits of the japonica subpanel was insignificant (Table 1). The broad-sense heritability (H^2) calculated for each trait with a significant genotypic effect was moderate to high, varying from 0.40 to 0.76, while high values were recorded for WC, Score and three ion content traits.

Table 1. Phenotypic variation and broad-sense trait heritability for the three panels.

Traits	n	Mean	SD	CV	Replication Effect	Genotype Effect		
					p-Value	*p*-Value	*F*-Value	H^2
Full panel								
WC	182	59.98	10.49	17.49	<0.001	<0.001	3.500	0.71
Score	182	4.41	1.16	26.30	<0.001	<0.001	2.576	0.61
Chl_total	182	1.4	0.65	46.43	<0.001	<0.001	1.924	0.48
Chla	182	0.82	0.35	42.68	<0.001	<0.001	1.902	0.47
Chlb	182	0.52	0.28	53.85	<0.001	<0.001	1.675	0.40
Chla_b	182	1.75	0.99	56.57	0.0560	0.3127	1.063	0.06
ConcK	182	43.39	7.61	17.54	<0.001	<0.001	4.226	0.76
ConcNa	182	18.63	11.97	64.25	<0.001	<0.001	3.470	0.71
Na_K	182	0.42	0.27	64.29	<0.001	<0.001	3.342	0.70
Indica subpanel								
WC	112	58.99	10.74	18.21	<0.001	<0.001	3.570	0.72
Score	112	4.51	1.1	24.39	<0.001	<0.001	2.644	0.62
Chl_total	112	1.39	0.72	51.80	<0.001	<0.001	1.849	0.46
Chla	112	0.82	0.39	47.56	<0.001	<0.001	2.105	0.53
Chlb	112	0.51	0.3	58.82	<0.001	<0.001	1.859	0.46
Chla_b	112	1.82	1.22	67.03	0.0413	0.3759	1.051	0.05
ConcK	112	45.86	7.04	15.35	<0.001	<0.001	2.855	0.65
ConcNa	112	19.48	11.63	59.70	<0.001	<0.001	3.414	0.71
Na_K	112	0.42	0.25	59.52	<0.001	<0.001	2.992	0.67
Japonica subpanel								
WC	64	61.7	9.75	15.80	<0.001	<0.001	2.858	0.65
Score	64	4.26	1.23	28.87	0.9317	<0.001	2.525	0.60
Chl_total	64	1.38	0.52	37.68	<0.001	<0.001	2.292	0.56
Chla	64	0.81	0.29	35.80	0.0026	0.0171	1.580	0.37
Chlb	64	0.52	0.24	46.15	<0.001	0.0414	1.455	0.31
Chla_b	64	1.65	0.43	26.06	0.0072	0.4666	1.014	0.01
ConcK	64	39.38	6.8	17.27	<0.001	<0.001	2.698	0.63
ConcNa	64	17.61	12.41	70.47	<0.001	<0.001	3.056	0.67
Na_K	64	0.44	0.3	68.18	<0.001	<0.001	2.879	0.65

n: number of accessions; Rep: replication; CV: coefficient of variations; H^2: broad-sense heritability; WC: leaf water content; Score: score of visual salt injury; Chl_total: total chlorophyll content in leaves; Chla: chlorophyll a content in leaves; Chlb: chlorophyll b content in leaves; Chla_b: chlorophyll a to chlorophyll b ratio in leaves; ConcK: K^+ concentration in leaves; ConcNa: Na^+ concentration in leaves; Na_K: ratio of Na^+/K^+ in leaves.

Significant phenotypic variation was observed for all of the traits, with "full name" (CVs) ranging from 15.35% to 70.47% (Table 1). Figure 1 shows statistically significant differences in the mean values of WC, Score and ConcK between the indica and japonica subpanels. Specifically, the indica subpanel displayed a lower WC and higher Score and ConcK than the japonica subpanel (Table 1, Figure 1). Consequently, for the Vietnamese rice landrace panel used in this study, indica accessions were considered less salt-tolerant than japonica accessions.

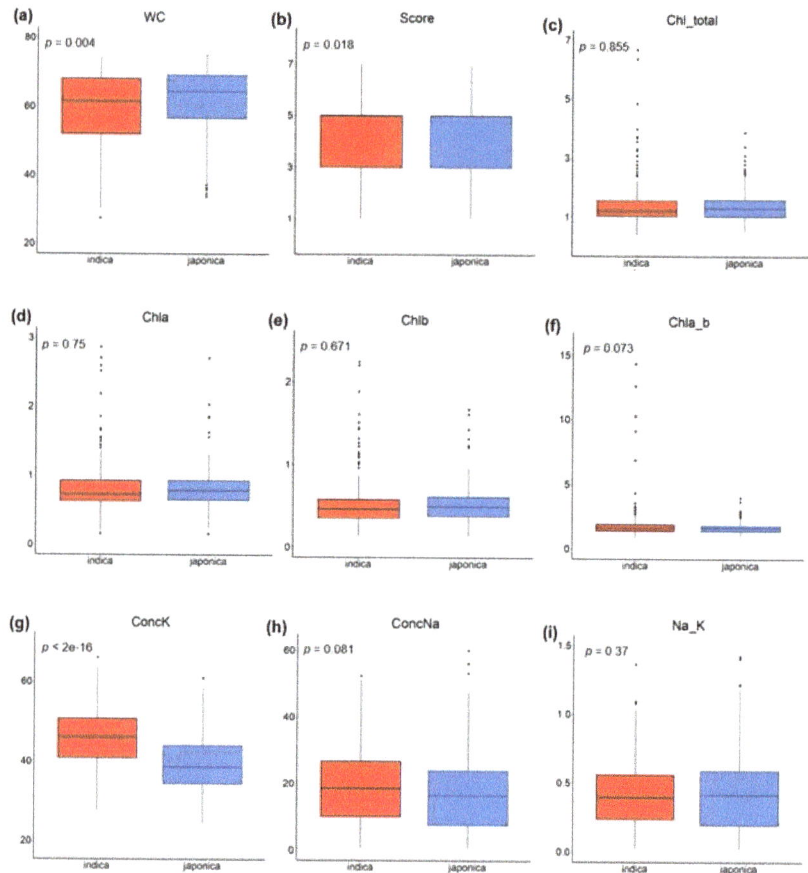

Figure 1. Boxplots of the distribution of salinity tolerance-related traits. Indica subpanel in red; japonica subpanel in blue; statistical significance (ANOVA *p*-values) between the two subpanels is indicated; (**a**) leaf water content (WC); (**b**) score of visual salt injury (Score); (**c**) total chlorophyll content in leaves (Chl_total); (**d**) chlorophyll a content in leaves (Chla); (**e**) chlorophyll b content in leaves (Chlb); (**f**) chlorophyll a to chlorophyll b ratio in leaves (Chla_b); (**g**) K^+ concentration in leaves (ConcK); (**h**) Na^+ concentration in leaves (ConcNa); (**i**) ratio of Na^+/K^+ in leaves (Na_K).

The correlations among the traits determined the same tendency within the full panel and the two subpanels (Figure S1). However, the correlation coefficients were largely variable between the traits (Table 2). For instance, Score, ConcNa and Na_K were strongly negatively correlated with WC. ConcNa and Na_K were also highly correlated with Score. In contrast, ConcK constituted weak correlations with the other traits, except for a moderate correlation with WC. Overall, higher correlations were observed among WC, Score, ConcNa and Na_K.

Table 2. Correlation coefficients between traits in the three panels (below the diagonal). Probabilities above the diagonal.

Traits	Panels	WC	Score	Chl_Total	Chla	Chlb	Chla_b	ConcK	ConcNa	Na_K
WC	F	1	<0.001	<0.001	<0.001	<0.001	0.608	<0.001	<0.001	<0.001
WC	I	1	<0.001	<0.001	<0.001	<0.001	0.709	<0.001	<0.001	<0.001
WC	J	1	<0.001	<0.001	0.004	0.012	0.469	<0.001	<0.001	<0.001
Score	F	−0.70	1	<0.001	<0.001	0.004	0.896	<0.001	<0.001	<0.001
Score	I	−0.71	1	<0.001	<0.001	0.004	0.803	<0.001	<0.001	<0.001
Score	J	−0.66	1	0.062	0.093	0.102	0.611	0.033	<0.001	<0.001
Chl_total	F	−0.34	0.17	1	<0.001	<0.001	<0.001	0.065	<0.001	<0.001
Chl_total	I	−0.44	0.24	1	<0.001	<0.001	<0.001	0.027	<0.001	<0.001
Chl_total	J	−0.26	0.13	1	<0.001	<0.001	<0.001	0.507	0.165	0.315
Chla	F	−0.31	0.18	0.82	1	<0.001	0.278	0.231	0.001	0.002
Chla	I	−0.39	0.24	0.85	1	<0.001	0.704	0.042	<0.001	<0.001
Chla	J	−0.21	0.12	0.78	1	<0.001	0.180	0.941	0.664	0.793
Chlb	F	−0.24	0.12	0.75	0.76	1	<0.001	0.845	0.007	0.008
Chlb	I	−0.29	0.16	0.73	0.72	1	<0.001	0.192	<0.001	<0.001
Chlb	J	−0.18	0.12	0.78	0.81	1	<0.001	0.865	0.556	0.739
Chla_b	F	0.02	0.01	−0.27	−0.05	−0.61	1	0.763	0.525	0.591
Chla_b	I	0.02	0.01	−0.24	−0.02	−0.62	1	0.835	0.586	0.638
Chla_b	J	0.05	-0.04	−0.32	−0.10	−0.62	1	0.438	0.611	0.817
ConcK	F	−0.41	0.32	0.08	0.05	0.01	0.01	1	<0.001	<0.001
ConcK	I	−0.43	0.39	0.12	0.11	0.07	0.01	1	<0.001	<0.001
ConcK	J	−0.33	0.15	0.05	0.01	0.01	−0.06	1	<0.001	0.005
ConcNa	F	−0.84	0.63	0.16	0.14	0.11	−0.03	0.42	1	<0.001
ConcNa	I	−0.83	0.63	0.25	0.24	0.19	−0.03	0.41	1	<0.001
ConcNa	J	−0.84	0.59	0.10	0.03	0.04	−0.04	0.41	1	<0.001
Na_K	F	−0.79	0.59	0.15	0.13	0.11	−0.02	0.20	0.96	1
Na_K	I	−0.79	0.61	0.23	0.22	0.18	−0.03	0.23	0.97	1
Na_K	J	−0.81	0.59	0.07	0.02	0.02	−0.02	0.20	0.97	1

F: full panel; I: indica subpanel; J: japonica subpanel; WC: leaf water content; Score: score of visual salt injury; Chl_total: total chlorophyll content in leaves; Chla: chlorophyll a content in leaves; Chlb: chlorophyll b content in leaves; Chla_b: chlorophyll a to chlorophyll b ratio in leaves; ConcK: K^+ concentration in leaves; ConcNa: Na^+ concentration in leaves; Na_K: ratio of Na^+/K^+ in leaves.

2.2. SNP-Trait Associations

GWAS analyses were conducted for the full panel and for the indica and japonica subpanels separately. The GWAS results are presented in the Q-Q and Manhattan plots in Figure 2 and Figure S2. Using the *p*-value threshold of 1×10^{-4}, we identified 64 associations between 58 SNPs and the studied traits, but no associations were detected in the japonica subpanel. These 58 significant SNPs were distributed in 26 QTL regions. Within the detected QTL regions, the number of significant SNPs increased to 119 when the threshold value was set at 1×10^{-3} (Table S1). Among these values, 110 SNPs were found in the full panel, 44 were identified in the indica subpanel, and 35 were common between the full panel and the indica subpanel.

A total of 16 QTLs were associated with Chla_b, 6 with WC, 6 with Score, 4 with ConcNa, 3 with Chl_total, 3 with Chlb, 2 with Chla, 3 with Na_K, and 1 with ConcK (Table 3). Ten of the 26 identified QTLs were associated with multiple traits, including QTL_25 on chromosome 11 associated with five traits (i.e., WC, Score, ConcK, ConcNa, and Na_K); QTL_21 on chromosome 9 associated with 4 traits (i.e., WC, Chla_b, ConcNa, and Na_K); three QTLs (QTL_9, QTL_20, and QTL_23) associated with three traits; and five other QTLs (i.e., QTL_13, QTL_16, QTL_17, QTL_19, and QTL_24) associated with two traits. Most of the individual trait-associated QTLs were detected for chlorophyll traits, except for QTL_1, which was related to Score. The number of significant SNPs within each QTL varied from 1 to 33, whereas QTL_25 was defined by 33 SNPs, QTL_21 by 14 SNPs, QTL_1 and QTL_4 by 8 SNPs, and QTL_16 and QTL_19 by 7 SNPs (Table 3).

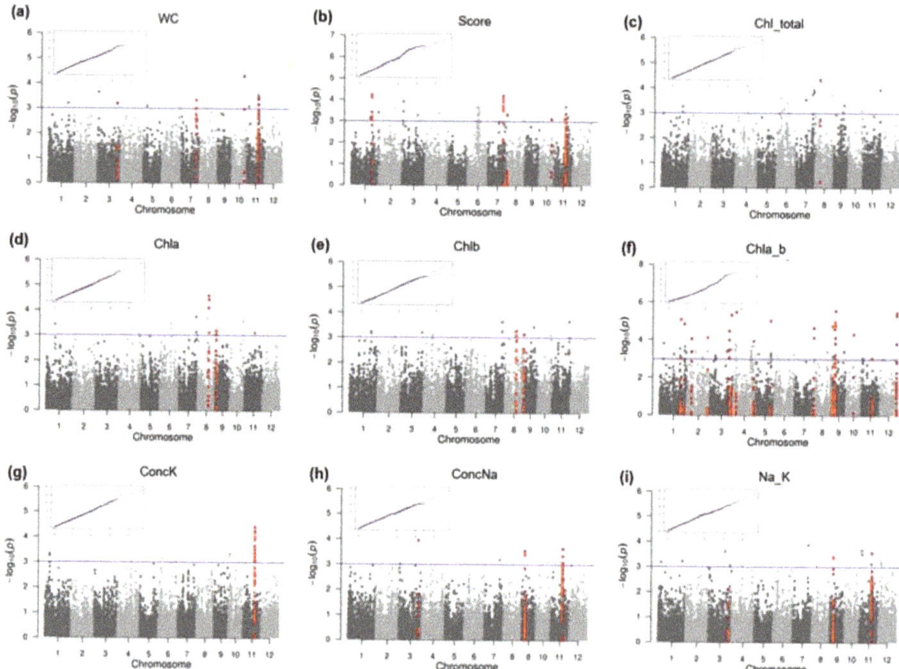

Figure 2. Manhattan plots and Q-Q plots for GWAS of salinity tolerance-related traits in the full panel. (**a**) Leaf water content (WC); (**b**) score of visual salt injury (Score); (**c**) total chlorophyll content in leaves (Chl_total); (**d**) chlorophyll a content in leaves (Chla); (**e**) chlorophyll b content in leaves (Chlb); (**f**) chlorophyll a to chlorophyll b ratio in leaves (Chla_b); (**g**) K^+ concentration in leaves (ConcK); (**h**) Na^+ concentration in leaves (ConcNa); (**i**) ratio of Na^+/K^+ in leaves (Na_K). In the Manhattan plots, significant SNPs are highlighted in red.

Therefore, among the 26 detected QTLs, QTL_25 was supposed to be the major QTL due to being mapped by the highest number (33) of significant SNPs and associated with the greatest number (5) of traits in both the full panel and the indica subpanel (Figure 3). The next was QTL_21, which was common to 4 traits and supported by 14 significant SNPs.

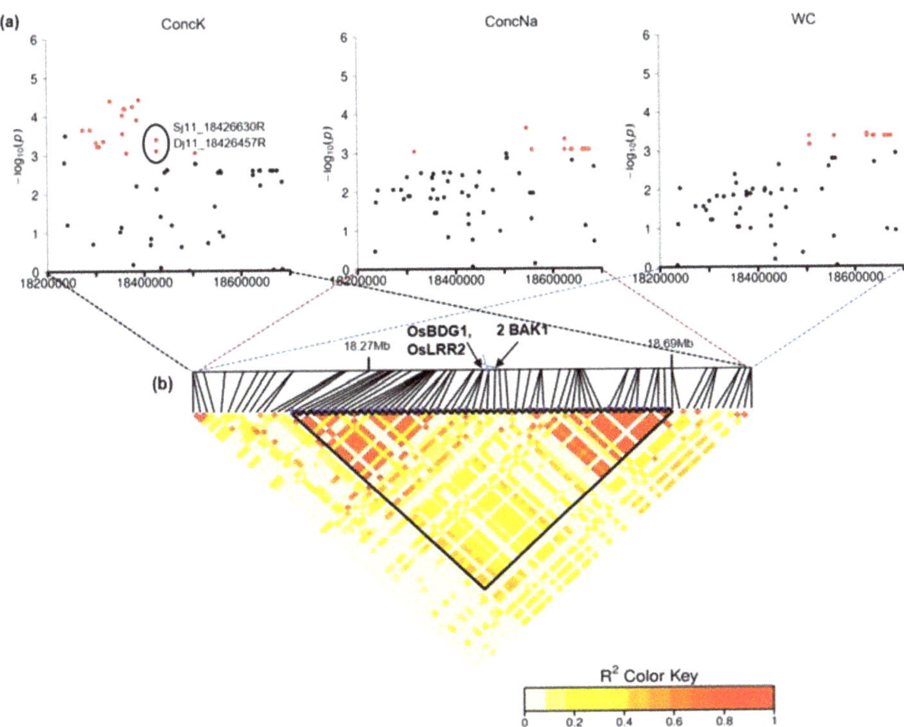

Figure 3. QTL_25. (**a**) Manhattan plot for K⁺, Na⁺ and water content in leaves in the full panel; (**b**) linkage disequilibrium (LD) heatmap. In the Manhattan plots, significant SNPs are highlighted in red, and genes of interest are mentioned. The genomic region of QTL_25 is specified in the boundary area in the LD heatmap.

2.3. Colocalizing QTLs and Candidate Genes Underlying the Detected QTLs Involved in Salinity Tolerance

The sites of the QTLs identified in this study were compared with QTLs detected in mapping populations and derived by other GWASs related to salinity tolerance. Consequently, most of our QTLs colocalized with already known QTLs, except for QTL_6, QTL_17 and QTL_22 (Table S2). We found a total of 100 colocalizations, of which 17 were detected by GWAS [22–26], and 83 other colocalizations were mapped in biparental populations [13,27–45]. In particular, 8 colocalizations shared similar traits (leaf chlorophyll content, K⁺ concentration, Na⁺ concentration, leaf water content). In addition, colocalizations with QTLs identified in previous studies for other traits using the same Vietnamese rice panel and genotyping data were observed (Table S3). For the latter, forty-nine overlapping associations were found that underlie QTL_3, QTL_6, QTL_8, QTL_17, QTL_20, and QTL_26.

In the region of almost all QTLs identified, a number of candidate genes related to the response of plants to salt or abiotic stress were found, with the exception of QTL_7, QTL_11, QTL_14, QTL_15, QTL-17, QTL_18 and QTL_22 (Table 3). These candidate genes encode different kinds of proteins including transcription factors, receptor-like kinases (RLKs), mitogen-activated kinase (MAPK), enzymes and transporters.

Table 3. List of candidate genes located within the identified QTLs.

QTL Name	Chr	QTL Position (bp)	Panel	Traits	No. of Signif. SNPs	Candidate Gene				References
						Locus	Gene	Protein	Description	
QTL_1	1	31,557,933–31,695,659	F	Score	8	Os01g54890	OsERF922	Ethylene-responsive transcription factor 2	Ethylene response transcription factor, negative regulation of salt resistance	[46]
						Os01g54930	OsVOZ1/EIP8		Vascular one zinc-finger 1/EBR1-interacting protein 8	
QTL_2	1	32,165,198–33,076,887	F	Chla_b	6	Os01g55940	OsGH3.2	IAA-amido synthetase	Modulation of free IAA and ABA homeostasis and drought and cold tolerance	[47]
						Os01g55974	STRIPE 2, ALR	Deoxycytidylate deaminase	Chloroplast development	[48,49]
						Os01g56040	OsSAP3	A20/AN1 zinc-finger protein 3	Inducibility to drought and salinity stress	[50]
						Os01g56070	OsRDCP3	RING finger protein 5	increase tolerance to drought stress in rice	[51]
						Os01g56400	OsABCI6	ABC transporter ATP-binding protein	response to abiotic stress	[52,53]
						Os01g56680	PsbW	Photosystem II reaction center W protein	photosynthesis regulation	[54]
QTL_3	1	38,515,041–38,722,651	F, I	Chla_b	1	Os01g66590	OsAS2	LOB domain protein	Regulation of shoot differentiation and leaf development	[55]
						Os01g66610		Lectin receptor-like kinase (LecRLK)	Regulation of plant growth and developmental processes in response to stress	[56]
						Os01g66420	OsPHD7	PHD finger protein (ZF-TF)	Up-regulated under drought stress	[57]

Table 3. *Cont.*

QTL Name	Chr	QTL Position (bp)	Panel	Traits	No. of Signif. SNPs	Candidate Gene				References
						Locus	Gene	Protein	Description	
QTL_4	2	6,668,466–6,853,020	F, I	Chla_b	8	*Os02g12750*	*OsTET2*	Tetraspanin domain containing protein	Response to heat, salt and water deficit stresses at seedling stage	[58]
						Os02g12790	*OsCga1*	GATA transcription factor (ZF-TF)	Regulation of chloroplast development and plant architecture, relating to natural variation in strong stay-green	[59,60]
						Os02g12794	*eEF-1B gamma*	Elongation factor 1-gamma	Salinity stress adaptation	[61]
						Os02g12800	*EF-1gamma*	Elongation factor 1-gamma	Salinity stress adaptation	[61]
QTL_5	2	32,011,340–32,679,510	F	Chla_b	1	*Os02g52290*	*OsFKBP12*	Peptidyl-prolyl cis-trans isomerase	Salinity stress response	[62,63]
						Os02g52650	*LhCa5*	Chlorophyll a/b-binding protein	Light-harvesting chlorophyll a/b-binding protein	[64]
						Os02g52670	*OsERF#103*	Ethylene-responsive transcription factor	Responsive to drought and salinity stress	[65]
						Os02g52744		DCL chloroplast precursor		
						Os02g52780	*OsbZIP23*	bZIP transcription factor	Regulation of ABA signaling and biosynthesis, salinity and drought tolerance	[66,67]
						Os02g53030	*OsRLCK84*	MAPK kinase	Response to salinity stress	[68]

Table 3. *Cont.*

QTL Name	Chr	QTL Position (bp)	Panel	Traits	No. of Signif. SNPs	Candidate Gene				References
						Locus	Gene	Protein	Description	
QTL_6	3	526,748–1,177,466	I	Chla_b	1	Os03g02010	OsDRM2	DNA methyl-transferase	Tissue- and genotype-dependent response to salinity stress	[69]
						Os03g02280	OsS40-4	S40-like protein	Response to leaf senescence and salinity stress	[70,71]
						Os03g02590	OsPEX11-1	Peroxisomal biogenesis factor 11	Relating to leaf senescence, salt responsive	[72,73]
QTL_7	3	7,197,414–7,297,414	I	Chlb	1					
QTL_8	3	29,719,291–29,898,084	F, I	Chla_b	5	Os03g51970	OsGRF6	Growth-regulating factor	Targeted by osa-miR396 and drought-up sRNA56202 responsive to salt and drought stress	[74–76]
						Os03g52090	OsECA2	Calcium-transporting ATPase 3	P-type Ca2+ ATPase IIA, harboring multiple stress-induced cis-acting elements	[77]
QTL_9	3	30,313,283–30,481,199	F	WC, Con-cNa, Na_K	1	Os03g53060				
QTL_10	3	33,128,341–33,501,467	F	Chla_b	1	Os03g58250	OsbZIP33	bZIP transcription factor	ABA-dependent enhancer of drought tolerance, responsive to high salinity, H2O2 and high temperature stress	[78]
						Os03g58300	OsIGL	Indole-3-glycerol phosphate lyase	Chloroplast precursor	
						Os03g58390	OsSIRP2	RING Ub E3 ligase	Salt and osmotic stress tolerance enhancer	[79]
						Os03g58540	TSV3/OsObgC2	Obg-like GTPase protein	Chloroplast development at the early leaf stage under cold stress	[80]

Table 3. *Cont.*

QTL Name	Chr	QTL Position (bp)	Panel	Traits	No. of Signif. SNPs	Candidate Gene				References
						Locus	Gene	Protein	Description	
QTL_11	4	4,254,414–4,354,414	F, I	Chla_b	1					
QTL_12	4	31,433,085–31,558,275	F, I	Chla_b	1	Os04g52960	OsNUC1	Nucleolin-like protein	Photosynthesis adaptation, reduction of oxidative stress and yield loss under salinity stress, enhancement of salt-stress tolerance	[81,82]
QTL_13	5	22,437,918–22,840,944	F, I	WC, Chla_b	2	Os05g38370	OsFKBP20-1a	Peptidyl-prolyl cis-trans isomerase FKBP-type	Drought and heat stress-response	[83]
						Os05g38290	OsPP2C49	Protein phosphatase 2C	Regulation of ABA-mediated signaling pathways	[84]
QTL_14	7	7,040,925–7,140,925	I	Chl_total	1					
QTL_15	7	21,360,003–21,460,003	I	Chl_total	1					
QTL_16	7	23,502,762–23,623,244	F	WC, Score	7	Os07g39270	OsGGPPS1	Geranylgeranyl pyrophosphate synthase	Chlorophyll biosynthesis	[85]
						Os07g39350		Sugar transporter	osmo protection	
						Os07g39360		Sugar transporter	osmo protection	
QTL_17	8	235,171–472,039	F, I	Score, Chla_b	4					
QTL_18	8	7,116,026–7,249,222	F	Chl_total	1					
QTL_19	8	17,191,665–17,648,853	F, I	Chla, Chlb	7	Os08g28710	OsRLCK253	Receptor-like kinase	Improvement of water-deficit and salinity stress tolerance	[86]
QTL_20	9	799,160–1,286,768	F	Chla, Chlb, Chla_b	5	Os09g02270	OsCYL4	Protein containing cyclase domain	Negative regulation of abiotic stress tolerance in relation to accumulation of ROS	[76]

Table 3. *Cont.*

QTL Name	Chr	QTL Position (bp)	Panel	Traits	No. of Signif. SNPs	Candidate Gene				References
						Locus	Gene	Protein	Description	
QTL_21	9	4,452,802–5,809,538	F, I	WC, Chla_b, Con-cNa, Na_K	14	*Os09g10600*		NADH-dependent enoyl-ACP reductase	Chloroplast precursor	
QTL_22	10	11,126,654–11,242,896	F	Chla_b	1					
QTL_23	10	18,944,166–19,070,983	F	WC, Score, Con-cNa	2	*Os10g35640*		Rf1 mitochondrial precursor (Nin-like)	Down-regulated salt-responsive, up-regulated cold-responsive	[87,88]
						Os10g35560	*OsSFR6*	Expressed protein	Osmotic stress and chilling tolerance	[89]
QTL_24	11	16,335,298–16,441,782	F, I	Score, Chla_b	1					
QTL_25	11	18,273,105–18,684,503	F, I	WC, Score, ConcK, Con-cNa, Na_K	33	*Os11g31530*	*OsBDG1*	BRASSINOSTEROID INSENSITIVE 1-associated receptor kinase 1, OsBri1	Salinity tolerance (upregulated in roots in response to salinity)	[90]
						Os11g31540	*OsLRR2*	BRASSINOSTEROID INSENSITIVE 1-associated receptor kinase 1, OsBri1	Stress tolerance (upregulated in leaves in response to cold and drought stress)	[91]
						Os11g31550		BRASSINOSTEROID INSENSITIVE 1-associated receptor kinase 1, OsBri1		
						Os11g31560		BRASSINOSTEROID INSENSITIVE 1-associated receptor kinase 1, OsBri1		

Table 3. *Cont.*

QTL Name	Chr	QTL Position (bp)	Panel	Traits	No. of Signif. SNPs	Candidate Gene				References
						Locus	Gene	Protein	Description	
QTL_26	12	25,841,227–26,215,713	F	Chla_b	5	*Os12g41860*	*OsHox33*	HDZIP III transcription factor	Targeted by a miRNA responsive to salinity stress, control of leaf senescence	[75,92]
						Os12g41950	*OsARF6b, OsARF25*	Auxin response factor	Candidate salinity tolerance-related gene at the seedling stage	[93]
						Os12g42060	*OsWAK128*	OsWAK receptor-like kinase	Candidate salinity tolerance-related gene at the seedling stage	[93]
						Os12g42070	*OsRLCK375, OsWAK129*	OsWAK receptor-like kinase	Down-regulated in cold, salt and drought stress conditions at the seedling stage	[94]
						Os12g42090		37 kDa inner envelope membrane protein	Chloroplast precursor, salinity-inducible	[93,95]
						Os12g42200	*OsCHX15*	ATCHX protein	Cation H+ antiporter, candidate salinity tolerance-related gene at the seedling stage	[93]
						Os12g42250	*OsZFP213, PINE1*	C2H2 transcription factor	Interacting with OsMAPK3 to enhance salinity tolerance by enhancing ROS-scavenging ability, regulating internode elongation and photoperiodic signals	[96,97]

3. Discussion

Rice is considered to be very sensitive to salinity [98,99]. Here, to determine the response of the Vietnamese rice landrace panel to salinity, a moderate salinity stress (100 mM NaCl) was applied at the seedling stage. We assessed a total of 9 phenotypic traits, all of which showed high variability within the panel in response to salinity stress. Of these 9 traits, WC, Score and three ion content traits (ConcNa, ConcK and Na_K) exhibited high heritability (0.60–0.76). Additionally, strong correlations (0.59–0.97) were observed among these traits with the exception of ConcK, indicating that WC, Score, ConcNa and Na_K were strongly associated with the response of rice plants to salinity stress, which is consistent with previous studies on rice salinity tolerance evaluation [100–102]. WC is a physiological parameter of the plant water status that expresses the response of plants to osmotic stress [103,104], ionic content traits reflect the level of ionic stress (ion homeostasis) [105], and the salt injury score is an indicator of plant damage/survival (growth performance) under salinity stress [100]. Previous studies reported that rice accessions tolerant to salinity stress have the ability to reduce the osmotic stress, prevent the excess accumulation of Na^+ and absorb greater K^+ to maintain a low shoot Na^+/K^+ ratio [106–109].

Correlations among the traits varied in the same direction in the full panel and the two subpanels. However, the japonica subpanel had, on average, greater WC and lower Score than the indica subpanel, indicating that japonica accessions are more salt-tolerant than indica accessions. This finding contradicts the results reported in a previous study [101] that used 4 japonica varieties and 6 indica varieties. This contradiction can be explained by the difference in the number of rice accessions included in the screening. In our study, 112 indica and 64 japonica accessions were evaluated.

In this study, GWAS analyses were applied for the full panel and for the two subpanels. Thus far, we succeeded in identifying 119 significant SNPs assigned to 26 QTLs. Twenty-two QTLs were detected in the full panel and 15 QTLs were detected only in the indica subpanel, but no japonica-specific QTLs were found, although japonica seems to have an average higher salinity tolerance than indica accessions. Similarly, in other studies using the same rice panel and genotyping data screened for water deficit tolerance and leaf traits, no japonica-specific QTLs were detected [17,18], likely because japonica accessions represent only one-third of the total accessions of the panel. In this study, of the 26 identified QTLs, 11 QTLs were detected in both the full panel and the indica subpanel, and 10 QTLs were associated with two or more traits (Table 3). Interestingly, all the QTLs that were detected for WC colocalized with QTLs associated with Score and/or ion content traits, except for QTL_13, which was found for WC and Chla_b. These results suggest that WC, Score and ion content traits have a shared genetic basis related to salinity stress responses. However, there was no overlap between QTLs detected for chlorophyll-related traits and ion content traits.

The QTLs discovered in this study were located on most chromosomes, apart from chromosome 6, and we found 3 QTLs (QTL_1, QTL_2 and QTL3) on chromosome 1, but none of these QTLs colocalized with *Saltol*, a well-known major QTL for rice salinity tolerance at the seedling stage [110,111]. A large number of QTLs for salinity tolerance detected in this study colocalized with QTLs detected in other studies and populations under conditions of salt stress at vegetative or reproductive stages, which validates our approach (Tables S2 and S3). Interestingly, of 26 QTLs identified in the present study, 3 QTLs (QTL_6, QTL_17 and QTL_22) did not colocalize with previously reported QTLs and thus constituted novel QTLs. They can be of high interest to bring new salinity tolerance sinks into breeding programs.

The major QTL identified in our study, QTL_25 at 18,273.1–18,684.5 kb on chromosome 11 was associated with WC, Score, ConcK, ConcNa and Na_K (Table 3, Figure 3). In particular, QTL_25 was mapped by 33 significant SNPs, and each of them contributed 5.75–14.09% to the phenotypic variation (Table S1). QTL_25 colocalized with previously identified QTLs under conditions of salinity stress using different mapping populations (Table S2), i.e., with 2 QTLs associated with leaf water content [42], with QTL qSHL11.1

for shoot length and QTL qRTL11.1 for root length [39], and with 4 GWAS-derived QTLs for the number of unfilled grains per plant [25] (Table S2), suggesting that this QTL has a pleiotropic effect on plant growth and reproduction under salinity stress and likely acts synergistically with other major salinity tolerance QTLs such as *Saltol*, in enhancing the salinity tolerance in rice.

Compared to the previous GWAS [14,17,18,20,112], using the same rice panel and genotyping data, we found 28 associations of 6 QTLs identified in this study colocalized with 23 associations of [14,17,20,112], but there was no colocalization between QTLs for salinity tolerance-related traits and for leaf mass traits [18] (Table S3). Remarkably, 9 associations for various drought tolerance-related traits, including relative water content after 2 and 3 weeks of drought stress, slope of relative water content after 2 weeks of drought stress, drought sensitivity score after 2, 3 and 4 weeks of drought stress, and recovery ability, belonging to QTL q9 of [17], were colocalized with all associations in QTL_17 for Chla_b and Score, suggesting that this genomic region contains important genetic determinants for rice adaptation to osmotic stresses.

Underlying 19 out of the 26 QTLs detected in this study, a high number of genes were annotated or functionally associated with salinity tolerance (Table 3).

Most candidate genes encode transcription factors reported to be involved in the rice response to salinity or abiotic stresses. Found in QTL_1 for Score, the *OsERF922* gene (*ETHYLENE RESPONSE FACTOR 922, Os01g54890*) negatively regulates tolerance to salinity stress through an ABA signaling pathway, since rice transgenic plants over-expressing *OsERF922* exhibited reduced salinity tolerance with increased shoot Na^+/K^+ ratio and ABA level, and knockdown of *OsERF922* expression reduced the ABA accumulation [46]. Additionally, as a member of the ERF gene family, *OsERF922*, the expression of *OsERF#103* (*ETHYLENE RESPONSE FACTOR 103, Os02g52670*) (in QTL_5), was reported to be upregulated under drought and salinity stress conditions at the seedling stage [65].

Furthermore, we found two potential genes encoding bZIP transcription factors. In plants, bZIP genes are involved in the response to abiotic stress [66,113]. One of these two genes is *OsbZIP23* (*b-ZIP TRANSCRIPTION FACTOR 23, Os02g52780*) (in QTL_5), which was functionally characterized as being an ABA-dependent enhancer of drought and salinity tolerance [66,67]. On the one hand, *OsbZIP23* overexpression significantly enhances tolerance to drought stress, especially to high salinity stress, compared with the wild type [66,67]. On the other hand, the *OsbZIP23* mutant displays significantly reduced tolerance to drought and salinity stress [67]. In addition, the SUMO protease *OsOSTS1* (*OVERLY TOLERANT TO SALT 1*), a gene involved in tolerance to high salinity [114], was reported to directly target OsbZIP23, which results in activation of OsbZIP23 and stimulation of OsbZIP23-dependent gene expression, which helps promote tolerance to drought stress [115]. Similar to *OsbZIP23*, *OsbZIP33* (*b-ZIP TRANSCRIPTION FACTOR 33, Os03g58250*), located in QTL_10, also plays a role as an ABA-dependent enhancer of drought and salinity tolerance. *OsbZIP33* is highly upregulated under drought and high salinity stress conditions. *OsbZIP33*-overexpressing transgenic plants exhibited significantly increased drought tolerance [78].

Three candidate genes belonging to the zinc-finger transcription factors were identified: *OsSAP3* (*STRESS-ASSOCIATED PROTEIN 3, Os01g56040*) in QTL_2, *OsPHD7* (*PHD FINGER PROTEIN 7, Os01g66420*) in QTL_3, and *OsCga1* (*CYTOKININ GATA TRANSCRIPTION FACTOR 1, Os02g12790*) in QTL_4. *OsSAP3* and *OsPHD7* are related to abiotic stress responses. In particular, the expression of *OsSAP3* is induced in response to drought and salinity stress [50], and *OsPHD7* is upregulated under drought stress [57]; moreover, *OsCga1* is associated with the development of chloroplasts [59] and stay-green [60]. Stay-green refers to the ability to maintain green leaves and photosynthetic capacity and is thus related to plant adaptation to osmotic stress [116]. Overexpression of *OsCga1* delays leaf senescence [59].

Underlying QTL_2, *OsRDCP3* (*RING DOMAIN-CONTAINING PROTEIN 3, Os01g56070*) was predicted to be involved in drought stress tolerance [51], and *OsABCI6* (*ABC TRANS-*

PORTER I FAMILY MEMBER 6, Os01g56400) was supposed to be involved in the response to abiotic stress [52,53]. Similarly, the expression of *OsTET2* (*TETRASPANIN 2, Os02g12750*), an integral membrane protein found in QTL_4, was increased in drought-stress seedlings; in addition, this gene was highly upregulated under heat and salinity stress [58].

Two other candidate transcription factor genes were found in QTL_26 on chromosome 11, including *OsHox33* (*HOMEOBOX GENE 33, Os12g41860*) and *OsARF25* (*AUXIN RESPONSE FACTOR 25, Os12g41950*). *OsHox33*, encoding an HDZIP transcription factor, is involved in leaf senescence because its knockdown accelerates leaf senescence [92] and is a target of a salinity stress-responsive miRNA [75]. *OsARF25* is also a salinity tolerance-related candidate gene discovered by GWAS, as reported by [93].

Another transcription factor gene identified, *OsAS2* (*ASYMMETRIC LEAVES 2, Os01g66590*) in QTL_3 [55], was associated with the development of plants. *LhCa5* (*PHOTOSYSTEM I LIGHT HARVESTING COMPLEX GENE 5, Os02g52650*) in QTL_5 was predicted to function in the photosystem [64].

Within the region of QTL_25, the strongest QTL found in this study, we detected a consecutive set of four *BRASSINOSTEROID INSENSITIVE 1-associated receptor kinase 1 (BAK1)*, including *Os11g31530 (OsBDG1)*, *Os11g31540 (OsLRR2)*, *Os11g31550*, and *Os11g31560* (Figure 3b). *BAK1*, encoding a leucine-rich repeat type II receptor-like kinase, functions as a coreceptor of BRI1 in brassinosteroid plant signaling [117]. Perception of brassinosteroids through the BRI1-BAK1 complex can influence the growth and development of rice plants [118], e.g., regulating the leaf angle and grain size [119] and regulating ABA-induced stomatal closure, which is critical for the survival of plants under water stress [120]. Among these four *BAK1* genes, *OsBDG1* and *OsLRR2* are considered to be involved in salt and/or abiotic stress responses [90,91]. Under salinity stress conditions, *OsBDG1* is significantly upregulated in roots of the rice-sensitive cultivar IR29, whereas *OsLRR2* is upregulated in roots of the rice-tolerant cultivar FL478 [90]. Additionally, the expression of *OsLRR2* is highly induced in leaves after cold and drought treatment; thus, *OsLRR2* is a supposed candidate gene involved in tolerance to abiotic stress [91]. Interestingly, two significant SNPs identified in this study, Sj11_18426630R and Dj11_18426457R, were located in the sequence of *OsBDG1* (Figure 3). Dj11_18426457R is intronic, while Sj11_18426630R is positioned within a coding sequence (i.e., exon 5) that changes the amino acid sequence in the LRR domain of the OsBDG1 protein. Thus, the perspective of a functional characterization of these *BAK1* candidate genes is opened.

Three other genes encoding receptor-like kinase (RLK) with enhanced abiotic stress tolerance are *Os08g28710 (OsRLCK253)* in QTL_19 and *Os12g42060 (OsWAK128)* and *Os12g42070 (OsRLCK375, OsWAK129)* in QTL_26. Functionally, *OsRLCK253* confers tolerance to salt and water deficits in transgenic *Arabidopsis thaliana* plants during different growth stages, resulting in yield protection against stress [86]. *OsWAK128* and *Os12g42070* were candidate genes near a GWAS-derived QTL related to salinity tolerance at the seedling stage [93]. In addition, a mitogen-activated protein kinase (MAPK) encoded by the *OsRLCK84* gene (*Os02g53030*) in QTL_5 was activated in response to salinity stress [68].

4. Materials and Methods

4.1. Plant Materials and Genotyping

This study included 179 Vietnamese rice landraces and 3 control genotypes (Nipponbare, Azucena and IR64). The Vietnamese rice accessions came from diverse locations throughout Vietnam and were originally provided by the Plant Resource Center (21°00′05″N and 105°43′33″E). All 182 accessions were genotyped by 21,623 single nucleotide polymorphism (SNP) markers using genotyping-by-sequencing with a minor allele frequency above 5% [14]. IR29 was used as a susceptibility check for phenotyping experiments. The names of the accessions, provinces of origin and ecosystem are described in Table S4. More detailed information on this panel can be found in [14].

4.2. Phenotyping Experiment

4.2.1. Salt Treatment

The experiment was conducted from August 26, 2019, to September 24, 2019, at the Agriculture Genetics Institute, Hanoi, Vietnam (21°02′55″ N and 105°46′58″ E). The accessions were grown in hydroponics following the IRRI standard protocol with three replicates [100]. Within each replicate, the accessions were randomly distributed in 5.2 individual plastic trays (36 × 31 × 15 cm) fitted with styrofoam float of 35 slots (2 mm diameter) filled with Peters solution composed of 1 g/L Peters water-soluble fertilizer (20-20-20 NPK) and 200 mg/L ferrous sulfate [21]. A total of 16 plastic trays were used.

The experiment was set under greenhouse conditions. After breaking dormancy at 50 °C for five days, seeds were soaked in water for 2–3 days. When germination began, seeds were incubated in a culture room (28 °C, photoperiod 12 h light/12 h dark) for 2 days. Once the primary root emerged well at a length of 2–3 cm, seedlings were cultured in styrofoam floats with a nylon net bottom according to the experimental design. Four seedlings were cultured per slot. Three days after seeding, seedlings were thinned to keep 3 well-developed plants per slot. The pH (5.2) and the level of nutrient solution were adjusted daily. The Peters solution was replaced weekly until the end of the experiment. Salinity stress was applied when plants reached the fourth leaf stage. Salt NaCl was gradually supplemented to the hydroponic medium to avoid osmotic shocks. Each time, 50 mM NaCl was separated by two days to obtain a final concentration of 100 mM NaCl. The experiment was stopped once all the plants exhibited drying in most leaves (average evaluation score of 7).

4.2.2. Scoring and Sampling

For each plant, salinity tolerance score was evaluated based on leaf injury symptoms using the modified standard evaluation score (SES) for rice [21], as follows: score 1—normal growth, no leaf symptoms; score 3—near normal growth, but leaf tips or few leaves whitish and rolled; score 5—growth severely retarded, most leaves rolled, only a few elongating; score 7—complete cessation of growth, most leaves dry, some plants dying; score 9—almost all plants dead or dying.

After scoring, the second fully expanded leaves of three plants in each hole were harvested. Quickly cut a 1.5 cm fragment from the leaf base, separately pack the material of each hole in aluminum foil, avoiding folding the leaves, and place on ice for chlorophyll determination. The rest of the cut leaves were immediately put into a small zip plastic bag of known weight for measuring the water content.

4.2.3. Chlorophyll Determination

The chlorophyll content was estimated as described in the protocol of [121] with some modifications. The harvested samples were weighed, put into 2-mL Eppendorf tubes, and ground in liquid nitrogen. The pellet was resuspended in 1.5 mL of 85% acetone solution and centrifuged at 12,000× g at 4 °C for 15 min. One milliliter of the supernatant was collected, and the absorbance was measured at wavelengths of 645 and 663 nm using a 7305 UV/visible spectrophotometer (Jenway, Staffordshire, UK). The chlorophyll content was calculated as follows: total chlorophyll (Chl_total, μg/mL) = 20.2 (A_{645}) + 8.02 (A_{663}), chlorophyll a (Chla, μg/mL) = 12.7 (A_{663}) − 2.69 (A_{645}), chlorophyll b (Chlb, μg/mL) = 22.9 (A_{645}) − 4.68 (A_{663}). The values were then converted to the amount of chlorophyll per milligram of fresh tissue (μg/mg). The ratio of chlorophyll a to chlorophyll b (Chla_b) was also determined.

4.2.4. Water Content Measurement

The bags containing samples were weighed to determine the sample fresh weight (FW). After being dried for 3 days at 70 °C in an oven, the sample dry weight (DW) was measured. The leaf water content of each sampling was calculated using the formula: WC (%) = (FW − DW) × 100/FW.

4.2.5. Ion Content Measurement

The above dried samples with known weight (DW, mg) were used for measurement of Na^+ and K^+ ion content. The samples were put into 15-mL Falcon tubes, and 10 mL of 0.1 N hydrochloric acid solution was added. After sample ion solubilization at room temperature overnight, 2 mL of sample solution at 200-fold dilution (10 μL of first sample solution + 2 mL 0,1 N hydrochloric acid solution) was used to measure Na^+ and K^+ concentrations (mg/L) by a SpectrAA 220FS atomic absorption spectrometer (Varian, US). The Na^+ and K^+ contents (ConcNa and ConcK) were then converted back to the quantity of Na^+ and K^+ ions per gram of dry weight (mg/gDW) by the following equations: ConcNa = [Na^+ measurement (mg/L) × dilution rate (200) × volume of first sample solution (10 mL)]/DW (mg); ConcK = [K^+ measurement (mg/L) × dilution rate (200) × volume of first sample solution (10 mL)]/DW (mg). The Na^+/K^+ ratio (Na_K) was calculated as the proportion of Na^+ content to K^+ content.

4.3. Statistical Analysis of Phenotypic Data

Statistical analysis of phenotypic data (means, standard deviations, coefficients of variation (CVs), graphs) was carried out in the R software v3.6.2. Analysis of variance (ANOVA) was performed to test the effect of genotype and replication using a linear model of the R function lm(). Broad-sense heritability (H^2) was used to estimate the genetic variance based on the variance among phenotypic measurements between three replicates of the panel. H^2 was computed using the following formula: $H^2 = (F\text{-value} - 1)/F\text{-value}$, where the F-value was derived from analysis of ANOVA for the genotype effect [18]. Phenotypic correlations between traits were evaluated by the Pearson method using the corrplot R package. The R function cor.test() was used to test the significance of the correlation coefficients.

4.4. Genome-Wide Association Study

The phenotypic data from the salt test and SNP genotypic data on the full panel and the indica and japonica subpanels were separately used to study the marker-trait associations by incorporating a kinship matrix along with population structure. In the Tassel software v.5.0, the structure matrix was determined with 6 axes on the SNP data of the population by running a principal component analysis (PCA). The kinship matrix was built by the pairwise identity-by-state method, to account for relatedness of individuals among 182 accessions. Q-Q and Manhattan plots of the negative $\log_{10}$-transformed observed p-values for each SNP-trait association were created to visualize the GWAS results. Markers with a p-value $\geq 5 \times 10^{-4}$ were declared significant.

The number of QTLs from the detected associations was determined based on linkage disequilibrium (LD) between SNPs surrounding the significant markers. The LD heatmaps were plotted by using the LDheatmap R package, and the genomic regions of QTLs were limited by LD blocks with r^2 values (squared allele frequency correlation) between SNPs > 0.4. For a low LD block (<50 kb), the interval of QTLs was enlarged by a distance of $+/- 50$ kb. The qqman package in R software was utilized to highlight the significant markers of strong QTLs in Manhattan plots. The genes in the genomic regions of strong QTLs were scanned in the MSU rice database.

5. Conclusions and Future Prospects

Our approach identified different QTLs characterized by the presence of a high number of genes associated with the response to salinity or abiotic stress. Interestingly, these genes are related to hormone transduction pathways or transcriptional modulation of gene expression in response to stress, suggesting that these QTLs act in complementary ways to control the salinity tolerance, which is of major interest for breeding programs. Pyramiding several favorable QTLs in a variety will ensure a better resilience of the plant to salinity stress under different environmental conditions and then a better sustainability of the variety. Therefore, it will be interesting to conduct introgression of the major QTLs identified in this study such as QTL25 in modern varieties cultivated in the Mekong or Red River Delta areas such as Bac Thom 7 and Khang Dan 18. The function of the four *BAK1* genes in QTL25 should be specified by generating single and multiple gene mutations using the CRISPR Cas9 system.

Supplementary Materials: The following are available online at https://www.mdpi.com/article/10.3390/plants10061088/s1, Figure S1: Correlation plots in the full panel and the indica and japonica subpanels, Figure S2: Manhattan plots and Q-Q plots for GWAS of salinity tolerance-related traits in the indica panel. In the Manhattan plots, significant SNPs are highlighted in red, Table S1: GWAS associations and significant SNPs at $p \leq 1 \times 10^{-3}$ in the full panel and the indica subpanel, Table S2: Colocalizations of the QTLs identified in this study with previous reports, Table S3: Colocalizations of the QTLs detected in this and previous studies using the same rice panel and genotyping data, Table S4: List of the 183 rice accessions used in the experiment.

Author Contributions: Conceptualization, P.G.; methodology, formal analysis, data curation, G.T.H., A.-A.V. and P.G.; investigation, T.D.L., H.T.V., K.L.N., F.G., L.H.T., H.T.T.V., F.L., X.H.P., A.-A.V., G.T.H.; writing—original draft preparation, G.T.H. and P.G.; writing—review and editing, T.D.L., H.T.V., K.L.N., H.T.T.V., T.X.D., X.H.P., A.-A.V.; visualization, supervision, project administration, G.T.H.; funding acquisition, G.T.H. and P.G. All authors have read and agreed to the published version of the manuscript.

Funding: This research was funded by the Ministry of Science and Technology of Vietnam and the French embassy in Vietnam in the frame of project "Application of functional genomics and association genetics to characterize genes involved in abiotic stresses tolerance in rice" (code: NDT.56.FRA/19). This research was also supported by the Global Rice Science Partnership (2011–2016) and by the CGIAR Research Program (CRP) on rice agri-food systems (RICE, 2017–2022).

Data Availability Statement: The GBS genotyping dataset supporting the results of this study has been deposited as a downloadable Excel file in TropGeneDB: http://tropgenedb.cirad.fr/tropgene/JSP/interface.jsp?module=RICE (accessed on 24 May 2021) tab "studies", study type "genotype", study "Vietnamese panel-GBS data". The seeds of the accessions are available in the National Key Laboratory for Plant Cell Biotechnology of Agricultural Genetics Institute, Hanoi, Vietnam.

Conflicts of Interest: The authors declare no conflict of interest.

References

1. Rengasamy, P. Soil Processes Affecting Crop Production in Salt-Affected Soils. *Funct. Plant Biol.* **2010**, *37*, 613. [CrossRef]
2. Wassmann, R.; Nelson, G.; Peng, S.; Sumfleth, K.; Jagadish, K.; Hosen, Y.; Rosegrant, M. Rice and global climate change. In *Rice in the Global Economy: Strategic Research and Policy Issues for Food Security*; IRRI: Los Banos, Philippines, 2010; pp. 411–433.
3. Timmer, C.P.; Block, S.; Dawe, D. Long-run dynamics of rice consumption, 1960–2050. In *Rice in the Global Economy: Strategic Research and Policy Issues for Food Security*; IRRI: Los Banos, Philippines, 2010; pp. 139–174.
4. Sarah, K.B. *System of Rice Intensification in Vietnam: Doing More with Less*; 52 Profiles on Agroecology; Oxfam: Food and Agriculture Organisation of the United States: Washington, DC, USA, 2016.
5. Ngoc Huyen Việt Nam chịu ảnh hưởng ra sao bởi biến đổi khí hậu? *Báo Đồng Khởi.* 2008. Available online: https://baodongkhoi.vn/viet-nam-chiu-anh-huong-ra-sao-boi-bien-doi-khi-hau-14012008-a543.html (accessed on 24 May 2021).
6. Ha, P.V.; Nguyen, H.T.M.; Kompas, T.; Che, T.N.; Trinh, B. Rice Production, Trade and the Poor: Regional Effects of Rice Export Policy on Households in Vietnam. *J. Agric. Econ.* **2015**, *66*, 280–307. [CrossRef]
7. Maitah, K.; Smutka, L.; Sahatqija, J.; Maitah, M.; Phuong Anh, N. Rice as a Determinant of Vietnamese Economic Sustainability. *Sustainability* **2020**, *12*, 5123. [CrossRef]
8. Loc, H.H.; Van Binh, D.; Park, E.; Shrestha, S.; Dung, T.D.; Son, V.H.; Truc, N.H.T.; Mai, N.P.; Seijger, C. Intensifying Saline Water Intrusion and Drought in the Mekong Delta: From Physical Evidence to Policy Outlooks. *Sci. Total Environ.* **2021**, *757*, 143919. [CrossRef] [PubMed]
9. Tin, H.Q.; Loi, N.H.; Labarosa, S.J.E.; McNally, K.L.; McCouch, S.; Kilian, B. Phenotypic Response of Farmer-selected CWR-derived Rice Lines to Salt Stress in the Mekong Delta. *Crop Sci.* **2021**, *61*, 201–218. [CrossRef]
10. Ganie, S.A.; Molla, K.A.; Henry, R.J.; Bhat, K.V.; Mondal, T.K. Advances in Understanding Salt Tolerance in Rice. *Theor. Appl. Genet.* **2019**, *132*, 851–870. [CrossRef]
11. Negrão, S.; Courtois, B.; Ahmadi, N.; Abreu, I.; Saibo, N.; Oliveira, M.M. Recent Updates on Salinity Stress in Rice: From Physiological to Molecular Responses. *Crit. Rev. Plant Sci.* **2011**, *30*, 329–377. [CrossRef]
12. Pang, Y.; Chen, K.; Wang, X.; Wang, W.; Xu, J.; Ali, J.; Li, Z. Simultaneous Improvement and Genetic Dissection of Salt Tolerance of Rice (*Oryza Sativa* L.) by Designed QTL Pyramiding. *Front. Plant Sci.* **2017**, *8*, 1275. [CrossRef]

13. Wang, Z.; Cheng, J.; Chen, Z.; Huang, J.; Bao, Y.; Wang, J.; Zhang, H. Identification of QTLs with Main, Epistatic and QTL × Environment Interaction Effects for Salt Tolerance in Rice Seedlings under Different Salinity Conditions. *Theor. Appl. Genet.* **2012**, *125*, 807–815. [CrossRef] [PubMed]

14. Phung, N.T.P.; Mai, C.D.; Mournet, P.; Frouin, J.; Droc, G.; Ta, N.K.; Jouannic, S.; Lê, L.T.; Do, V.N.; Gantet, P.; et al. Characterization of a Panel of Vietnamese Rice Varieties Using DArT and SNP Markers for Association Mapping Purposes. *BMC Plant Biol.* **2014**, *14*, 371. [CrossRef]

15. Fukuoka, S.; Alpatyeva, N.V.; Ebana, K.; Luu, N.T.; Nagamine, T. Analysis of Vietnamese Rice Germplasm Provides an Insight into Japonica Rice Differentiation. *Plant Breed.* **2003**, *122*, 497–502. [CrossRef]

16. Li, J.-Y.; Wang, J.; Zeigler, R.S. The 3000 Rice Genomes Project: New Opportunities and Challenges for Future Rice Research. *GigaScience* **2014**, *3*, 8. [CrossRef]

17. Hoang, G.T.; Van Dinh, L.; Nguyen, T.T.; Ta, N.K.; Gathignol, F.; Mai, C.D.; Jouannic, S.; Tran, K.D.; Khuat, T.H.; Do, V.N.; et al. Genome-Wide Association Study of a Panel of Vietnamese Rice Landraces Reveals New QTLs for Tolerance to Water Deficit during the Vegetative Phase. *Rice* **2019**, *12*, 4. [CrossRef] [PubMed]

18. Hoang, G.T.; Gantet, P.; Nguyen, K.H.; Phung, N.T.P.; Ha, L.T.; Nguyen, T.T.; Lebrun, M.; Courtois, B.; Pham, X.H. Genome-Wide Association Mapping of Leaf Mass Traits in a Vietnamese Rice Landrace Panel. *PLoS ONE* **2019**, *14*, e0219274. [CrossRef] [PubMed]

19. Phung, N.T.P.; Mai, C.D.; Hoang, G.T.; Truong, H.T.M.; Lavarenne, J.; Gonin, M.; Nguyen, K.L.; Ha, T.T.; Do, V.N.; Gantet, P.; et al. Genome-Wide Association Mapping for Root Traits in a Panel of Rice Accessions from Vietnam. *BMC Plant Biol.* **2016**, *16*, 64. [CrossRef] [PubMed]

20. Ta, K.N.; Khong, N.G.; Ha, T.L.; Nguyen, D.T.; Mai, D.C.; Hoang, T.G.; Phung, T.P.N.; Bourrie, I.; Courtois, B.; Tran, T.T.H.; et al. A Genome-Wide Association Study Using a Vietnamese Landrace Panel of Rice (*Oryza Sativa*) Reveals New QTLs Controlling Panicle Morphological Traits. *BMC Plant Biol.* **2018**, *18*, 282. [CrossRef]

21. Egdane, J.; Vispo, N.; Mohammadi, R.; Amas, J.; Katimbang, M.; Platten, J.; Ismail, A.; Gregorio, G. *Phenotyping Protocols for Salinity and Other Problem Soils*; International Rice Research Institute: Los Banos, Philippines, 2007.

22. Batayeva, D.; Labaco, B.; Ye, C.; Li, X.; Usenbekov, B.; Rysbekova, A.; Dyuskalieva, G.; Vergara, G.; Reinke, R.; Leung, H. Genome-Wide Association Study of Seedling Stage Salinity Tolerance in Temperate Japonica Rice Germplasm. *BMC Genet.* **2018**, *19*, 2. [CrossRef]

23. Frouin, J.; Languillaume, A.; Mas, J.; Mieulet, D.; Boisnard, A.; Labeyrie, A.; Bettembourg, M.; Bureau, C.; Lorenzini, E.; Portefaix, M.; et al. Tolerance to Mild Salinity Stress in Japonica Rice: Agenome-Wide Association Mapping Study Highlights Calcium Signaling and Metabolism Genes. *PLoS ONE* **2018**, *13*, e0190964. [CrossRef]

24. Kumar, V.; Singh, A.; Mithra, S.V.A.; Krishnamurthy, S.L.; Parida, S.K.; Jain, S.; Tiwari, K.K.; Kumar, P.; Rao, A.R.; Sharma, S.K.; et al. Genome-Wide Association Mapping of Salinity Tolerance in Rice (*Oryza Sativa*). *DNA Res.* **2015**, *22*, 133–145. [CrossRef]

25. Lekklar, C.; Pongpanich, M.; Suriya-arunroj, D.; Chinpongpanich, A.; Tsai, H.; Comai, L.; Chadchawan, S.; Buaboocha, T. Genome-Wide Association Study for Salinity Tolerance at the Flowering Stage in a Panel of Rice Accessions from Thailand. *BMC Genom.* **2019**, *20*, 76. [CrossRef]

26. Yuan, J.; Wang, X.; Zhao, Y.; Khan, N.U.; Zhao, Z.; Zhang, Y.; Wen, X.; Tang, F.; Wang, F.; Li, Z. Genetic Basis and Identification of Candidate Genes for Salt Tolerance in Rice by GWAS. *Sci. Rep.* **2020**, *10*, 9958. [CrossRef]

27. Ammar, M.H.M.; Pandit, A.; Singh, R.K.; Sameena, S.; Chauhan, M.S.; Singh, A.K.; Sharma, P.C.; Gaikwad, K.; Sharma, T.R.; Mohapatra, T.; et al. Mapping of QTLs Controlling Na$^+$, K$^+$ and CI$^-$ Ion Concentrations in Salt Tolerant Indica Rice Variety CSR27. *J. Plant Biochem. Biotechnol.* **2009**, *18*, 139–150. [CrossRef]

28. Cheng, L.; Wang, Y.; Meng, L.; Hu, X.; Cui, Y.; Sun, Y.; Zhu, L.; Ali, J.; Xu, J.; Li, Z. Identification of Salt-Tolerant QTLs with Strong Genetic Background Effect Using Two Sets of Reciprocal Introgression Lines in Rice. *Genome* **2012**, *55*, 45–55. [CrossRef] [PubMed]

29. De Leon, T.B.; Linscombe, S.; Subudhi, P.K. Molecular Dissection of Seedling Salinity Tolerance in Rice (*Oryza Sativa* L.) Using a High-Density GBS-Based SNP Linkage Map. *Rice* **2016**, *9*, 52. [CrossRef]

30. Ghomi, K.; Rabiei, B.; Sabouri, H.; Sabouri, A. Mapping QTLs for Traits Related to Salinity Tolerance at Seedling Stage of Rice (*Oryza Sativa* L.): An Agrigenomics Study of an Iranian Rice Population. *OMICS* **2013**, *17*, 242–251. [CrossRef]

31. Gimhani, D.R.; Gregorio, G.B.; Kottearachchi, N.S.; Samarasinghe, W.L.G. SNP-Based Discovery of Salinity-Tolerant QTLs in a Bi-Parental Population of Rice (*Oryza Sativa*). *Mol. Genet. Genom.* **2016**, *291*, 2081–2099. [CrossRef] [PubMed]

32. Hemamalini, G.S.; Shashidhar, H.E.; Hittalmani, S. Molecular Marker Assisted Tagging of Morphological and Physiological Traits under Two Contrasting Moisture Regimes at Peak Vegetative Stage in Rice (*Oryza Sativa* L.). *Euphytica* **2000**, *112*, 69–78. [CrossRef]

33. Hossain, H.; Rahman, M.A.; Alam, M.S.; Singh, R.K. Mapping of Quantitative Trait Loci Associated with Reproductive-Stage Salt Tolerance in Rice. *J. Agron. Crop Sci.* **2015**, *201*, 17–31. [CrossRef]

34. Koyama, M.L.; Levesley, A.; Koebner, R.M.D.; Flowers, T.J.; Yeo, A.R. Quantitative Trait Loci for Component Physiological Traits Determining Salt Tolerance in Rice. *Plant Physiol.* **2001**, *125*, 406–422. [CrossRef]

35. Lafitte, R.H.; Courtois, B. Genetic Variation in Performance under Reproductive-Stage Water Deficit in a Doubled Haploid Rice Population in Upland Fields. In Proceedings of the Molecular Approaches for the Genetic Improvement of Cereals for Stable Production in Water-Limited Environments, CIMMYT, El Batan, Mexico, 21 June 1999.

36. Mohammadi, R.; Mendioro, M.S.; Diaz, G.Q.; Gregorio, G.B.; Singh, R.K. Mapping Quantitative Trait Loci Associated with Yield and Yield Components under Reproductive Stage Salinity Stress in Rice (*Oryza Sativa* L.). *J. Genet.* **2013**, *92*, 433–443. [CrossRef]

37. Prasad, S.R.; Bagali, P.G.; Hittalmani, S.; Shashidhar, H.E. Molecular Mapping of Quantitative Trait Loci Associated with Seedling Tolerance to Salt Stress in Rice (*Oryza Sativa* L.). *Curr. Sci.* **2000**, *78*, 162–164.

38. Price, A.H.; Steele, K.A.; Moore, B.J.; Jones, R.G.W. Upland Rice Grown in Soil-Filled Chambers and Exposed to Contrasting Water-Deficit Regimes II. Mapping Quantitative Trait Loci for Root Morphology and Distribution. *Field Crops Res.* **2002**, *76*, 25–43. [CrossRef]

39. Puram, V.R.R.; Ontoy, J.; Linscombe, S.; Subudhi, P.K. Genetic Dissection of Seedling Stage Salinity Tolerance in Rice Using Introgression Lines of a Salt Tolerant Landrace Nona Bokra. *J. Hered.* **2017**, *108*, 658–670. [CrossRef] [PubMed]

40. Sabouri, H.; Rezai, A.M.; Moumeni, A.; Kavousi, A.; Katouzi, M.; Sabouri, A. QTLs Mapping of Physiological Traits Related to Salt Tolerance in Young Rice Seedlings. *Biol. Plant.* **2009**, *53*, 657–662. [CrossRef]

41. Thomson, M.J.; de Ocampo, M.; Egdane, J.; Rahman, M.A.; Sajise, A.G.; Adorada, D.L.; Tumimbang-Raiz, E.; Blumwald, E.; Seraj, Z.I.; Singh, R.K.; et al. Characterizing the Saltol Quantitative Trait Locus for Salinity Tolerance in Rice. *Rice* **2010**, *3*, 148–160. [CrossRef]

42. Ul Haq, T.; Gorham, J.; Akhtar, J.; Akhtar, N.; Steele, K.A. Dynamic Quantitative Trait Loci for Salt Stress Components on Chromosome 1 of Rice. *Funct. Plant Biol.* **2010**, *37*, 634–645. [CrossRef]

43. Wang, Z.; Chen, Z.; Cheng, J.; Lai, Y.; Wang, J.; Bao, Y.; Huang, J.; Zhang, H. QTL Analysis of Na^+ and K^+ Concentrations in Roots and Shoots under Different Levels of NaCl Stress in Rice (*Oryza Sativa* L.). *PLoS ONE* **2012**, *7*, e51202. [CrossRef] [PubMed]

44. Wang, Z.; Wang, J.; Bao, Y.; Wu, Y.; Zhang, H. Quantitative Trait Loci Controlling Rice Seed Germination under Salt Stress. *Euphytica* **2011**, *178*, 297–307. [CrossRef]

45. Yue, B.; Xue, W.; Xiong, L.; Yu, X.; Luo, L.; Cui, K.; Jin, D.; Xing, Y.; Zhang, Q. Genetic Basis of Drought Resistance at Reproductive Stage in Rice: Separation of Drought Tolerance from Drought Avoidance. *Genetics* **2006**, *172*, 1213–1228. [CrossRef]

46. Liu, D.; Chen, X.; Liu, J.; Guo, Z. The Rice ERF Transcription Factor OsERF922 Negatively Regulates Resistance to Magnaporthe Oryzae and Salt Tolerance. *J. Exp. Bot.* **2012**, *63*. [CrossRef]

47. Du, H.; Wu, N.; Fu, J.; Wang, S.; Li, X.; Xiao, J.; Xiong, L. A GH3 Family Member, OsGH3-2, Modulates Auxin and Abscisic Acid Levels and Differentially Affects Drought and Cold Tolerance in Rice. *J. Exp. Bot.* **2012**, *63*, 6467–6480. [CrossRef]

48. Niu, M.; Wang, Y.; Wang, C.; Lyu, J.; Wang, Y.; Dong, H.; Long, W.; Wang, D.; Kong, W.; Wang, L.; et al. ALR Encoding DCMP Deaminase Is Critical for DNA Damage Repair, Cell Cycle Progression and Plant Development in Rice. *J. Exp. Bot.* **2017**, *68*, 5773–5786. [CrossRef] [PubMed]

49. Xu, J.; Deng, Y.; Li, Q.; Zhu, X.; He, Z. STRIPE2 Encodes a Putative DCMP Deaminase That Plays an Important Role in Chloroplast Development in Rice. *J. Genet. Genom.* **2014**, *41*, 539–548. [CrossRef] [PubMed]

50. Vij, S.; Tyagi, A.K. Genome-Wide Analysis of the Stress Associated Protein (SAP) Gene Family Containing A20/AN1 Zinc-Finger(s) in Rice and Their Phylogenetic Relationship with Arabidopsis. *Mol. Genet. Genom.* **2006**, *276*, 565–575. [CrossRef]

51. Bae, H.; Kim, S.K.; Cho, S.K.; Kang, B.G.; Kim, W.T. Overexpression of OsRDCP1, a Rice RING Domain-Containing E3 Ubiquitin Ligase, Increased Tolerance to Drought Stress in Rice (*Oryza Sativa* L.). *Plant Sci.* **2011**, *180*, 775–782. [CrossRef] [PubMed]

52. Nguyen, V.N.T.; Moon, S.; Jung, K.-H. Genome-Wide Expression Analysis of Rice ABC Transporter Family across Spatio-Temporal Samples and in Response to Abiotic Stresses. *J. Plant Physiol.* **2014**, *171*, 1276–1288. [CrossRef]

53. Saha, J.; Sengupta, A.; Gupta, K.; Gupta, B. Molecular Phylogenetic Study and Expression Analysis of ATP-Binding Cassette Transporter Gene Family in Oryza Sativa in Response to Salt Stress. *Comput. Biol. Chem.* **2015**, *54*, 18–32. [CrossRef]

54. Wang, Y.; Jiang, Q.; Liu, J.; Zeng, W.; Zeng, Y.; Li, R.; Luo, J. Comparative Transcriptome Profiling of Chilling Tolerant Rice Chromosome Segment Substitution Line in Response to Early Chilling Stress. *Genes Genom.* **2017**, *39*, 127–141. [CrossRef]

55. Ma, Y.; Wang, F.; Guo, J.; Zhang, X.S. Rice OsAS2 Gene, a Member of LOB Domain Family, Functions in the Regulation of Shoot Differentiation and Leaf Development. *J. Plant Biol.* **2009**, *52*, 374–381. [CrossRef]

56. Passricha, N.; Saifi, S.; Ansari, M.W.; Tuteja, N. Prediction and Validation of Cis-Regulatory Elements in 5′ Upstream Regulatory Regions of Lectin Receptor-like Kinase Gene Family in Rice. *Protoplasma* **2017**, *254*, 669–684. [CrossRef] [PubMed]

57. Sun, M.; Jia, B.; Yang, J.; Cui, N.; Zhu, Y.; Sun, X. Genome-Wide Identification of the PHD-Finger Family Genes and Their Responses to Environmental Stresses in *Oryza Sativa* L. *Int. J. Mol. Sci.* **2017**, *18*, 2005. [CrossRef] [PubMed]

58. Mani, B.; Agarwal, M.; Katiyar-Agarwal, S. Comprehensive Expression Profiling of Rice Tetraspanin Genes Reveals Diverse Roles during Development and Abiotic Stress. *Front. Plant Sci.* **2015**, *6*. [CrossRef]

59. Hudson, D.; Guevara, D.R.; Hand, A.J.; Xu, Z.; Hao, L.; Chen, X.; Zhu, T.; Bi, Y.-M.; Rothstein, S.J. Rice Cytokinin GATA Transcription Factor1 Regulates Chloroplast Development and Plant Architecture. *Plant Physiol.* **2013**, *162*, 132–144. [CrossRef]

60. Zhao, Y.; Qiang, C.; Wang, X.; Chen, Y.; Deng, J.; Jiang, C.; Sun, X.; Chen, H.; Li, J.; Piao, W.; et al. New Alleles for Chlorophyll Content and Stay-Green Traits Revealed by a Genome Wide Association Study in Rice (*Oryza Sativa*). *Sci. Rep.* **2019**, *9*, 2541. [CrossRef] [PubMed]

61. Hossain, M.R.; Bassel, G.W.; Pritchard, J.; Sharma, G.P.; Ford-Lloyd, B.V. Trait Specific Expression Profiling of Salt Stress Responsive Genes in Diverse Rice Genotypes as Determined by Modified Significance Analysis of Microarrays. *Front. Plant Sci.* **2016**, *7*. [CrossRef] [PubMed]

62. Lakra, N.; Kaur, C.; Singla-Pareek, S.L.; Pareek, A. Mapping the 'Early Salinity Response' Triggered Proteome Adaptation in Contrasting Rice Genotypes Using ITRAQ Approach. *Rice* **2019**, *12*, 3. [CrossRef] [PubMed]

63. Lim, S.D.; Jung, C.G.; Park, Y.C.; Lee, S.C.; Lee, C.; Lim, C.W.; Kim, D.S.; Jang, C.S. Molecular Dissection of a Rice Microtubule-Associated RING Finger Protein and Its Potential Role in Salt Tolerance in Arabidopsis. *Plant Mol. Biol.* **2015**, *89*, 365–384. [CrossRef] [PubMed]

64. Umate, P. Genome-Wide Analysis of the Family of Light-Harvesting Chlorophyll a/b-Binding Proteins in Arabidopsis and Rice. *Plant Signal. Behav.* **2010**, *5*, 1537–1542. [CrossRef]

65. Mishra, M.; Kanwar, P.; Singh, A.; Pandey, A.; Kapoor, S.; Pandey, G.K. Plant Omics: Genome-Wide Analysis of ABA Repressor1 (ABR1) Related Genes in Rice during Abiotic Stress and Development. *OMICS* **2013**, *17*, 439–450. [CrossRef]

66. Park, S.-H.; Jeong, J.S.; Lee, K.H.; Kim, Y.S.; Do Choi, Y.; Kim, J.-K. OsbZIP23 and OsbZIP45, Members of the Rice Basic Leucine Zipper Transcription Factor Family, Are Involved in Drought Tolerance. *Plant Biotechnol. Rep.* **2015**, *9*, 89–96. [CrossRef]

67. Xiang, Y.; Tang, N.; Du, H.; Ye, H.; Xiong, L. Characterization of OsbZIP23 as a Key Player of the Basic Leucine Zipper Transcription Factor Family for Conferring Abscisic Acid Sensitivity and Salinity and Drought Tolerance in Rice. *Plant Physiol.* **2008**, *148*, 1938–1952. [CrossRef]

68. Mirdar Mansuri, R.; Shobbar, Z.-S.; Babaeian Jelodar, N.; Ghaffari, M.R.; Nematzadeh, G.-A.; Asari, S. Dissecting Molecular Mechanisms Underlying Salt Tolerance in Rice: A Comparative Transcriptional Profiling of the Contrasting Genotypes. *Rice* **2019**, *12*, 13. [CrossRef]

69. Ferreira, L.J.; Azevedo, V.; Maroco, J.; Oliveira, M.M.; Santos, A.P. Salt Tolerant and Sensitive Rice Varieties Display Differential Methylome Flexibility under Salt Stress. *PLoS ONE* **2015**, *10*, e0124060. [CrossRef] [PubMed]

70. Jehanzeb, M.; Zheng, X.; Miao, Y. The Role of the S40 Gene Family in Leaf Senescence. *Int. J. Mol. Sci.* **2017**, *18*, 2152. [CrossRef] [PubMed]

71. Zheng, X.; Jehanzeb, M.; Habiba; Zhang, Y.; Li, L.; Miao, Y. Characterization of S40-like Proteins and Their Roles in Response to Environmental Cues and Leaf Senescence in Rice. *BMC Plant Biol.* **2019**, *19*, 174. [CrossRef] [PubMed]

72. Liu, L.; Zhou, Y.; Zhou, G.; Ye, R.; Zhao, L.; Li, X.; Lin, Y. Identification of Early Senescence-Associated Genes in Rice Flag Leaves. *Plant Mol. Biol.* **2008**, *67*, 37–55. [CrossRef] [PubMed]

73. Molla, K.A.; Debnath, A.B.; Ganie, S.A.; Mondal, T.K. Identification and Analysis of Novel Salt Responsive Candidate Gene Based SSRs (CgSSRs) from Rice (*Oryza Sativa* L.). *BMC Plant Biol.* **2015**, *15*, 122. [CrossRef] [PubMed]

74. Gao, P.; Bai, X.; Yang, L.; Lv, D.; Li, Y.; Cai, H.; Ji, W.; Guo, D.; Zhu, Y. Over-Expression of Osa-MIR396c Decreases Salt and Alkali Stress Tolerance. *Planta* **2010**, *231*, 991–1001. [CrossRef]

75. Huang, X.; Feng, J.; Wang, R.; Zhang, H.; Huang, J. Comparative Analysis of MicroRNAs and Their Targets in the Roots of Two Cultivars with Contrasting Salt Tolerance in Rice (*Oryza Sativa* L.). *Plant Growth Regul.* **2019**, *87*, 139–148. [CrossRef]

76. Qin, J.; Ma, X.; Tang, Z.; Meng, Y. Construction of Regulatory Networks Mediated by Small RNAs Responsive to Abiotic Stresses in Rice (*Oryza Sativa*). *Comput. Biol. Chem.* **2015**, *58*, 69–80. [CrossRef]

77. Huda, K.K.; Yadav, S.; Banu, M.S.A.; Trivedi, D.K.; Tuteja, N. Genome-Wide Analysis of Plant-Type II Ca^{2+} ATPases Gene Family from Rice and Arabidopsis: Potential Role in Abiotic Stresses. *Plant Physiol. Biochem.* **2013**, *65*, 32–47. [CrossRef]

78. Chen, H.; Dai, X.J.; Gu, Z.Y. OsbZIP33 Is an ABA-Dependent Enhancer of Drought Tolerance in Rice. *Crop Sci.* **2015**, *55*, 1673–1685. [CrossRef]

79. Chapagain, S.; Park, Y.C.; Kim, J.H.; Jang, C.S. Oryza Sativa Salt-Induced RING E3 Ligase 2 (OsSIRP2) Acts as a Positive Regulator of Transketolase in Plant Response to Salinity and Osmotic Stress. *Planta* **2018**, *247*, 925–939. [CrossRef] [PubMed]

80. Lin, D.; Jiang, Q.; Ma, X.; Zheng, K.; Gong, X.; Teng, S.; Xu, J.; Dong, Y. Rice TSV3 Encoding Obg-Like GTPase Protein Is Essential for Chloroplast Development during the Early Leaf Stage under Cold Stress. *Genes Genomes Genet.* **2018**, *8*, 253–263. [CrossRef] [PubMed]

81. Boonchai, C.; Udomchalothorn, T.; Sripinyowanich, S.; Comai, L.; Buaboocha, T.; Chadchawan, S. Rice Overexpressing OsNUC1-S Reveals Differential Gene Expression Leading to Yield Loss Reduction after Salt Stress at the Booting Stage. *Int. J. Mol. Sci.* **2018**, *19*, 3936. [CrossRef]

82. Sripinyowanich, S.; Chamnanmanoontham, N.; Udomchalothorn, T.; Maneeprasopsuk, S.; Santawee, P.; Buaboocha, T.; Qu, L.-J.; Gu, H.; Chadchawan, S. Overexpression of a Partial Fragment of the Salt-Responsive Gene OsNUC1 Enhances Salt Adaptation in Transgenic Arabidopsis Thaliana and Rice (*Oryza Sativa* L.) during Salt Stress. *Plant Sci.* **2013**, *213*, 67–78. [CrossRef]

83. Ahn, J.C.; Kim, D.-W.; You, Y.N.; Seok, M.S.; Park, J.M.; Hwang, H.; Kim, B.-G.; Luan, S.; Park, H.-S.; Cho, H.S. Classification of Rice (Oryza Sativa l. Japonica Nipponbare) Immunophilins (Fkbps, Cyps) and Expression Patterns under Water Stress. *BMC Plant Biol.* **2010**, *10*, 253. [CrossRef] [PubMed]

84. Xue, T.; Wang, D.; Zhang, S.; Ehlting, J.; Ni, F.; Jakab, S.; Zheng, C.; Zhong, Y. Genome-Wide and Expression Analysis of Protein Phosphatase 2C in Rice and Arabidopsis. *BMC Genomics* **2008**, *9*, 550. [CrossRef] [PubMed]

85. Zhou, F.; Wang, C.-Y.; Gutensohn, M.; Jiang, L.; Zhang, P.; Zhang, D.; Dudareva, N.; Lu, S. A Recruiting Protein of Geranylgeranyl Diphosphate Synthase Controls Metabolic Flux toward Chlorophyll Biosynthesis in Rice. *Proc. Natl. Acad. Sci. USA* **2017**, *114*, 6866–6871. [CrossRef] [PubMed]

86. Giri, J.; Vij, S.; Dansana, P.K.; Tyagi, A.K. Rice A20/AN1 Zinc-Finger Containing Stress-Associated Proteins (SAP1/11) and a Receptor-like Cytoplasmic Kinase (OsRLCK253) Interact via A20 Zinc-Finger and Confer Abiotic Stress Tolerance in Transgenic Arabidopsis Plants. *New Phytol.* **2011**, *191*, 721–732. [CrossRef] [PubMed]

87. Goswami, K.; Tripathi, A.; Sanan-Mishra, N. Comparative MiRomics of Salt-Tolerant and Salt-Sensitive Rice. *J. Integr. Bioinform.* **2017**, *14*. [CrossRef] [PubMed]

88. Pradhan, S.K.; Pandit, E.; Nayak, D.K.; Behera, L.; Mohapatra, T. Genes, Pathways and Transcription Factors Involved in Seedling Stage Chilling Stress Tolerance in Indica Rice through RNA-Seq Analysis. *BMC Plant Biol.* **2019**, *19*, 352. [CrossRef]

89. Lakmini, W.G. The Role of SENSITIVE TO FREEZING6 (SFR6) in Plant Tolerance to Stress. In *Durham Theses*; Durham University: Durham, UK, 2010.

90. Senadheera, P.; Maathuis, F.J.M. Differentially Regulated Kinases and Phosphatases in Roots May Contribute to Inter-Cultivar Difference in Rice Salinity Tolerance. *Plant Signal. Behav.* **2009**, *4*, 1163–1165. [CrossRef]

91. Liao, Y.; Hu, C.; Zhang, X.; Cao, X.; Xu, Z.; Gao, X.; Li, L.; Zhu, J.; Chen, R. Isolation of a Novel Leucine-Rich Repeat Receptor-like Kinase (OsLRR2) Gene from Rice and Analysis of Its Relation to Abiotic Stress Responses. *Biotechnol. Biotechnol. Equip.* **2017**, *31*, 51–57. [CrossRef]

92. Luan, W.; Shen, A.; Jin, Z.; Song, S.; Li, Z.; Sha, A. Knockdown of OsHox33, a Member of the Class III Homeodomain-Leucine Zipper Gene Family, Accelerates Leaf Senescence in Rice. *Sci. China Life Sci.* **2013**, *56*, 1113–1123. [CrossRef] [PubMed]

93. Yu, J.; Zao, W.; He, Q.; Kim, T.-S.; Park, Y.-J. Genome-Wide Association Study and Gene Set Analysis for Understanding Candidate Genes Involved in Salt Tolerance at the Rice Seedling Stage. *Mol. Genet. Genom.* **2017**, *292*, 1391–1403. [CrossRef] [PubMed]

94. Vij, S.; Giri, J.; Dansana, P.K.; Kapoor, S.; Tyagi, A.K. The Receptor-like Cytoplasmic Kinase (OsRLCK) Gene Family in Rice: Organization, Phylogenetic Relationship, and Expression during Development and Stress. *Mol. Plant* **2008**, *1*, 732–750. [CrossRef]

95. Hedayati, P.; Monfard, H.H.; Isa, N.M.; Hwang, D.J.; Zain, C.R.C.M.; Uddin, M.I.; Zuraida, A.R.; Ismail, I.; Zainal, Z. Construction and Analysis of an Oryza Sativa (Cv. MR219) Salinity-Related CDNA Library. *Acta Physiol. Plant.* **2015**, *37*, 91. [CrossRef]

96. Gómez-Ariza, J.; Brambilla, V.; Vicentini, G.; Landini, M.; Cerise, M.; Carrera, E.; Shrestha, R.; Chiozzotto, R.; Galbiati, F.; Caporali, E.; et al. A Transcription Factor Coordinating Internode Elongation and Photoperiodic Signals in Rice. *Nat. Plants* **2019**, *5*, 358–362. [CrossRef]

97. Zhang, Z.; Liu, H.; Sun, C.; Ma, Q.; Bu, H.; Chong, K.; Xu, Y. A C2H2 Zinc-Finger Protein OsZFP213 Interacts with OsMAPK3 to Enhance Salt Tolerance in Rice. *J. Plant Physiol.* **2018**, *229*, 100–110. [CrossRef]

98. Flowers, T.J.; Yeo, A.R. Variability in the Resistance of Sodium Chloride Salinity within Rice (*Oryza Sativa* L.) Varieties. *New Phytol.* **1981**, *88*, 363–373. [CrossRef]

99. Lutts, S.; Kinet, J.M.; Bouharmont, J. Changes in plant response to NaCl during development of rice (Oryza sativa L.) varieties differing in salinity resistance. *J. Exp. Bot.* **1995**, *46*, 1843–1852. [CrossRef]

100. Gregorio, G.B.; Senadhira, D.; Mendoza, R.D. *Screening Rice for Salinity Tolerance*; IRRl Discussion Paper Series; IRRI: Los Banos, Philippines, 1997; pp. 2–23.

101. Lee, K.-S.; Choi, W.-Y.; Ko, J.-C.; Kim, T.-S.; Gregorio, G.B. Salinity Tolerance of Japonica and Indica Rice (*Oryza Sativa* L.) at the Seedling Stage. *Planta* **2003**, *216*, 1043–1046. [CrossRef] [PubMed]

102. Neang, S.; de Ocampo, M.; Egdane, J.A.; Platten, J.D.; Ismail, A.M.; Seki, M.; Suzuki, Y.; Skoulding, N.S.; Kano-Nakata, M.; Yamauchi, A.; et al. A GWAS Approach to Find SNPs Associated with Salt Removal in Rice Leaf Sheath. *Ann. Bot.* **2020**, *126*, 1193–1202. [CrossRef] [PubMed]

103. Negrão, S.; Schmöckel, S.M.; Tester, M. Evaluating Physiological Responses of Plants to Salinity Stress. *Ann. Bot.* **2017**, *119*, 1–11. [CrossRef] [PubMed]

104. Verslues, P.E.; Agarwal, M.; Katiyar-Agarwal, S.; Zhu, J.; Zhu, J.-K. Methods and Concepts in Quantifying Resistance to Drought, Salt and Freezing, Abiotic Stresses That Affect Plant Water Status. *Plant J.* **2006**, *45*, 523–539. [CrossRef]

105. Zhu, J.-K. Regulation of Ion Homeostasis under Salt Stress. *Curr. Opin. Plant Biol.* **2003**, *6*, 441–445. [CrossRef]

106. Chakraborty, K.; Chattaopadhyay, K.; Nayak, L.; Ray, S.; Yeasmin, L.; Jena, P.; Gupta, S.; Mohanty, S.K.; Swain, P.; Sarkar, R.K. Ionic Selectivity and Coordinated Transport of Na^+ and K^+ in Flag Leaves Render Differential Salt Tolerance in Rice at the Reproductive Stage. *Planta* **2019**, *250*, 1637–1653. [CrossRef]

107. Gregorio, G.B.; Senadhira, D. Genetic Analysis of Salinity Tolerance in Rice (*Oryza Sativa* L.). *Theoret. Appl. Genet.* **1993**, *86*, 333–338. [CrossRef] [PubMed]

108. Moradi, F.; Ismail, A.M. Responses of Photosynthesis, Chlorophyll Fluorescence and ROS-Scavenging Systems to Salt Stress during Seedling and Reproductive Stages in Rice. *Ann. Bot.* **2007**, *99*, 1161–1173. [CrossRef]

109. Yeo, A.R.; Yeo, M.E.; Flowers, S.A.; Flowers, T.J. Screening of Rice (*Oryza Sativa* L.) Genotypes for Physiological Characters Contributing to Salinity Resistance, and Their Relationship to Overall Performance. *Theoret. Appl. Genet.* **1990**, *79*, 377–384. [CrossRef] [PubMed]

110. Bonilla, P.; Dvorak, J.; Mackill, D.; Deal, K.; Gregorio, G. RFLP and SSLP Mapping of Salinity Tolerance Genes in Chromosome 1 of Rice (*Oryza Sativa* L.) Using Recombinant Inbred Lines. *Philipp. Agric. Sci.* **2002**, *85*, 68–76.

111. Gregorio, G.B. *Tagging Salinity Tolerance Genes in Rice Using Amplified Fragment Length Polymorphism (AFLP)*; University of the Philippines: Los Baños, Philippines, 1997.

112. To, H.T.M.; Nguyen, H.T.; Dang, N.T.M.; Nguyen, N.H.; Bui, T.X.; Lavarenne, J.; Phung, N.T.P.; Gantet, P.; Lebrun, M.; Bellafiore, S.; et al. Unraveling the Genetic Elements Involved in Shoot and Root Growth Regulation by Jasmonate in Rice Using a Genome-Wide Association Study. *Rice* **2019**, *12*, 69. [CrossRef]

113. Baillo, E.H.; Kimotho, R.N.; Zhang, Z.; Xu, P. Transcription Factors Associated with Abiotic and Biotic Stress Tolerance and Their Potential for Crops Improvement. *Genes* **2019**, *10*, 771. [CrossRef]

114. Srivastava, A.K.; Zhang, C.; Yates, G.; Bailey, M.; Brown, A.; Sadanandom, A. SUMO Is a Critical Regulator of Salt Stress Responses in Rice. *Plant Physiol.* **2016**, *170*, 2378–2391. [CrossRef]

115. Srivastava, A.K.; Zhang, C.; Caine, R.S.; Gray, J.; Sadanandom, A. Rice SUMO Protease Overly Tolerant to Salt 1 Targets the Transcription Factor, OsbZIP23 to Promote Drought Tolerance in Rice. *Plant J.* **2017**, *92*, 1031–1043. [CrossRef]
116. Kamal; Alnor Gorafi; Abdelrahman; Abdellatef; Tsujimoto Stay-Green Trait: A Prospective Approach for Yield Potential, and Drought and Heat Stress Adaptation in Globally Important Cereals. *IJMS* **2019**, *20*, 5837. [CrossRef]
117. Li, J.; Wen, J.; Lease, K.A.; Doke, J.T.; Tax, F.E.; Walker, J.C. BAK1, an Arabidopsis LRR Receptor-like Protein Kinase, Interacts with BRI1 and Modulates Brassinosteroid Signaling. *Cell* **2002**, *110*, 213–222. [CrossRef]
118. Wang, X. Brassinosteroids Regulate Dissociation of BKI1, a Negative Regulator of BRI1 Signaling, from the Plasma Membrane. *Science* **2006**, *313*, 1118–1122. [CrossRef]
119. Jang, S.; Li, H.-Y. Oryza Sativa BRASSINOSTEROID UPREGULATED1 LIKE1 Induces the Expression of a Gene Encoding a Small Leucine-Rich-Repeat Protein to Positively Regulate Lamina Inclination and Grain Size in Rice. *Front. Plant Sci.* **2017**, *8*, 1253. [CrossRef]
120. Shang, Y.; Dai, C.; Lee, M.M.; Kwak, J.M.; Nam, K.H. BRI1-Associated Receptor Kinase 1 Regulates Guard Cell ABA Signaling Mediated by Open Stomata 1 in Arabidopsis. *Mol. Plant* **2016**, *9*, 447–460. [CrossRef] [PubMed]
121. Arnon, D.I. Copper Enzymes in Isolated Chloroplasts. Polyphenoloxidase in Beta Vulgaris. *Plant Physiol.* **1949**, *24*, 1–15. [CrossRef] [PubMed]

Article

Morphological Analysis, Protein Profiling and Expression Analysis of Auxin Homeostasis Genes of Roots of Two Contrasting Cultivars of Rice Provide Inputs on Mechanisms Involved in Rice Adaptation towards Salinity Stress

Shivani Saini [1], Navdeep Kaur [1], Deeksha Marothia [1], Baldev Singh [1], Varinder Singh [1], Pascal Gantet [2,3,*] and Pratap Kumar Pati [1,*]

[1] Department of Biotechnology, Guru Nanak Dev University, Amritsar 143005, Punjab, India; shivani.saini23@gmail.com (S.S.); kaurnavdeep302@yahoo.com (N.K.); deekshamarothia@yahoo.in (D.M.); bsbt89@gmail.com (B.S.); varinder75@gmail.com (V.S.)

[2] Université de Montpellier, UMR DIADE, Centre de Recherche de l'IRD, Avenue Agropolis, BP 64501, CEDEX 5, 34394 Montpellier, France

[3] Centre of the Region Haná for Biotechnological and Agricultural Research, Department of Molecular Biology, Palacký University Olomouc, Šlechtitelů 27, 783 71 Olomouc, Czech Republic

* Correspondence: pascal.gantet@umontpellier.fr (P.G.); pkpati@yahoo.com (P.K.P.)

Citation: Saini, S.; Kaur, N.; Marothia, D.; Singh, B.; Singh, V.; Gantet, P.; Pati, P.K. Morphological Analysis, Protein Profiling and Expression Analysis of Auxin Homeostasis Genes of Roots of Two Contrasting Cultivars of Rice Provide Inputs on Mechanisms Involved in Rice Adaptation towards Salinity Stress. Plants 2021, 10, 1544. https://doi.org/10.3390/plants10081544

Academic Editors: Igor G. Loskutov and Masayuki Fujita

Received: 15 April 2021
Accepted: 24 July 2021
Published: 28 July 2021

Publisher's Note: MDPI stays neutral with regard to jurisdictional claims in published maps and institutional affiliations.

Abstract: Plants remodel their root architecture in response to a salinity stress stimulus. This process is regulated by an array of factors including phytohormones, particularly auxin. In the present study, in order to better understand the mechanisms involved in salinity stress adaptation in rice, we compared two contrasting rice cultivars—Luna Suvarna, a salt tolerant, and IR64, a salt sensitive cultivar. Phenotypic investigations suggested that Luna Suvarna in comparison with IR64 presented stress adaptive root traits which correlated with a higher accumulation of auxin in its roots. The expression level investigation of auxin signaling pathway genes revealed an increase in several auxin homeostasis genes transcript levels in Luna Suvarna compared with IR64 under salinity stress. Furthermore, protein profiling showed 18 proteins that were differentially regulated between the roots of two cultivars, and some of them were salinity stress responsive proteins found exclusively in the proteome of Luna Suvarna roots, revealing the critical role of these proteins in imparting salinity stress tolerance. This included proteins related to the salt overly sensitive pathway, root growth, the reactive oxygen species scavenging system, and abscisic acid activation. Taken together, our results highlight that Luna Suvarna involves a combination of morphological and molecular traits of the root system that could prime the plant to better tolerate salinity stress.

Keywords: rice; abiotic stress; salinity; root; auxin; *YUCCA*; *PIN*; proteomics; mass spectrometry

1. Introduction

The plant root is the vital organ that serves a wide range of functions and regulates crop productivity. As roots are in direct interface with the soil, they act as the primary site for perceiving environmental stress-related signals for plants [1,2]. Among various environmental stresses, salinity has emerged as one of the most serious threats limiting global crop production and yield [3]. Currently, almost 20% of the world's total irrigated land is estimated to be affected by salinity stress and it is expected that by the end of the year 2050, more than 50% of the world's arable land will become saline [4–7]. High soil salinity induces undesirable changes at phenotypic, biochemical, physiological, cellular, genetic and molecular levels, which are detrimental to plant growth and survival [8]. The root system responds to abiotic stresses by triggering stress adaptive mechanisms, which are supposed to be regulated by a number of factors [2,9,10].

The potential of several phytohormones to ameliorate the damaging effects of salinity stress has attracted the attention of researchers in the recent past [11,12]. Among different phytohormones, auxin is an important plant hormone well-known for controlling the different aspects of plant growth and development including tropistic growth, vascular tissue differentiation, auxiliary bud formation, cell elongation, flower organ development and abiotic stress tolerance [13–16]. It has also been regarded as a master player in triggering salinity stress-induced changes in root system architecture [12]. Auxin regulates the root growth rates by promoting lateral root formation and mediating the size of root meristem by controlling the transition from cell division to cell differentiation processes [17,18]. The processes that determine the spatiotemporal distribution of auxin and the maintenance of auxin homeostasis required for root growth and development include local auxin biosynthesis, transport, perception, signaling, conjugation and degradation [19,20].

Although roots are the critical site for the perception of salinity stress signals and are responsible for triggering stress-related mechanisms in plants, very little attention has been paid to analyzing this underground part of the plant in the context of understanding salinity tolerance. Physiological, biochemical and genetic studies have provided ample evidence in support of the key role of auxin in triggering abiotic stress-mediated differential modifications in the root system architecture of plants [21]. The key role of the maintenance of auxin homeostasis in regulating salinity stress tolerance is emerging in plant biology [22]. In the present study, in order to better understand the mechanisms conferring rice adaptation to salinity stress, we conducted a comparative analysis of various auxin-related genes (which regulate auxin homeostasis) in the roots of salt sensitive IR64 and salt tolerant Luna Suvarna (LS) cultivars of rice under optimal as well as salinity stress conditions. Further, the endogenous content of indole-3-acetic acid (IAA) has also been estimated in the roots of two contrasting salinity stress cultivars of rice, and an analysis of their root morphology has been performed.

For finding significant clues on the adaptive behavior of plants to salinity stresses, studies at the protein level might be a better option compared to at the transcript level since many post-transcriptional and post-translational changes often take place in plants and hence, the rate of transcription and the translation will not necessarily always correlate. Hence, proteome-based approaches involving two-dimensional gel electrophoresis (2-DE) and mass spectrometry (MS) are often utilized for unraveling proteins associated with induced changes in plants as they are very reliable, sensitive and powerful technologies [9,23]. For example, a comparative study of the leaf proteome profiles of the wild salt tolerant Poaceae species *Porteresia coarctata* with two rice cultivars variable in salt sensitivity—IR64 (salt-sensitive) and Pokkali (salt-tolerant)—suggested that, in the leaves of *Porteresia coarctata*, several proteins exhibited up-regulation that could provide it a physiological advantage under salinity stress [24]. However, there are limited reports on a comparative proteome analysis of the root of contrasting salt-responsive cultivars of rice. Therefore, herein a comparison of the root proteome of two rice cultivars differing in salt tolerance has been conducted using 2-DE and MS. Our results showed that salt tolerant rice cultivar LS has better stress adaptive root traits, elevated expression of auxin homeostasis genes and more endogenous IAA content than IR64 cultivar, which could be linked to the acquisition of natural salinity stress tolerance in LS. Further, several salinity stress responsive proteins were detected exclusively in the roots of LS, which might be providing a peculiar property for attaining salinity stress adaptation and tolerance in rice.

2. Results

2.1. Analysis of Morphological Parameters and IAA Quantification in IR64 and LS

The differences in the morphological parameters of the two cultivars were clearly observable when cultivated in normal conditions (Figure 1A,B). An approximately 40% increase in the length of shoot and a 70% longer roots were observed in the salt-tolerant cultivar, LS, as compared to the sensitive cultivar IR64 (Figure 1A–C,F). The number of roots (primary root and crown roots) was also found to be 169% more in LS (Figure 1G). Moreover, an increase in the fresh weight of shoots and roots by 101% and 137% respectively, was noticed in LS (Figure 1D,H). Similarly, an approximately 122% and 170% enhancement in the dry weight of shoots and roots in the tolerant cultivar, respectively, was observed in LS with respect to IR64 (Figure 1E,I). The amount of endogenous IAA was also quantified in the roots of IR64 and LS. It was observed that in LS roots, the endogenous IAA concentration was significantly higher (1.086 µg/gFW) as compared to IR64 roots (0.6608 µg/gFW) (Figure 1J).

2.2. Expression Analysis of Genes Involved in Auxin Homeostasis in IR64 and LS Roots

To better understand the cause of the differences observed in the IAA content in the roots of both cultivars, the transcript levels of various genes involved in auxin homeostasis were measured by qRT-PCR under optimal and salinity stress conditions. Among various auxin biosynthesis genes, the transcript levels of *OsYUCCA5*, *OsYUCCA7*, and *OsYUCCA8* exhibited significant up-regulation of 2.79, 3.53, and 2.58 fold, respectively, in the roots of the salt-tolerant cultivar LS as compared to the salt-sensitive cultivar IR64 of rice under normal conditions (Figure 2A). In response to salinity stress, significant down-regulation of *OsYUCCA3*, *OsYUCCA4*, *OsYUCCA6*, *OsYUCCA7*, and *OsYUCCA9* by 0.43, 0.15, 0.18, 0.27, and 0.39 fold was observed in IR64 with respect to the control (Figure 2A). In the LS cultivar, auxin biosynthesis genes *OsYUCCA3*, *OsYUCCA4*, *OsYUCCA5*, *OsYUCCA6*, *OsYUCCA7*, and *OsYUCCA9* exhibited significant up-regulation by 2.83, 2.79, 7.88, 6.75, 4.53 and 3.07 fold, respectively, in the roots upon salinity stress with respect to the IR64 (control) (Figure 2A). On the contrary, the expression level of *OsYUCCA1* and *OsYUCCA8* exhibited significant down-regulation by 0.41 and 0.75 fold, respectively, in the LS root compared to the control (Figure 2A). The transcript levels of *OsYUCCA2* did not show any significant difference in two contrasting salinity stress responsive cultivars of rice under normal and salinity stress conditions.

Among different auxin efflux transporter *OsPIN* genes, *OsPIN2*, *OsPIN5a*, and *Os-PIN5b* exhibited higher transcript level accumulation of 1.95, 2.36, and 2.46 fold, respectively, in the roots of untreated LS as compared to the IR64 (Figure 2B). On the contrary, significant down-regulation in the expression of *OsPIN1b*, that is, 0.48, was found in the roots of LS as compared to the salt-sensitive cultivar IR64 of rice (Figure 2B). In response to salinity stress in the IR64 cultivar, significant down-regulation of *OsPIN1a* and *OsPIN2* by 0.28 and 0.23 fold, respectively, was observed as compared to the control. The expression of auxin efflux transporters *OsPIN1a*, *OsPIN1b*, *OsPIN2*, *OsPIN3a*, and *OsPIN5b* was up-regulated by 4.51, 1.62, 5.46, 1.42, and 1.54 fold, respectively, in the roots of the LS cultivar under salinity stress as compared to the untreated IR64 control (Figure 2B). However, the transcript levels of *OsPIN1c* and *OsPIN1d* did not show any significant differences between the two cultivars in response to control or salinity stress conditions (Figure 2B). Under the effect of salinity stress, *OsPIN5a* and *OsPIN9* exhibited down-regulation by 0.85 and 0.33 fold, respectively, in LS than the roots of the control (Figure 2B).

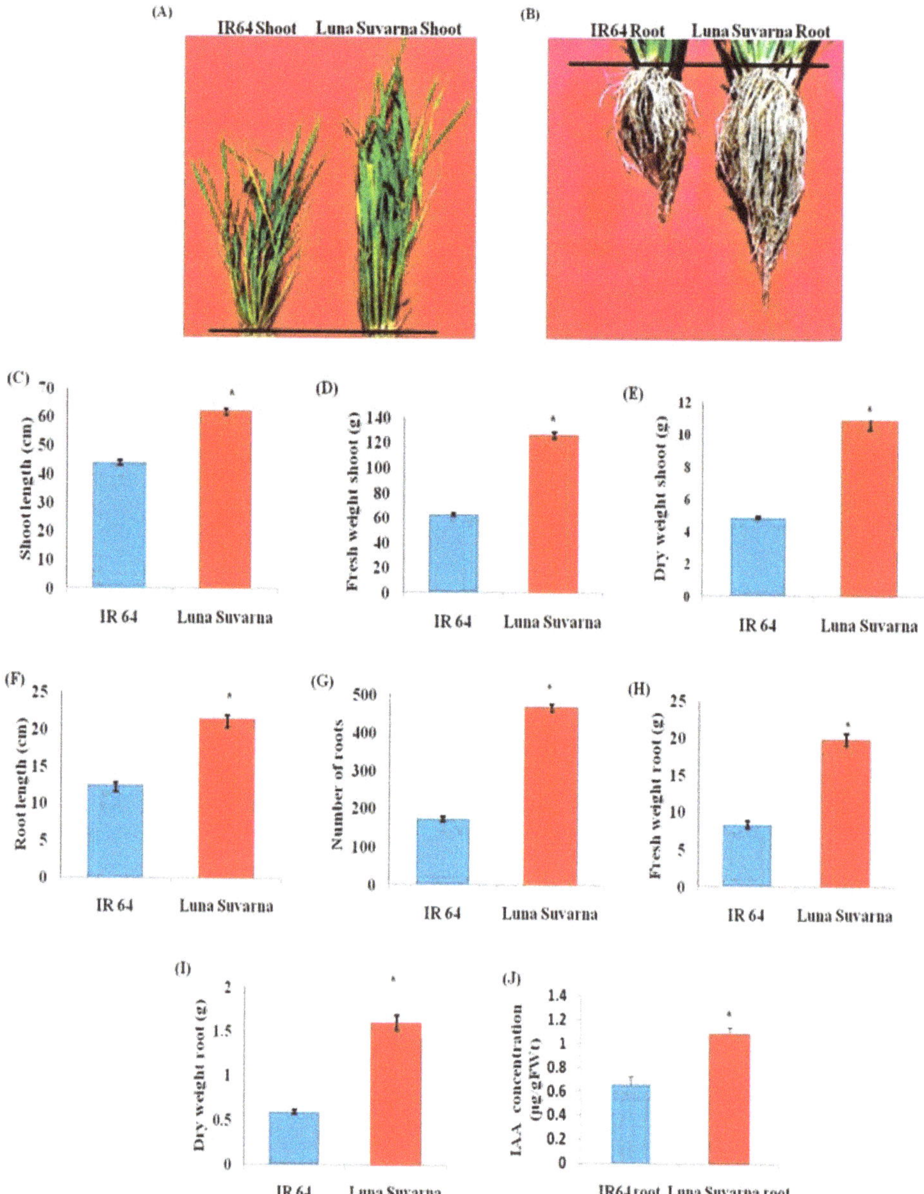

Figure 1. (**A**) Growth response of IR64 and Luna Suvarna (LS) shoot under control conditions; (**B**) Growth response of IR64 and Luna Suvarna (LS) roots under control conditions; (**C**) Comparative analysis of shoot length in IR64 and Luna Suvarna; (**D**) Comparative analysis of the fresh weight of shoot in IR64 and Luna Suvarna; (**E**) Comparative analysis of the dry weight of shoot in IR64 and Luna Suvarna; (**F**) Comparative analysis of root length in IR64 and Luna Suvarna; (**G**) Comparative analysis of the number of roots in IR64 and Luna Suvarna; (**H**) Comparative analysis of the fresh weight of roots in IR64 and Luna Suvarna; (**I**) Comparative analysis of the dry weight of roots in IR64 and Luna Suvarna; (**J**) IAA estimation in the roots of IR64 and LS. Data represent mean $\pm$ SE (n = 15) for the analysis of growth parameters while for IAA quantification mean $\pm$ SE (n = 3). Asterisks signs (*) represent values which were significantly different among different samples (Fisher LSD, $p \leq 0.05$). Blue color represents IR64 while red color represents LS cultivar.

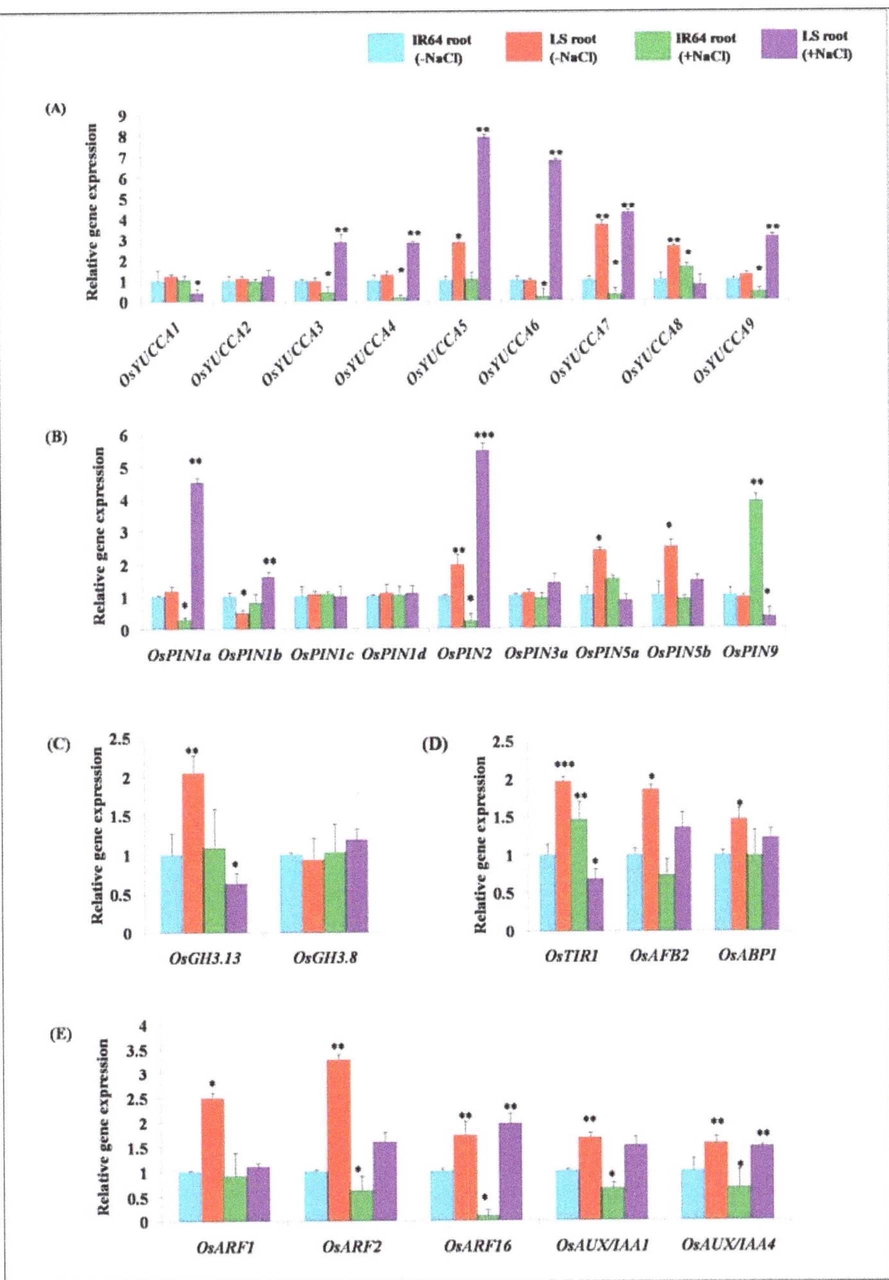

Figure 2. Real-time gene expression analysis of auxin homeostasis genes under control and salinity stress conditions in IR64 and LS roots. (**A**) Real-time gene expression of auxin biosynthesis genes; (**B**) Real-time gene expression of auxin transport genes; (**C**) Real-time gene expression of auxin conjugation and degradation genes; (**D**) Real-time gene expression of auxin receptor genes; (**E**) Real-time gene expression of auxin signaling genes. Three biological replicates were taken and bars represent mean ± SE. Asterisks signs (*, **, ***) represent values which were significantly different among different samples (Fisher LSD, $p \leq 0.05$). The transcript levels of LS under normal condition and, LS and IR64 upon salinity stress treatment were compared with IR64 (control), whose expression was assumed as 1. (-NaCl) refers to untreated samples, (+NaCl) refers to salinity treated samples. Blue color refers IR64 root (-NaCl), red color refers to LS root (-NaCl), green color refers to IR64 root (+NaCl), and purple color refers to LS root (+NaCl).

The transcript levels of auxin conjugation and degradation gene *OsGH3.13* were found to be higher by 2.04 fold in the LS root under normal conditions (Figure 2C). In response to the salinity stress, no significant change was observed in the gene expression of *OsGH3.13* in IR64, whereas *OsGH3.13* displayed significant down-regulation by 0.63 fold in the LS root with respect to the control (Figure 2C). The gene expression of *OsGH3.8* did not exhibit a significant change in the roots of LS under the normal conditions with respect to IR64. *OsGH3.8* displayed 1.21 fold higher accumulations in the tolerant cultivar LS, and its expression remained unaltered in IR64 in response to salinity stress (Figure 2C). The expression of auxin receptor genes, particularly *OsTIR1*, *OsAFB2* and *OsABP1*, was studied in the roots of salt-sensitive cultivar IR64 and salt-tolerant cultivar LS of rice under normal and salinity stress (Figure 2D). It was observed that the expression of *OsTIR1* and *OsAFB2* displayed up-regulation by 1.96 and 1.86 fold in the roots of the LS under the normal conditions as compared to IR64. Similarly, the transcript level accumulation of *OsABP1*, the auxin receptor of the proteasome independent pathway was also found to be higher by 1.46 fold in the roots of LS than IR64 (Figure 2D). Under the effect of salinity stress, *OsTIR1* exhibited 1.46 fold up-regulation, while *OsAFB2* displayed 0.73 fold down-regulation in IR64 than in the control (Figure 2D). In LS roots, it was observed that the expression of *OsAFB2* and *OsABP1* displayed up-regulation by 1.36 and 1.22 fold, respectively, in response to salinity stress. On the contrary, *OsTIR1* showed 0.68 fold down-regulation in the roots of LS (Figure 2D). Interestingly, various auxin signaling genes, such as *OsARF1*, *OsARF2*, *OsARF16*, *OsAUX/IAA1* and *OsAUX/IAA4*, also exhibited higher gene expression of 2.49, 3.26, 1.72, 1.66, and 1.54 fold, respectively, in the LS root as compared to the IR64 under control conditions (Figure 2E). In response to salinity stress, the transcript levels of *OsARF1*, *OsARF2*, *OsARF16*, *OsAUX/IAA1*, and *OsAUX/IAA4* displayed down-regulation by 0.91, 0.62, 0.09, 0.6, and 0.67 fold, respectively, in IR64 with respect to the control. On the contrary, *OsARF2*, *OsARF16*, *OsAUX/IAA1* and *OsAUX/IAA4* exhibited up-regulation by 1.59, 1.96, 1.52, and 1.48 fold, respectively, in the LS root under salinity stress stimuli (Figure 2E). The expression level of *OsARF1* did not show any significant differences in the two contrasting salt tolerant cultivars of rice under salinity stress (Figure 2E).

2.3. 2-DE Analysis of Root Proteins in IR64 and LS

In complement to auxin-related gene expression analysis, the proteins isolated from roots of 14-days old seedlings of salt-sensitive IR64 and salt-tolerant LS cultivars of rice were subjected to 2-DE analysis. A total of 146 spots were detected in IR64 while 166 spots were observed in LS (Figure 3), using PDQuest 8.0.2 software. Among these protein spots, 27 were observed to be either present/absent and few were of altered intensity in the roots of IR64 and LS. These spots were excised and sent for MALDI TOF/TOF MS/MS analysis. Out of 27 spots, only 18 proteins were successfully identified (Tables 1 and 2). The plant intracellular Ras group related LRR protein 2, B3 domain-containing protein, and Ubiquitin fold modifier protein 1 displayed higher protein expression by 1.76, 1.1, and 3.75 fold in the roots of LS in comparison to IR64, respectively (Table 2). Among the 18 identified proteins, 13 proteins were implicated in abiotic stress responses and two of them exhibited enzymatic activity. One protein constituted the core nucleosome component, however, another protein was involved in autophagy and protein transport and the remaining one protein function in the DNA methylation process.

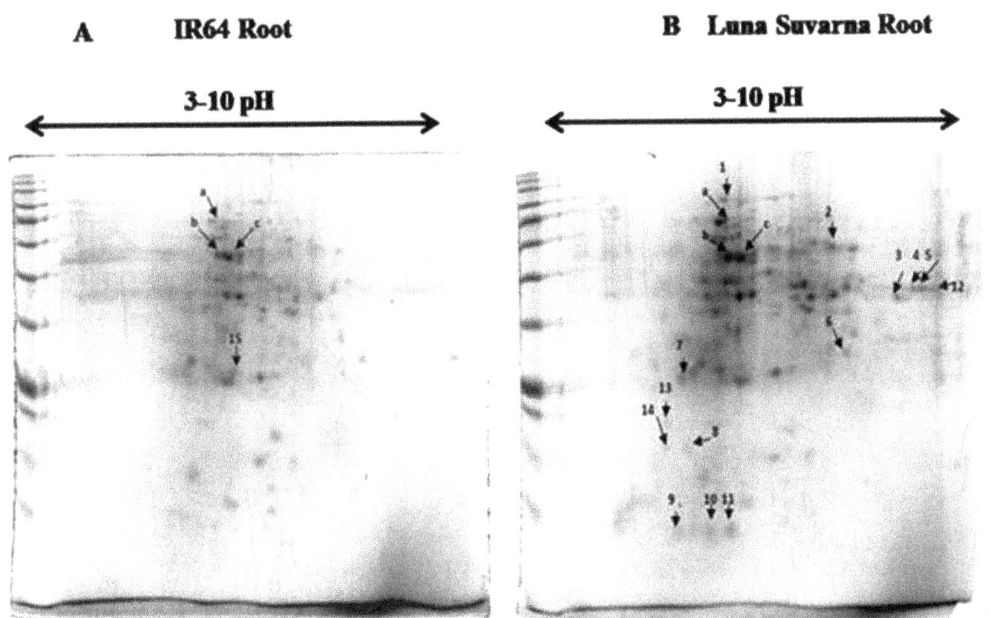

Figure 3. Two-dimensional gel electrophoretic analysis of the protein profiles of rice root proteome under control condition (**A**) IR64 root (**B**) Luna Suvarna (LS) root.

Table 1. Specifically expressed proteins in Luna Suvarna (LS) versus IR64 roots identified using MALDI-ToF/ToF mass spectrometry.

Spot	Protein Name	Accession Number	MSU Number	Reference Organism	Mrpi (Theoretical)	Score	Coverage	Function
1	Histone H2B.10	H2B10_ORYSI	LOC_Os01g05610.1	*Oryza sativa* subsp. Indica	16522 (10.02)	41	36%	Core nucleosome component
2	DEAD-box ATP-dependent RNA helicase 53	RH53_ORYSJ	LOC_Os07g05050.1	*Oryza sativa* subsp. Japonica	65429 (9.44)	47	15%	Abiotic stress responses
3	Glyceraldehyde-3-phosphate dehydrogenase 3 cytosolic	G3PC3_ORYSJ	LOC_Os02g38920	*Oryza sativa* subsp. Japonica	36716 (7.68)	37	38%	Glycolysis enzyme
4	Calcineurin B-like interacting protein kinase 21	CIPKL_ORYSJ	LOC_Os07g44290	*Oryza sativa* subsp. Japonica	59592 (9.26)	39	15%	Abiotic stress tolerance
5	4-Coumarate-CoA ligase-like 9	4CLL9_ORYSJ	LOC_Os04g24530	*Oryza sativa* subsp. Japonica	59782 (5.69)	46	22%	Abiotic stress tolerance

Table 1. Cont.

Spot	Protein Name	Accession Number	MSU Number	Reference Organism	Mrpi (Theoretical)	Score	Coverage	Function
6	β-glucosidases 12	BGL12_ORYSJ	LOC_Os04g39880	*Oryza sativa* subsp. Japonica	57713 (8.75)	47	11%	Abiotic stress tolerance
7	Ubiquitin-like protein autophagy-related	ATG12_ORYSI	LOC_Os06g10340	*Oryza sativa* subsp. Indica	10454/9.05	34	32%	Autophagy and protein transport
8	Phytosulfokines 3	PSK3_ORYSJ	LOC_Os03g47230	*Oryza sativa* subsp. Japonica	8341/5.83	34	29%	Abiotic stress tolerance
9	Cyanate hydratase	CYNS_ORYSI	LOC_Os10g33270	*Oryza sativa* subsp. Indica	18653/5.61	33	44%	Abiotic stress tolerance
10	Probable protein arginine N-methyltransferase	ANM3_ORYSI	LOC_Os07g44640	*Oryza sativa* subsp. Indica	69127/4.49	29	13%	Mediates methylation process
11	Heat stress transcription factor B 1	HSFB1_ORYSJ	LOC_Os09g28354	*Oryza sativa* subsp. Japonica	33121/9.35	33	43%	Abiotic stress responses
12	Delta 1-pyrroline-5-carboxylate synthase	P5CS_ORYSJ	LOC_Os05g38150	*Oryza sativa* subsp. Japonica	78153/6.37	53	14%	Abiotic stress tolerance
13	Calcineurin B-like-interacting protein kinase 29	CIPKT_ORYSJ	LOC_Os07g48090	*Oryza sativa* subsp. Japonica	48581/8.69	38	44%	Abiotic stress tolerance
14	Minichromosome maintenance 6	MCM6_ORYSI	LOC_Os05g14590	*Oryza sativa* subsp. Indica	93168/5.55	43	17%	Abiotic stress tolerance
15	UDP-glucose 6-dehydrogenase 1	UGDH1_ORYSJ	LOC_Os03g31210	*Oryza sativa* subsp. Japonica	52834 (5.75)	39	13%	Enzymatic function

Spots 1–14 were exclusively observed in the case of Luna Suvarna root, whereas spot 15 was differentially less expressed in the IR64 root.

Table 2. Highly expressed proteins in Luna Suvarna (LS) roots as compared to IR64 identified using MALDI-ToF/ToF mass spectrometry.

Spot	Protein Name	Accession Number	MSU Number	Reference Organism	Mrpi (Theoretical)	Mrpi (Experimental)	Score	Coverage	Function
a	Plant intracellular Ras group related LRR protein 2	PIRL2_ORYSJ	LOC_Os02g38040	*Oryza sativa* subsp. Japonica	55345/5.51	114.7/5.6	30	27%	Abiotic stress tolerant
b	Ubiquitin-fold modifier 1	UFM1_ORYSJ	LOC_Os01g73140	*Oryza sativa* subsp. Japonica	10356/9.60	81.67/5.98	30	44%	Abiotic stress tolerant
c	B3 domain-containing protein	Y1237_ORYSJ	LOC_Os01g52540	*Oryza sativa* subsp. Japonica	83749/5.81	118.37/5.71	38	34%	Abiotic stress tolerant

Comparative expression analysis of identified stress marker genes (which are known for exhibiting tolerance against salinity stress) (Table 3) was performed in the roots of IR64 and LS under the native condition to complete the results obtained with MS at the transcript level. The transcript levels of *4-coumarate CoA ligase 9 (CCoA), β-glucosidases 12 (β-gluc), phytosulfokines 3 (PSK),* and *B3 domain-containing protein (B3D)* were observed to be 1.09, 1.02, 3.64, and 1.78 fold higher, respectively, in LS roots compared to IR64 under normal conditions (Figure 4). However, the gene expression of *calcineurin B-like interacting protein kinase 21 (CIPK), delta-pyrroline-5-carboxylate synthase (P5CS), cyanate hydratase (CHS), DEAD-box ATP-dependent RNA helicases 53 (DEAD), plant intracellular Ras group related LRR protein 2 (RAS-LRR),* and *minichromosome maintenance 6 (MCM6)* was 0.69, 0.38, 0.30, 0.21, 0.64, and 0.22 fold lower, respectively, in the roots of tolerant cultivar LS (Figure 4) with respect to the salt-sensitive IR64. Under salinity stress, *β-gluc* and *B3D* displayed up-regulation of 1.22 and 2.44 fold, respectively, in IR64 roots compared to the control (Figure 4). On the contrary, *CIPK, CCoA, PSK, P5CS, CHS,* and *MCM6* showed lower transcript levels by 0.54, 0.53, 0.066, 0.15, 0.75, and 0.72 fold, respectively, in IR64, while *DEAD* exhibited no significant change in its expression. In response to salinity stress, the expression of *CIPK, CCoA, PSK, CHS, DEAD* and *MCM6* was 1.85, 1.86, 15.26, 1.32, 1.5, and 1.37 fold higher respectively, in LS roots compared to control IR64 (Figure 4). However, the transcript level of *β-gluc, P5CS, B3D* and *RAS-LRR* exhibited 0.83, 0.78, 0.41, and 0.65 fold (Figure 4) lower transcript level accumulation, respectively, in the roots of the tolerant cultivar (LS) under salinity stress stimuli.

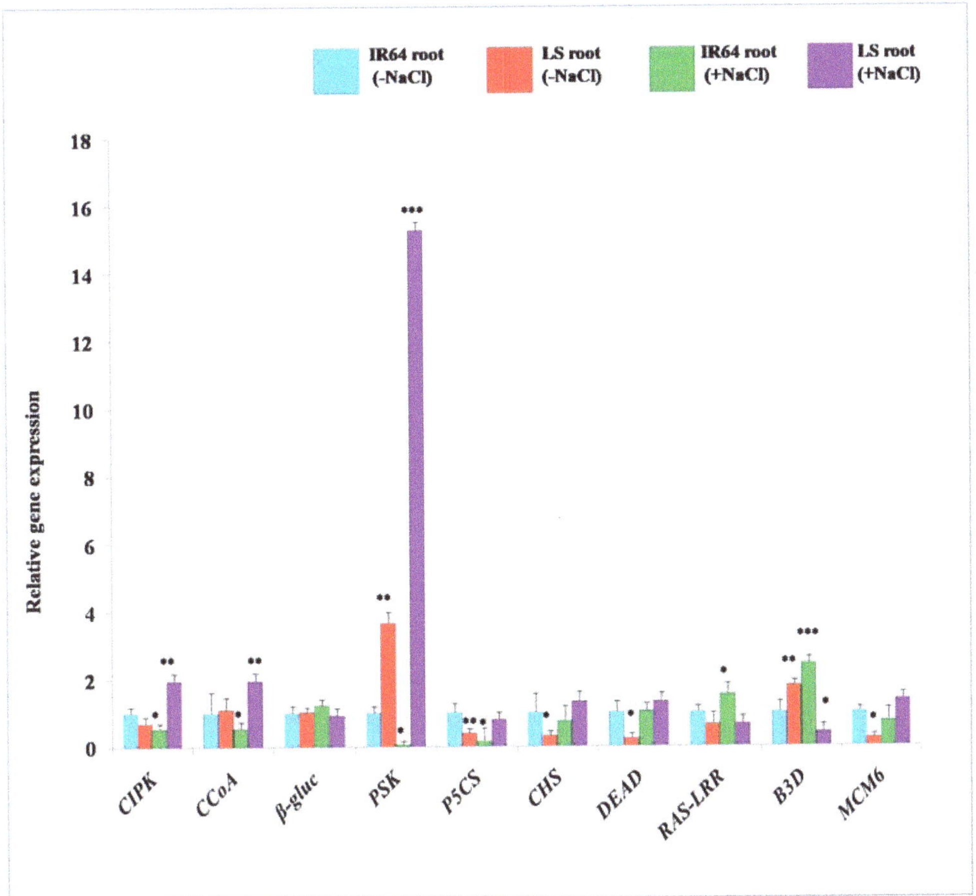

Figure 4. The relative expression levels of genes encoding for proteins differentially accumulated in IR64 and Luna Suvarna (LS) roots under control and salinity stress conditions. Three biological replicates were taken and bars represent mean ± SE. Asterisks signs (*, **, ***) represent values which were significantly different among different samples (Fisher LSD, $p \leq 0.05$). The transcript levels of LS under normal condition and, LS and IR64 upon salinity stress treatment were compared with IR64 (control), whose expression was assumed as 1. (-NaCl) refers to untreated samples, (+NaCl) refers to salinity treated samples. Blue color refers to IR64 root (-NaCl), red color refers to LS root (-NaCl), green color refers to IR64 root (+NaCl) and purple color refers to LS root (+NaCl). CIPK: Calcineurin B-like interacting protein kinase 21; CCoA: 4-coumarate CoA ligase like 9; β-gluc: Beta-glucosidase 12; PSK: Phytosulfokines 3; P5CS: Delta-1-pyrroline-5-carboxylate synthase; CHS: Cyanate hydratase; DEAD: DEAD-box ATP-dependent RNA helicase 53; RAS-LRR: Plant intracellular Ras group related LRR protein 2; B3D: B3 domain-containing protein; MCM6: minichromosome maintenance 6.

3. Discussion

Plant roots perceive the salinity stress signals and promptly pass them to the shoot to activate various stress-responsive pathways [1,2,9,10]. Although roots are the important site for the perception of salinity stress-related signals, not much attention has been paid to exploring this underground part of the plant in the context of understanding salinity tolerance attributes. In the present study, a positive correlation has been observed between the root system architecture, auxin content, stress marker proteins, and salinity stress adaptation. It was observed that the salinity stress tolerant LS cultivar of rice has a longer primary root, a larger number of roots and a higher fresh weight and dry weight in comparison to IR64. It is realized that plants acquire deeper roots, more lateral roots, more root hair length, and a larger number of roots and its biomass for achieving natural defense against stress conditions including salinity [25–27]. Moreover, it has been demonstrated that the presence of a larger root/shoot length ratio and a higher root biomass promoted the adaptation of plants towards environmental stresses [24,28]. Thus, the present observation of distinct differences in the root phenotype of LS compared to IR64 could be extrapolated to the acquisition of adaptive morphological traits that enable LS plants to mitigate salinity stress when exposed.

It is believed that phytohormones are critical signaling molecules that function downstream of environmental stimuli and regulate various stress adaptive pathways [29]. In previous studies, it has been demonstrated that high salinity stress greatly affects root architecture by inhibiting primary and lateral root growth through altering the accumulation and distribution of the critical phytohormone, auxin [30–33]. Among different auxins, the role of primary auxin IAA is thought to be fundamental as it is the key player in regulating root development [34,35]. Thus, in the present work, the endogenous levels of IAA have been estimated. Upon analysis, it was observed that LS exhibited significantly higher IAA content compared to IR64 roots. Earlier, it was reported that *iaam-OX* transgenic lines (with higher endogenous IAA level) and wild-type plants of Arabidopsis pretreated with IAA exhibited resistance towards drought stress [15]. However, the triple mutants, *yuc1yuc2yuc6*, which were deficient in endogenous IAA content, showed decreased resistance towards drought stress [15]. Moreover, augmented levels of indole-3-butyric acid (IBA) in growing leaves and higher IAA content in the roots of the highly salt-resistant maize variety, SR03, were observed in response to salinity stress [32]. It was revealed that the increased IAA concentration enhanced the accumulation of cell growth-related agents, such as β-expansins (involved in cell wall extension), under salinity stress [32,36]. In Arabidopsis *iar4* mutants, reduced root meristem activity and root growth were reported due to diminished auxin distribution in root tips, indicating the key role of auxin in root growth and development [30]. The exogenous application of auxin is also well known to positively modulate root architecture, especially the lateral root number [15,33,37,38]. Thus, the higher endogenous levels of IAA observed in salt tolerant rice cultivar (LS) could be considered as one of the prominent reasons for the acquisition of salinity stress adaptive root traits observed in the LS cultivar.

It is well realized that the process of auxin-mediated root development is regulated by a complex interplay between auxin metabolism, its signaling and transport leading to the spatio-temporal distribution of auxin [12,39,40]. Thus, to get insights into the molecular dynamics of auxin homeostasis, the transcript-level expression of different genes involved in the auxin pathway has been analyzed in roots in the present work. Recent studies suggest that the local biosynthesis of auxin by YUCCA flavin monooxygenases in the roots is the primary source for normal root development and root gravitropic responses [35]. Moreover, it has been demonstrated that five *YUCCA* genes—*YUCCA3*, *YUCCA5*, *YUCCA7*, *YUCCA8*, and *YUCCA9*—express highly in Arabidopsis roots, playing an essential role in the root development [35]. However, the link of *YUCCA* genes and salinity stress adaptation has never been evaluated in rice. Interestingly, in the present study, it was observed that the transcript level accumulation of different *OsYUCCA* genes was higher in the roots of LS. It might be the primary reason for more auxin biosynthesis and its accumulation in LS roots.

Further, in response to salinity stress, the transcript level accumulation of *OsYUCCA3*, *OsYUCCA4*, *OsYUCCA5*, *OsYUCCA6*, *OsYUCCA7*, and *OsYUCCA9* was enhanced in LS roots. In Arabidopsis, *YUCCA8* and *YUCCA9* have been linked with the development of lateral roots, while their mutants develop shorter primary roots suggesting their key role in the development of root system architecture [41]. Hence, the present study hints towards a link between *OsYUCCA* genes mediated enhanced auxin accumulation and subsequently better developed root system architecture for the acquisition of salinity stress tolerance in rice.

Once IAA is biosynthesized, it is transported to the area of its requirement with the help of cell-to-cell auxin transport mediated by *OsPINs* in rice [42]. Earlier, it was found that polar auxin transport is affected by osmotic stress caused by increased salinity or drought [31]. Moreover, flavonoids and phenolic compounds that are accumulated in response to stress exposure also inhibit polar auxin transport [43,44]. Interestingly, in the present study, the transcript levels of auxin efflux carrier genes, such as *OsPIN1a*, *OsPIN2*, *OsPIN3a*, *OsPIN5a*, and *OsPIN5b*, were found to be higher in the LS root. Moreover, under salinity stress, the expression of auxin transport genes, particularly *OsPIN1a*, *OsPIN1b*, *OsPIN2*, *OsPIN3a*, and *OsPIN5b*, was up-regulated in the roots of LS. *OsPIN1b* and *OsPIN9* have been suggested to participate in root development in rice, by regulating auxin-cytokinin interaction [45]. Further, *OsPIN2* expresses highly in roots and enhances shoot to root auxin transport [46]. Thus, the increased expression of such transporter genes in LS roots suggests that salt-tolerant rice cultivar has better capability to maintain auxin homeostasis under salinity stress; however, further investigations are necessary to consolidate these findings.

The optimum concentration of IAA is maintained in a cell through their conjugation and degradation by *OsGH3* genes [47]. Hence, the expression of auxin conjugation and degradation gene *OsGH3.13* was analyzed in the roots of salt-sensitive (IR64) and salt-tolerant cultivar (LS) of rice. It was observed that the expression of *OsGH3.13* was significantly higher in the roots of LS, which is contradictory to the observed higher IAA content in LS. Thus, it can be inferred that the role of the *OsYUCCA* genes is probably more critical in regulating auxin content in rice as compared to *OsGH3*. Further, the analysis of *OsGH3.8* at transcript level suggests no difference in expression level between IR64 and LS roots. However, under salinity stress, the transcript level accumulation of *OsGH3.13* was down-regulated in the roots of LS compared to IR64. On the contrary, *OsGH3.8* exhibited up-regulation in the root but not to a significant level. It indicates that the lower expression of *OsGH3.13* under NaCl application in LS root might be responsible for providing tolerance against salinity stress, probably by enhancing IAA levels.

Various findings have suggested that *OsAFB2* and *OsTIR1* are the auxin signaling receptors affected by salinity stress [31,48]. However, their probable role in providing a natural defense against salinity stress has never been evaluated. In the present study, the transcript level accumulation of auxin receptor genes *OsTIR1*, *OsAFB2*, and *OsABP1* were found to be elevated in the roots of LS with respect to IR64 under normal conditions. This identification of the enhanced expression of auxin receptors can be linked to higher auxin content in the roots of LS compared to IR64. Under salinity stress, the expression of *OsTIR1* showed down-regulation in LS roots with respect to control IR64. On the contrary, the transcript levels of *OsAFB2* and *OsABP1* exhibited up-regulation in the roots of LS compared to IR64 under the salt application. The expression of various auxin signaling genes, *OsARF1*, *OsARF2*, *OsARF16*, *OsAUX/IAA1*, and *OsAUX/IAA4*, was also found to be higher in the roots of LS compared to IR64 under both control as well as salinity stress conditions, demonstrating elevated endogenous IAA level in LS root compared to IR64. Previous studies have also linked elevated auxin concentration with increased auxin transport and downstream signaling genes [49], thus promoting auxin-mediated root development. In rice, 31 auxin repressor (*OsAUX/IAAs*) and 25 auxin activator (*OsARFs*) genes that participate in auxin signaling were observed to be suppressed by cold, heat and drought stress [31]. On the contrary, some *OsAUX/IAA*, such as *OsAUX/IAA 6,9,18,19,20* and *28* and *OsARF 4,11,13,14,16,18* and *19*, were induced by at least one among cold, heat, and drought stress [31]. Hence, various auxin signaling genes respond differentially to abiotic stresses such as cold, heat and drought [31]. However, to the best of our knowledge, there is no report on the effect of salinity stress on the auxin signaling genes. It has been suggested that, in Arabidopsis Aux/IAA protein, IAA14 participates in the early stages of lateral root development [50,51]. Hence, the observed elevated expression of *AUX/IAA* in LS root compared to IR64 might be extrapolated to the salt tolerance and a higher number of roots detected in the tolerant cultivar.

The comparative proteomics study revealed that some proteins were specifically present in the roots of LS compared to IR64. Among these, CIPK21 and CIPK29, the Ca^{2+} sensing proteins of the CIPK gene family, have been previously linked with the enhanced tolerance against salinity stress conditions in Arabidopsis [52,53]. It was suggested that *salt overly sensitive 3 (SOS3)* encodes for CBL, which functions in sensing the cytosolic Ca^{2+} concentration by directly binding to it [52,54]. The Ca^{2+} bound CBL proteins directly activate their interacting partners, such as CIPK6, which are involved in auxin transport, regulation of root length and lateral root development [52,54]. CIPK6 also enhances the transcript levels of *NAC, PIN2,* and *P5CS* genes, which promote salinity stress resistance in plants [52]. P5CS protein, which is involved in proline biosynthesis, was also observed to show increased levels in LS roots in the current study. Proline is involved in the maintenance of cell turgor or osmotic balance, stabilizing membranes to prevent the leakage of electrolytes, and regulates reactive oxygen species (ROS) homeostasis [55,56]. Thus, it can be supposed that the increased levels of an enzyme involved in proline biosynthesis could be responsible for enhancing the proline content in the LS roots that has been previously linked to salinity tolerance [2]. In LS roots, a higher abundance of lignin biosynthesis protein, CCoA, was also observed. There is evidence that salinity stress causes increased lignification of the cell wall through maintaining the structural rigidity and durability of desert poplar plants [57]. Thus, the current finding indicates the possible role of lignin deposition in enhancing salinity stress tolerance in rice [57–59]. Further, higher protein accumulation of detoxification enzyme, CHS (which detoxifies cytotoxic compounds such as cyanate), was observed exclusively in LS roots. CHS also supplies salinity-stressed plants with alternative sources of nitrogen and carbon for better adaptation [60,61]. PSK, which was found exclusively in LS roots, is linked to plant immunity and the maintenance of cellular homeostasis, and is also involved in normal root growth and development [62]. PSK also decreases ethylene production which hinders the primary root growth by inhibiting cell proliferation in the meristematic zone and cell elongation in the elongation zone [62–64]. Hence, the augmented root growth and salinity stress tolerance observed in LS compared to IR64 could also be linked to the higher PSK content. Another stress marker protein, DEAD, was observed exclusively in the roots of LS. In earlier reports, DEAD has been shown to provide salinity stress tolerance in transgenic tobacco by reducing oxidative stress through activating the ROS scavenging system [65,66]. It also improves a plant's photosynthesis machinery, enhances plant growth and development, and mitigates salinity stress [65]. Hence, the higher accumulation of DEAD in LS roots might be involved in scavenging excess ROS, leading to the promotion of salinity stress tolerance. Moreover, β-gluc, which enhances the ABA pool, was also found exclusively in LS roots. The key role of β-gluc in releasing active and free forms of abscisic acid (ABA) from physiologically inactive ABA-glucose conjugate pool, resulting in the alleviation of salinity stress, has already been reported [67,68]. Therefore, the higher accumulation of β-gluc in LS roots might promote ABA accumulation, thus enhancing salinity stress tolerance. The content of MCM6 protein was found exclusively in the roots of the salt tolerant cultivar (LS), which plays an important role in the initiation and elongation steps of eukaryotic DNA replication [69]. In one of the previous studies, the role of MCM6 in providing resistance against high salinity and cold stress has already been elucidated [69]. It was also suggested that the ectopic over-expression of *Pisum sativum PsMCM6* in tobacco confers salinity stress tolerance without affecting yield [65,69]. Further, there were some proteins, such as RAS-LRR, B3D and ubiquitin fold modifier 1, that expressed relatively higher in LS as compared to IR64 roots. The enhanced protein accumulation of RAS-LRR (which encodes polygalacturonase inhibitor proteins, PGIPs) plays a critical role in mitigating salinity stress [68,70]. Further, the roots of LS also exhibited higher protein accumulation of B3D (which triggers various stress-responsive genes) with respect to IR64 roots. RAV (related to ABI3/VP1) protein contains AP2 domain at N-terminal region and B3D in its C-terminal region, which also confer salinity stress resistance through regulating various stress-related genes (*RD29A, RD29B, RAB18, ABI1, ERD15, KIN, ERD10,* and *COR15a*) [71]. The protein content of ubiquitin fold modifier 1 (UFM1) was found to be several-fold higher in the roots of the tolerant cultivar (LS), which could prevent oxidative damage caused by free radicals. It has been suggested that, in addition to ubiquitin, plants utilize a number of ubiquitin-like proteins, such as those related to ubiquitin 1 (RUB1), small ubiquitin-like modifier (SUMO), UFM1, and homology to ubiquitin (HUB), which participates in providing abiotic stress tolerance [72,73]. These proteins confer resistance against salinity stress by prohibiting the damage caused by free radicals and also prevent endoplasmic reticulum-induced apoptosis in protein secretory cells [73,74]. In IR64, upon the application of salinity stress, the expression of *CIPK, CCoA, P5CS, PSK, CHS,* and *MCM6* genes exhibited down-regulations with respect to the control IR64, which might lead to salinity stress susceptibility in the sensitive cultivar. The present finding also indicates that, in IR64, the protein turnover rate might be high, probably leading to targeting of the salinity stress responsive proteins towards degradation, leading to salinity stress sensitivity. On the contrary, in LS upon salinity stress application, higher transcript accumulation of *CIPK, CCoA, PSK, CHS, DEAD* and *MCM6* was observed, which could be linked to its acquisition of the salinity stress resistance property.

Table 3. Role of identified stress marker proteins in salinity stress tolerance.

Protein	Functions	Reference
Calcineurin B-like interacting protein kinases	(1) Enhances shoot to root auxin transport. (2) Mediates root development and lateral root formation.(3) Promotes salt stress tolerance.	[52–54]
4-Coumarate CoA ligase	(1) Enhances salt stress tolerance. (2) Increases lignification of salt stress-tolerant varieties.	[57,58]
β-glucosidases	(1) In vacuoles, it converts ABA glucopyranosides into free ABA. (2) It leads to the adaptation of plants to salt stress conditions.	[67,68]
Phytosulfokines	(1) Emerged as a novel kind of plant hormone recently and involved in immunity and homeostasis. (2) Involved in root growth and development; and inhibits ethylene production.	[62–64]
Plant intracellular Ras-group-related LRR protein 2	(1) Promotes root development. (2) Promotes salt stress tolerance by encoding PGIPs.	[68,70]
Minichromosome maintenance 6 (MCM6)	(1) It confers salinity stress tolerance in pea by additional uptake of Na^+	[65,69]
B3 Domain containing protein	(1) RAV (Related to ABI3 and VP1) has AP2 and B3 domain. (2) Promotes salinity tolerance, enhances stress marker genes.	[71]
Cyanate hydratase	(1) Involved in detoxification of cyanate. (2) Provide alternative sources of nitrogen and carbon for enhancing salt stress tolerance.	[60,61]
Delta pyrroline-5-carboxylate synthase	(1) Maintenance of cell turgor, osmotic balance and lipid synthesis. (2) Salt stress resistance.	[55,56]
DEAD-box ATP-dependent RNA helicase 53 (DEAD)	(1) ReduceS oxidative stress through activation of ROS scavenging system. (2) Improves plant's photosynthesis machinery, enhancing plant growth and development and mitigates salt stress.	[65,66]

4. Materials and Methods

4.1. Plant Material

The certified and disease-free seeds of salinity stress-sensitive IR64 and salinity stress-tolerant LS rice (*Oryza sativa* L.) cultivars were procured from Punjab Agricultural University, Ludhiana, India, and the Central Rice Research Institute (CRRI), Cuttack, India, respectively. LS can tolerate the salt stress up to 8 dSm^{-1}. The seeds were surface sterilized with 70% ethanol (*v/v*) for 1 min and treated with 0.4% sodium hypochlorite solution containing a drop of tween-20 for 30 min. The seeds were washed thrice with autoclaved distilled water and were then dried on autoclaved Whatman paper (3 mm) for 5 min. After surface sterilization, the seeds were inoculated in the plastic tray containing autoclaved sand moistened with sterile distilled water and were incubated in the culture room at 25 °C (day/night), 70–80% relative humidity (day/night), and 14 h photoperiod for 14 days. After 14 days, IR64 and LS seedlings were treated with 100 mM NaCl for 8 h for imposing salinity stress. The roots were later separated for protein and RNA extraction to conduct 2-dimensional gel electrophoresis and gene expression studies, respectively.

4.2. Study of Morphological Parameters

The seedlings of IR64 and LS were harvested after 2 months and were dipped in water to remove the adhering sand particles. A representative sample of 15 seedlings of both IR64 and LS were selected to study the morphological parameters. Root and shoot length were measured using a meter scale and observations for fresh weight were taken in grams. The root and shoot of each sample were then dried in an oven at 70 °C until a constant weight was achieved, and then the observations for dry weight were recorded. The number of roots for each seedling of IR64 and LS was also counted.

4.3. IAA Estimation

To estimate the content of IAA, 5 g fresh roots of IR64 and LS were crushed finely in liquid nitrogen and extracted in chilled 80% ethanol (15 mL/g) containing butylated hydroxytoluene (BHT) (100 mg/L) [75]. The homogenized material was kept in the dark at 4 °C for 24 h and was filtered. The solid residues were re-extracted thrice with 80% ethanol for 4 h without adding BHT. The BHT containing filtrate and the filtrate without BHT were combined and were centrifuged at 8000 rpm for 20 min. The supernatant was concentrated by drying at 30 °C in a rotavapor in the dark and was used for further processing while the pellet was discarded. The concentrate was resuspended in 2.5 mL of 0.1 M potassium phosphate buffer (pH 8) and was applied to the PVP column after adding 3-bed volumes of potassium phosphate buffer into the PVP column. After elution, a 3-bed volume of potassium phosphate buffer in the PVP column was added again. The elute was concentrated by drying in the rotavapor at 30 °C to obtain 10 mL of elute and its pH was adjusted to 2.5 with 1N HCl. The concentrated 10 mL elute was dissolved in diethyl ether (30 mL) containing BHT (100 mg/L). It was vortexed and kept for 10 min and then the supernatant was taken in a fresh flask (approx. 30 mL). This step was repeated four times. The obtained elute was mixed with 1.5 g of Na_2SO_4 and kept for 30 min. After 30 min, it was evaporated and dried completely at 30 °C using the rotavapor. Then 5 mL of distilled water was added and evaporated on rotavapor at 30 °C. The step was repeated twice and a dried pellet was obtained. The pellet was then dissolved in 1.5 mL of methanol (HPLC grade) for IAA estimation. Further, the elution was carried out with 100 % methanol (HPLC grade): Water (Formic acid 0.05% *v/v*), 35:65, at a flow rate of 1 mL·min^{-1}. The column elutes were passed through a UV detector at 254 nm, and IAA was estimated with reference to an authentic standard of IAA (1 mM) (Sigma Chemical Co., St. Louis, MO, USA). The readings were taken in the replicates of three and the average of peaks was obtained.

4.4. RNA Extraction and cDNA Synthesis

A total of 150 mg of root sample of IR64 and LS was homogenized in liquid nitrogen using pestle and mortar. Total RNA was isolated using Trizol reagent (Invitrogen, http://www.invitrogen.com, last accessed on 20 July 2021), as per the manufacturer's instructions. RNase-free DNase (Sigma-Aldrich, USA) was used to remove the genomic DNA and 2 μg of RNA was used to synthesize cDNA in a total volume of 10 μL reaction using the iScript cDNA synthesis kit (Bio-Rad, Hercules, CA, USA) as per the manufacturer's recommendations.

4.5. Quantitative Real-Time (qRT) PCR Analysis

qRT-PCR was performed to study the differential expression of auxin homeostasis genes in the roots of IR64 and LS. The nucleotide sequences of different genes involved in auxin homeostasis were retrieved from the rice annotation project database (RAP-DB) and the gene-specific primers were designed using Integrated DNA Technologies, USA (http://www.idtdna.com/primerquest/Home/Index, last accessed on 20 July 2021) (Supporting Information-Table S1). The qRT-PCR reaction was performed in 96 well plates using SYBR Green detection chemistry in the StepOne Plus Realtime PCR machine (Applied Biosystems, Waltham, MA, USA). A 10 μL reaction was prepared using 5 μL of 2X Fast SYBR Green (Applied Biosystem), 7.5 ng of each cDNA, 5 μmol each of forward and reverse gene-specific primers and the final volume was raised to 10 μL using sterile nuclease-free water. No template control (NTC) was also set for each primer pair. The thermal cycling was carried out using the following parameters: initial denaturation step at 95 °C for the 20 s to activate the Taq DNA polymerase, followed by the 40 cycles of denaturation at 95 °C for 3 s and finally annealing at 60 °C for 30 s. The melting curve was generated by heating the amplicon from 60 to 90 °C. Baseline and threshold cycles (Ct) were automatically determined using the StepOne Plus Software version 2.3 (Applied Biosystems, USA). Fold changes were calculated using C_T ($\Delta\Delta C_T$) and normalized against *OsUBQ5* (LOC_Os1g328400) used as an endogenous control.

4.6. Protein Extraction

A phenol-based method was used for extracting proteins from 1 g roots of IR64 and LS as reported previously [76]. The samples were homogenized with 6 mL of extraction buffer containing 100 mM KCl, 700 mM sucrose, 50 mM EDTA, and 500 mM Tris-HCl pH 8.0. Further, 2% β-mercaptoethanol, 1 mM PMSF, and a 10 mM protease inhibitor cocktail were added to the extraction buffer just before use. The mixture was vortexed and incubated by agitating on ice for 10 min. After incubation, 6 mL of tris-buffered phenol was added to it and the mixture was again vortexed and incubated on a shaker on ice for 10 min. The homogeneous mixture was centrifuged at 12,000 rpm at 4 °C for 20 min. The upper phenolic phase was collected carefully in a fresh tube. Again, 3 mL of the extraction buffer was added to the tris-buffered phenol and the extraction process was repeated and the upper phenolic phase was collected. Further, 5 volumes of 0.1 M ammonium acetate in 100% cold methanol were added to the phenolic phase and the tube was shaken gently. The mixture was incubated at −20 °C for protein precipitation overnight. The protein pellet was recovered after 24 h by centrifugation at 12,000 rpm at 4 °C for 10 min and the supernatant was discarded. The pellet so obtained was washed thrice with 0.1 M ammonium acetate in cold methanol and then with a mixture containing 80% methanol and 20% acetone, followed by washing with 100% cold methanol. The final washing was given with 100% chilled acetone and the washed pellet was air-dried and stored at −80 °C for 2-DE.

4.7. Protein Solubilization and Quantification

The protein pellets were suspended thoroughly in rehydration buffer (ReadyPrep™ 2-D Starter Kit Rehydration/Sample Buffer #1632106, Bio-Rad, USA). Protein concentration was quantified with a Bradford protein estimation assay [77] using bovine serum albumin (BSA) taken as standard.

4.8. Two-Dimensional Gel Electrophoretic (2-DE) Analysis

For isoelectric focusing (IEF), 150 μg of protein was dissolved in a total of 130 μL of rehydration buffer containing 8 M urea, 2% CHAPS, 50 mM DTT, 0.2% Bio-Lyte® 3/10 ampholyte, 0.001% bromophenol blue (Bio-Rad, USA) and passively rehydrated over IPG strips (7 cm, pH 3–10, Readystrips, Cat. No. 163-200, Bio-Rad, USA) overnight at 20 °C. After rehydration, the strips were focused at 250V for 40 min, 4000 V for 2 h with linear voltage amplification, and finally to 10,000 V h with rapid amplification. After IEF, the strips were incubated with equilibration buffer I, containing 6 M urea, 375 mM Tris-HCL pH 8.8, 2% SDS and 2% DTT for 15 min for reduction (ReadyPrep 2-D Starter Kit Equilibration Buffer I #1632107, Bio-Rad USA). For alkylation of the proteins, the strip was further incubated with 2.5% iodoacetamide dissolved in equilibration buffer II containing 6 M urea, 375 mM Tris-HCL pH 8.8 and 2% SDS (ReadyPrep 2-D Starter Kit Equilibration Buffer I #1632108) for 15 min. The second-dimensional electrophoresis was performed using 12% polyacrylamide gel. After mounting the strip on the gel, it was sealed with 0.5% agarose containing 0.1% bromophenol blue, and the protein molecular marker was also loaded. Electrophoresis was performed at a constant voltage of 100 V for 2 h in tris-glycine-SDS containing running buffer.

4.9. Gel Staining, Imaging, and Analysis

After 2-DE, gels were stained with Coomassie brilliant blue and were stored in 5% acetic acid until further analysis. Gel imaging was conducted using the Molecular Imager Gel Doc XR system (Bio-Rad, USA) and the images were analyzed using PDQuest 8.0.2 software.

4.10. Protein in-Gel Digestion and Mass Spectrometry (MS) Analysis

Proteins spots showing variations in their intensities, presence and absence were manually excised from Coomassie brilliant blue-stained gels and were subjected to mass spectrometric analysis [78]. The excised gel pieces were destained properly using 100 mM NH_4HCO_3/50% ACN solution and washed twice with 200 μL of Milli-Q water for 5 min each and were dehydrated using 100 μL of acetonitrile. The samples were subjected to trypsinolysis in 25 μL of trypsin solution (Sigma, USA) with a concentration of 20 μg/mL in 25 mmol/L NH_4HCO_3, and were incubated overnight at 37 °C. Each digested peptide was further extracted from the gels using 50% trifluoroacetic acid/ 50% acetonitrile, twice at room temperature. The extracted peptides were mixed with 0.5 μL of α-cyano-4-hydroxy-cinnamic acid (Bruker) of a concentration of 20 mg/mL prepared in 0.1% trifluoroacetic acid, 30% (*v/v*) acetonitrile and dried at room temperature. The trypsin digested protein samples were subjected to mass spectrometric analysis using an UltrafleXtreme™ mass spectrometer (Bruker Daltonics Inc. Germany). The instrument was calibrated and fine-tuned with a mass standard starter kit (Bruker) and standard tryptic digested BSA (Bruker, Germany). TOF spectra were recorded in positive ion reflector mode between mass ranges of 700–3500 Da. For protein characterization, the obtained MS spectra were searched against a non-redundant database (SwissProt database) using a MASCOT search engine with these parameters: taxonomy: *Oryza sativa* (rice); parent ion mass tolerance was set at ± 1.2 Da and MS/MS tolerance at 100ppm; variable modifications, oxidation of methionine (M) and carbadomethylation of cysteine (C) and trypsin enzyme.

4.11. Statistical Analysis

All the data obtained from different experiments were evaluated using statistical analysis. An unpaired t-test and a one-way analysis of variance (ANOVA) (the Fischer LSD, Waltham, MA, USA) test were conducted to compare the mean differences using Sigma Stat version 3.5. Comparisons with $p < 0.05$ were considered significantly different.

5. Conclusions

The present study shows that salt tolerant rice cultivars present salinity stress adaptive root traits, likely due to an elevated endogenous auxin content and augmented levels of key salinity stress providing proteins in its roots. Salt tolerant rice LS cultivars exhibited higher transcript-level expression of different genes involved in auxin homeostasis both under control and salinity stress conditions. Thus, our study suggests that an elevated level of auxin and a higher buffering capacity of the auxin homeostasis process may be critical for the acquisition of salinity stress adaptation in rice. Upon 2-DE and MS analysis, several salinity stress tolerance providing proteins were detected that exhibited higher constitutive expression in the roots of LS with respect to IR64. In LS roots, the transcript level of some identified stress marker proteins exhibited lower expression; on the contrary, their protein accumulation was higher in the tolerant cultivar, LS. It indicates that their protein turnover rate might be low. Taken together; these results highlight morphological and molecular features that are critical for rice adaptation towards salinity stress and reveal that this process is multifactorial. Moreover, our results pinpoint several candidate genes that could be artificially overexpressed to increase salinity stress tolerance in rice.

Supplementary Materials: The following are available online at https://www.mdpi.com/article/10.3390/plants10081544/s1, Table S1: List of primers used for the quantitative real-time polymerase chain reaction.

Author Contributions: Conceptualization, P.K.P.; methodology, P.K.P. and S.S.; validation, P.K.P., S.S. and N.K.; formal analysis, S.S., N.K., D.M., B.S. and V.S.; investigation, S.S.; resources, P.K.P., S.S., N.K. and D.M.; data curation, P.K.P., S.S. and N.K.; writing—original draft preparation, S.S. and N.K.; writing—review and editing, P.K.P., P.G., S.S. and N.K.; visualization, S.S., N.K. and D.M.; supervision, P.K.P. and P.G.; project administration, P.K.P.; funding acquisition, P.K.P. and P.G. All authors have read and agreed to the published version of the manuscript.

Funding: We are thankful to the funding agencies, the Department of Science and Technology (DST), Government of India, and the Department of Biotechnology (DBT), Government of India, and to CGIAR Research Program (CRP) on rice agri-food systems (RICE, 2017–2022) for supporting this research work.

Data Availability Statement: The available data are presented in the manuscript.

Conflicts of Interest: The authors declare no conflict of interest.

References

1. Jackson, M. Hormones from roots as signals for the shoots of stressed plants. *Trends Plant Sci.* **1997**, *2*, 22–28. [CrossRef]
2. Saini, S.; Kaur, N.; Pati, P.K. Reactive oxygen species dynamics in roots of salt sensitive and salt tolerant cultivars of rice. *Anal. Biochem.* **2018**, *550*, 99–108. [CrossRef]
3. Gerona, M.E.B.; Deocampo, M.P.; Egdane, J.A.; Ismail, A.M.; Dionisio-Sese, M.L. Physiological Responses of Contrasting Rice Genotypes to Salt Stress at Reproductive Stage. *Rice Sci.* **2019**, *26*, 207–219. [CrossRef]
4. Qadir, M.; Quillérou, E.; Nangia, V.; Murtaza, G.; Singh, M.; Thomas, R.J.; Drechsel, P.; Noble, A.D. Economics of salt-induced land degradation and restoration. *Nat. Resour. Forum.* **2014**, *38*, 282–295. [CrossRef]
5. Shrivastava, P.; Kumar, R. Soil salinity: A serious environmental issue and plant growth promoting bacteria as one of the tools for its alleviation. *Saudi. J. Biol. Sci.* **2015**, *22*, 123–131. [CrossRef]
6. Egamberdieva, D.; Wirth, S.; Bellingrath-Kimura, S.D.; Mishra, J.; Naveen, K.; Arora, N.K. Salt-Tolerant Plant Growth Promoting Rhizobacteria for Enhancing Crop Productivity of Saline Soils. *Front. Microbiol.* **2019**, *10*, 2791. [CrossRef]
7. Solis, C.A.; Yong, M.T.; Vinarao, R.; Jena, K.; Holford, P.; Shabala, L.; Zhou, M.; Shabala, S.; Chen, Z. Back to the Wild: On a Quest for Donors toward Salinity Tolerant Rice. *Front. Plant Sci.* **2020**, *11*, 323. [CrossRef]
8. AbdElgawad, H.; Zinta, G.; Hegab, M.M.; Pandey, R.; Asard, H.; Abuelsoud, W. High salinity induces different oxidative stress and antioxidant responses in maize seedlings organs. *Front. Plant Sci.* **2016**, *7*, 276. [CrossRef] [PubMed]
9. Zhao, Q.; Zhang, H.; Wang, T.; Chen, S.; Dai, S. Proteomics-based investigation of salt-responsive mechanisms in plant roots. *J. Proteomics* **2013**, *2*, 230–253. [CrossRef] [PubMed]
10. Ghosh, D.; Xu, J. Abiotic stress responses in plant roots: A proteomics perspective. *Front. Plant Sci.* **2014**, *5*, 6. [CrossRef] [PubMed]
11. Ryu, H.; Cho, Y. Plant hormones in salt stress tolerance. *J. Plant Biol.* **2015**, *58*, 147–155. [CrossRef]

12. Korver, R.A.; Koevoets, I.T.; Testerink, C. Out of Shape during Stress: A Key Role for Auxin. *Trends Plant Sci.* **2018**, *23*, 783–793. [CrossRef]
13. Quint, M.; Gray, W.M. Auxin signaling. *Curr. Opin. Plant Biol.* **2006**, *9*, 448–453. [CrossRef]
14. Davies, P.J. Plant hormones: Their nature, occurrence, and functions. In *Plant Hormones*; Davies, P.J., Ed.; Springer: Dordrecht, The Netherlands, 2010; pp. 1–15.
15. Shi, H.; Chen, L.; Ye, T.; Liu, X.; Ding, K.; Chan, Z. Modulation of auxin content in Arabidopsis confers improved drought stress resistance. *Plant Physiol. Biochem.* **2014**, *82*, 209–217. [CrossRef] [PubMed]
16. Jadamba, C.; Kang, K.; Paek, N.; Lee, S.I.; Yoo, S. Overexpression of Rice *Expansin7* (*Osexpa7*) Confers Enhanced Tolerance to Salt Stress in Rice. *Int. J. Mol. Sci.* **2020**, *21*, 454. [CrossRef]
17. Fukaki, H.; Okushima, Y.; Tasaka, M. Auxin-mediated lateral root formation in higher plants. *Int. Rev. Cytol.* **2007**, *256*, 111–137.
18. Di Mambro, R.; De Ruvo, M.; Pacifici, E.; Salvi, E.; Sozzani, R.; Benfey, P.N.; Busch, W.; Novak, O.; Ljung, K.; Di Paola, L.; et al. Auxin minimum triggers the developmental switch from cell division to cell differentiation in the Arabidopsis root. *Proc. Natl. Acad. Sci. USA* **2017**, *114*, 641–649. [CrossRef]
19. Rosquete, M.R.; Barbez, E.; Kleine-Vehn, J. Cellular auxin homeostasis: Gatekeeping is housekeeping. *Mol. Plant* **2012**, *5*, 772–786. [CrossRef] [PubMed]
20. Saini, S.; Sharma, I.; Pati, P.K. Integrating the Knowledge of Auxin Homeostasis with Stress Tolerance in Plants. In *Mechanism of Plant Hormone Signaling under Stress*; Pandey, G.K., Ed.; John Wiley and Sons, Inc.: Hoboken, NJ, USA, 2017; pp. 53–70.
21. Ribba, T.; Garrido-Vargas, F.; O'Brien, J.A. Auxin-mediated responses under salt stress: From developmental regulation to biotechnological applications. *J. Exp. Bot.* **2020**, *71*, 3843–3853. [CrossRef] [PubMed]
22. Vaseva, I.I.; Mishev, K.; Depaepe, T.; Vassileva, V.; Van Der Straeten, D. The Diverse Salt-Stress Response of Arabidopsis ctr1-1 and ein2-1 Ethylene Signaling Mutants is Linked to Altered Root Auxin Homeostasis. *Plants* **2021**, *10*, 452. [CrossRef]
23. Zhang, H.; Han, B.; Wang, T.; Chen, S.; Li, H.; Zhang, Y.; Dai, S. Mechanisms of plant salt response: Insights from proteomics. *J. Proteome Res.* **2012**, *11*, 49–67. [CrossRef]
24. Sengupta, S.; Majumder, A.L. Insight into the salt tolerance factors of a wild halophytic rice, *Porteresia coarctata*: A physiological and proteomic approach. *Planta* **2009**, *29*, 911–929. [CrossRef]
25. Comas, L.H.; Becker, S.R.; Cruz, V.M.V.; Byrne, P.F.; Dierig, D.A. Root traits contributing to plant productivity under drought. *Front. Plant Sci.* **2013**, *4*, 442. [CrossRef]
26. Kim, Y.; Chung, Y.S.; Lee, E.; Tripathi, P.; Heo, S.; Kim, K.H. Root response to drought stress in rice (*Oryza sativa* L.). *Int. J. Mol. Sci.* **2020**, *21*, 1513. [CrossRef] [PubMed]
27. Arif, M.R.; Islam, M.T.; Robin, A.H.K. Salinity stress alters root morphology and root hair traits in *Brassica napus*. *Plants* **2019**, *8*, 192. [CrossRef]
28. Polania, J.; Poschenrieder, C.; Rao, I.; Beebe, S. Root traits and their potential links to plant ideotypes to improve drought resistance in common bean. *Theor. Exp. Plant Physiol.* **2017**, *29*, 143–154. [CrossRef]
29. Ku, Y.; Sintaha, M.; Cheung, M.; Lam, H. Plant Hormone Signaling Crosstalks between Biotic and Abiotic Stress Responses. *Int. J. Mol. Sci.* **2018**, *19*, 3206. [CrossRef] [PubMed]
30. Fu, Y.; Yang, Y.; Chen, S.; Ning, N.; Hu, H. Arabidopsis IAR4 modulates primary root growth under salt stress through ROS-mediated modulation of auxin distribution. *Front. Plant Sci.* **2019**, *10*, 522. [CrossRef] [PubMed]
31. Du, H.; Liu, H.; Xiong, L. Endogenous auxin and jasmonic acid levels are differentially modulated by abiotic stresses in rice. *Front. Plant Sci.* **2013**, *4*, 397. [CrossRef] [PubMed]
32. Zörb, C.; Geilfus, C.M.; Mühling, K.H.; Ludwig-Müller, J. The influence of salt stress on ABA and auxin concentrations in two maize cultivars differing in salt resistance. *J. Plant Physiol.* **2013**, *170*, 220–224. [CrossRef] [PubMed]
33. Koevoets, I.T.; Venema, J.H.; Elzenga, J.T.M.; Testerink, C. Roots Withstanding their Environment: Exploiting Root System Architecture Responses to Abiotic Stress to Improve Crop Tolerance. *Front. Plant Sci.* **2016**, *7*, 1335. [CrossRef]
34. Saini, S.; Sharma, I.; Kaur, N.; Pati, P.K. Auxin: A master regulator in plant root development. *Plant. Cell Rep.* **2013**, *32*, 741–757. [CrossRef]
35. Chen, Q.; Dai, X.; De-Paoli, H.; Cheng, Y.; Takebayashi, Y.; Kasahara, H.; Kamiya, Y.; Zhao, Y. Auxin overproduction in shoots cannot rescue auxin deficiencies in Arabidopsis roots. *Plant. Cell Physiol.* **2014**, *55*, 1072–1079. [CrossRef]
36. Kao, C. Mechanisms of Salt Tolerance in Rice Plants: Cell Wall-Related Genes and Expansins. *J. Taiwan Agric. Res.* **2017**, *66*, 87–93.
37. Cheng, Y.; Dai, X.; Zhao, Y. Auxin biosynthesis by the YUCCA flavin monooxygenases controls the formation of floral organs and vascular tissues in Arabidopsis. *Genes Dev.* **2006**, *20*, 1790–1799. [CrossRef] [PubMed]
38. Alarcón, M.; Salguero, J.; Lloret, P.G. Auxin modulated initiation of lateral roots is linked to pericycle cell length in maize. *Front. Plant Sci.* **2019**, *10*, 11. [CrossRef] [PubMed]
39. Ljung, K. Auxin metabolism and homeostasis during plant development. *Development* **2013**, *140*, 943–950. [CrossRef]
40. Fukui, K.; Hayashi, K.I. Manipulation and sensing of auxin metabolism, transport and signaling. *Plant Cell Physiol.* **2018**, *59*, 1500–1510. [CrossRef]
41. Hentrich, M.; Böttcher, C.; Düchting, P.; Cheng, Y.; Zhao, Y.; Berkowitz, O.; Masle, J.; Medina, J.; Pollmann, S. The jasmonic acid signaling pathway is linked to auxin homeostasis through the modulation of YUCCA8 and YUCCA9 gene expression. *Plant J.* **2013**, *74*, 626–637. [CrossRef] [PubMed]

42. Zwiewka, M.; Bilanovičová, V.; Seifu, Y.W.; Nodzyński, T. The Nuts and Bolts of PIN Auxin Efflux Carriers. *Front. Plant Sci.* **2019**, *10*, 985. [CrossRef] [PubMed]

43. Bielach, A.; Hrtyan, M.; Tognetti, V.B. Plants under stress: Involvement of auxin and cytokinin. *Int. J. Mol. Sci.* **2017**, *18*, 1427. [CrossRef]

44. Potters, G.; Pasternak, T.P.; Guisez, Y.; Jansen, M.A. Different stresses, similar morphogenic responses: Integrating a plethora of pathways. *Plant Cell Environ.* **2009**, *32*, 158–169. [CrossRef] [PubMed]

45. Singh, P.; Mohanta, T.K.; Sinha, A.K. Unraveling the Intricate Nexus of Molecular Mechanisms Governing Rice Root Development: OsMPK3/6 and Auxin-Cytokinin Interplay. *PLoS ONE* **2015**, *10*, e0123620. [CrossRef] [PubMed]

46. Balzan, S.; Johal, G.S.; Carraro, N. The role of auxin transporters in monocots development. *Front. Plant Sci.* **2014**, *5*, 393. [CrossRef]

47. Kong, W.; Zhong, H.; Deng, X.; Gautam, M.; Gong, Z.; Zhang, Y.; Zhao, G.; Liu, C.; Li, Y. Evolutionary Analysis of *GH3* Genes in Six *Oryza* Species/Subspecies and Their Expression Under Salinity Stress in *Oryza sativa* ssp. *Japonica*. *Plants* **2019**, *8*, 30. [CrossRef]

48. Xia, K.; Wang, R.; Ou, X.; Fang, Z.; Tian, C.; Duan, J.; Wang, Y.; Zhang, M. OsTIR1 and OsAFB2 downregulation via OsmiR393 overexpression leads to more tillers, early flowering and lesstolerance to salt and drought in rice. *PLoS ONE* **2012**, *7*, e30039.

49. Xu, D.; Miao, J.; Yumoto, E.; Yokota, T.; Asahina, M.; Watahiki, M. YUCCA9 Mediated Auxin Biosynthesis and Polar Auxin Transport Synergistically Regulate Regeneration of Root Systems Following Root Cutting. *Plant Cell Physiol.* **2017**, *58*, 1710–1723. [CrossRef]

50. Guseman, J.M.; Hellmuth, A.; Lanctot, A.; Feldman, T.P.; Moss, B.L.; Klavins, E.; Calderón Villalobos, L.I.; Nemhauser, J.L. Auxin-induced degradation dynamics set the pace for lateral root development. *Development* **2015**, *142*, 905–909.

51. Leyser, O. Auxin Signaling. *Plant Physiol.* **2018**, *176*, 465–479. [CrossRef] [PubMed]

52. Tripathi, V.; Parasuraman, B.; Laxmi, A.; Chattopadhyay, D. CIPK6, a CBL-interacting protein kinase is required for development and salt tolerance in plants. *Plant J.* **2009**, *58*, 778–790. [CrossRef]

53. Chen, L.; Wang, Q.Q.; Zhou, L.; Ren, F.; Li, D.D.; Li, X.B. Arabidopsis CBL-interacting protein kinase (CIPK6) is involved in plant response to salt/osmotic stress and ABA. *Mol. Biol. Rep.* **2013**, *40*, 4759–4767. [CrossRef]

54. Hu, W.; Xia, Z.; Yan, Y.; Ding, Z.; Tie, W.; Wang, L.; Zou, M.; Wei, Y.; Lu, C.; Hou, X.; et al. Genome-wide gene phylogeny of CIPK family in cassava and expression analysis of partial drought-induced genes. *Front. Plant Sci.* **2015**, *6*, 914. [CrossRef]

55. Hayat, S.; Hayat, Q.; Alyemeni, M.N.; Wani, A.S.; Pichtel, J.; Ahmad, A. Role of proline under changing environments: A review. *Plant Signal. Behav.* **2012**, *7*, 1456–1466. [CrossRef] [PubMed]

56. Gharsallah, C.; Fakhfakh, H.; Grubb, D.; Gorsane, F. Effect of salt stress on ion concentration, proline content, antioxidant enzyme activities and gene expression in tomato cultivars. *AoB Plants* **2016**, *8*, plw055. [CrossRef] [PubMed]

57. Zhang, C.H.; Ma, T.; Luo, W.C.; Xu, J.M.; Liu, J.Q.; Wan, D.S. Identification of 4CL Genes in Desert Poplars and Their Changes in Expression in Response to Salt Stress. *Genes* **2015**, *6*, 901–917. [CrossRef] [PubMed]

58. Shafi, A.; Chauhan, R.; Gill, T.; Swarnkar, M.K.; Sreenivasulu, Y.; Kumar, S.; Kumar, N.; Shankar, R.; Ahuja, P.S.; Singh, A.K. Expression of SOD and APX genes positively regulates secondary cell wall biosynthesis and promotes plant growth and yield in Arabidopsis under salt stress. *Plant Mol. Biol.* **2015**, *87*, 615–631. [CrossRef]

59. Le Gall, H.; Philippe, F.; Domon, J.M.; Gillet, F.; Pelloux, J.; Rayon, C. Cell Wall Metabolism in Response to Abiotic Stress. *Plants* **2015**, *4*, 112–166. [CrossRef] [PubMed]

60. Nveawiah-Yoho, P.; Zhou, J.; Palmer, M.; Sauve, R.; Zhou, S. Identification of Proteins for Salt Tolerance Using a Comparative Proteomics Analysis of Tomato Accessions with Contrasting Salt Tolerance. *J. Am. Soc. Hortic. Sci.* **2013**, *138*, 382–394. [CrossRef]

61. Maršálová, L.; Vítámvás, P.; Hynek, R.; Prášil, I.T.; Kosová, K. Proteomic Response of *Hordeum vulgare* cv. Tadmor and *Hordeum marinum* to Salinity Stress: Similarities and Differences between a Glycophyte and a Halophyte. *Front. Plant Sci.* **2016**, *7*, 1154. [CrossRef] [PubMed]

62. Lorbiecke, R.; Steffens, M.; Tomm, J.M.; Scholten, S.; Wiegen, P.; Kranz, E.; Wienand, U.; Sauter, M. Phytosulphokine gene regulation during maize (*Zea mays* L.) reproduction. *J. Exp. Bot.* **2005**, *56*, 1805–1819. [CrossRef]

63. Sauter, M. Phytosulfokine peptide signaling. *J. Exp. Bot.* **2015**, *66*, 5161–5169. [CrossRef]

64. Kutschmar, A.; Rzewuski, G.; Stührwohldt, N.; Beemster, G.T.; Inzé, D.; Sauter, M. PSK-α promotes root growth in Arabidopsis. *New Phytol.* **2009**, *181*, 820–831. [CrossRef]

65. Macovei, A.; Vaid, N.; Tula, S.; Tuteja, N. A new DEAD-box helicase ATP-binding protein (OsABP) from rice is responsive to abiotic stress. *Plant Signal. Behav.* **2012**, *7*, 1138–1143. [CrossRef] [PubMed]

66. Tuteja, N.; Banu, M.S.; Huda, K.M.; Gill, S.S.; Jain, P.; Pham, X.H.; Tuteja, R. Pea p68, a DEAD-box helicase, provides salinity stress tolerance in transgenic tobacco by reducing oxidative stress and improving photosynthesis machinery. *PLoS ONE* **2014**, *30*, e98287. [CrossRef]

67. Dietz, K.J.; Sauter, A.; Wichert, K.; Messdaghi, D.; Hartung, W. Extracellular beta-glucosidase activity in barley involved in the hydrolysis of ABA glucose conjugate in leaves. *J. Exp. Bot.* **2000**, *51*, 937–944. [CrossRef] [PubMed]

68. Li, W.; Zhao, F.; Fang, W.; Xie, D.; Hou, J.; Yang, X.; Zhao, Y.; Tang, Z.; Nie, L.; Lv, S. Identification of early salt stress responsive proteins in seedling roots of upland cotton (*Gossypium hirsutum* L.) employing iTRAQ-based proteomic technique. *Front. Plant Sci.* **2015**, *6*, 732. [CrossRef]

69. Dang, H.Q.; Tran, N.Q.; Gill, S.S.; Tuteja, R.; Tuteja, N. A single subunit MCM6 from pea promotes salinity stress tolerance without affecting yield. *Plant Mol. Biol.* **2011**, *76*, 19–34. [CrossRef] [PubMed]
70. Xu, Z.S.; Xiong, T.F.; Ni, Z.Y.; Chen., X.P.; Chen, M.; Li, L.C.; Gao, D.Y.; Yu, X.D.; Liu, P.; Ma, Y.Z. Isolation and identification of two genes encoding leucine-rich repeat (LRR) proteins differentially responsive to pathogen attack and salt stress in tobacco. *Plant Sci.* **2009**, *176*, 38–45. [CrossRef]
71. Li, X.J.; Li, M.; Zhou, Y.; Hu, S.; Hu, R.; Chen, Y.; Li, X.B. Overexpression of cotton RAV1 gene in Arabidopsis confers transgenic plants high salinity and drought sensitivity. *PLoS ONE* **2015**, *10*, e0118056. [CrossRef] [PubMed]
72. Miura, K.; Hasegawa, P.M. Sumoylation and other ubiquitin-like post-translational modifications in plants. *Trends Cell Biol.* **2010**, *20*, 223–232. [CrossRef]
73. Srivastava, A.K.; Zhang, C.; Yates, G.; Bailey, M.; Brown, A.; Sadanandom, A. SUMO is a Critical Regulator of Salt Stress Responses in Rice. *Plant Physiol.* **2016**, *170*, 2378–2391. [CrossRef] [PubMed]
74. Lemaire, K.; Moura, R.F.; Granvik, M.; Igoillo-Esteve, M.; Hohmeier, H.E.; Hendrickx, N.; Newgard, C.B.; Waelkens, E.; Cnop, M.; Schuit, F. Ubiquitin fold modifier 1 (UFM1) and its target UFBP1 protect pancreatic beta cells from ER stress-induced apoptosis. *PLoS ONE* **2011**, *6*, e18517. [CrossRef]
75. Nagar, P.K.; Sood, S. Changes in endogenous auxins during winter dormancy in tea (*Camellia sinensis* L.) O. Kuntze. *Acta Physiol. Plant* **2006**, *28*, 165–169. [CrossRef]
76. Faurobert, M.; Pelpoir, E.; Chaïb, J. Phenol extraction of proteins for proteomic studies of recalcitrant plant tissues. *Methods Mol. Biol.* **2007**, *355*, 9–14. [PubMed]
77. Bradford, M. A rapid and sensitive method for the quantitation of microgram quantities of protein utilizing the principle of protein-dye binding. *Anal. Biochem.* **1976**, *72*, 248–254. [CrossRef]
78. Kaur, D.; Dogra, V.; Thapa, P.; Bhattacharya, A.; Sood, A.; Sreenivasulu, Y. In vitro flowering associated protein changes in *Dendrocalamus hamiltonii*. *Proteomics* **2015**, *15*, 1291–1306. [CrossRef] [PubMed]

MDPI
St. Alban-Anlage 66
4052 Basel
Switzerland
Tel. +41 61 683 77 34
Fax +41 61 302 89 18
www.mdpi.com

Plants Editorial Office
E-mail: plants@mdpi.com
www.mdpi.com/journal/plants